GW00836550

10 764 7305

Biology and Ecology of the Brown and Sea Trout

Springer
London
Berlin
Heidelberg
New York
Barcelona
Hong Kong
Milan
Paris
Santa Clara
Singapore
Tokyo

J. L. Baglinière, G. Maisse

Biology and Ecology of the Brown and Sea Trout

Published in association with
Praxis Publishing
Chichester, UK

Editeurs/Editors: J. L. Baglinière, G. Maisse
INRA-ENSA, Station de Physiologie et Ecologie des Poissons
65, rue de St Brieuc, 35042 Rennes Cedex
Translator: Dr. Jenny Watson, Department of Zoology, University of Aberdeen.
Translation Editor: Dr. Lindsay Laird, Department of Zoology, University of Aberdeen.

Original French edition, *La Truite: biologie et écologie*
Published by © INRA, Paris, 1991

This work has been published with the help of the French Ministère de la Culture.

SPRINGER–PRAXIS SERIES IN AQUACULTURE & FISHERIES
SERIES EDITOR Dr. Lindsay Laird, M.A., Ph.D., University of Aberdeen, UK
CONSULTANT EDITOR Dr. Selina Stead, B.Sc., M.Sc., Ph.D., Scottish Agricultural College, Aberdeen, UK

ISBN 1-85233-117-8 Sprınger-Verlag Berlin Heidelberg New York

British Library Cataloging in Publication Data
Bagliniere, J. L.
Biology and ecology of the brown and sea trout. -
(Springer–Praxis series in aquaculture and fisheries)
1.Sea-run brown trout 2.Sea-run brown trout - Ecology
I.Title II.Maisse, G.
597.5'7

ISBN 1852331178

Library of Congress Cataloging-in-Publication Data
Truite, biologie et écologie. English
Biology and ecology of the brown sea trout/[edited by] J.L. Baglinière, G. Maisse:
[translator, Jenny Watson].
p. cm. -- (Spring-Praxis series in acquaculture & fisheries)

ISBN 1-85233-117-8 (alk. paper)
1. Brown trout. I. Baglinière, Jean-Luc. II. Maise, G.
III. Title. IV. Series.
QL638.S2T8213 1999
597.5'7--dc21 98-51613
CIP

Printed by MPG Books Ltd, Bodmin, Cornwall, UK

Cover design: Jim Wilkie
Typesetting: Heather FitzGibbon, Christchurch, Dorset, UK

Printed on paper supplied by Precision Publishing Papers Ltd, UK

Contents

Part III. The management of natural populations of brown trout

The colour illustration of brown trout habitats is facing page 42.

Acknowledgements

This book was conceived by the scientific committee of the Colloquium on Trout, organized from 6 to 8 September 1988 at the Centre du Paraclet (Conseil Superieur de la Pêche) by INRA, the National Agronomic Research Institute (Physiology and Ecology of Fishes Station) and the Conseil Superieur de la Pêche (French Bulletin of Fishing and Fish Culture). The colloquium was financed by SPRETIE (the Secrétariat de l'Etat a l'Environnement), the Conseil Superior de la Pêche and INRA. Other articles presented during this colloquium have been published in two editions of the Bulletin Français de Pêche et de Pisciculture (nos 318 and 319 and special Trout Colloquium edition).

The scientific committee consisted of: J. Allardi (CEMAGREF, Paris), J. Arrignon (Union Nationale des Fédérations de Pêche), J. L. Baglinière (INRA, Rennes), B. Buttiker (Conservatoire de la Faune, Switzerland), A. Champigneulle (INRA, Thonon-les-Bains), B. Chevassus (INRA, Paris), Y. Coté (Minister for Leisure, Hunting and Fishing, Québec), P. Dumont (Minister for Leisure, Hunting and Fishing), Françoise Fournel (CSP, Compiègne), M. Héland (INRA, St Pée/Nivelle), G. Maisse (INRA, Rennes), A. Neveu (INRA, Rennes), A. Nihouarn (CSP, Rennes), J. C. Phillipart (University of Liège, Belgium), A Richard (CSP, Rennes), E. Vigneux (CSP, Paraclet).

We wish to thank all the people who, through their remarks and suggestions, helped to improve the quality of the articles presented in this book: J. Arrignon (Union Nationale des Fédérations de Pêche), P. Bergot (INRA, St Pée/Nivelle), B. Chevassus (INRA, Jouy-en-Josas), Brigitte Dasaigues (University of Paris I), P. Gaudin (University of Claude Bernard Lyon I), J. Y. Gautier (University of Rennes), J. Genermont (University of Paris South), D. Gerdeaux (INRA, Thonon-les-Bains), R. Guyomard (INRA, Jouy-en-Josas), M. Héland (INRA, St Pée/Nivelle), C. Lagier (University of Lyon II), P. Y. Lebail (INRA, Rennes), J. Lecomte (INRA, Jouy-en-Josas), A. Neveu (INRA, Rennes), Dominique Ombredane (INRA, Rennes), E. Prevost (INRA, Rennes), M. Thibault (INRA, Rennes).

Introduction: The brown trout (*Salmo trutta* L)—its origin, distribution and economic and scientific significance

J. L. Baglinière

I. INTRODUCTION

The brown trout is a species of salmonid with a facultative migratory character (Hoar, 1976) and has a great capacity for adaptation to different environments. This has led to a major degree of polymorphism in this species, which has been classified in the past under different scientific names (Melhaoui, 1985; Elliott, 1989, 1994). However, the interpretation of the polymorphy of the trout remains delicate in terms of genetic differences (Krieg, 1984), but the idea of the existence of a single species, *Salmo trutta* Linnaeus, actually remains the most likely.

II. PHYLOGENY

The genus *Salmo* constitutes, with six others (*Brachymystax*, *Salmothymus*, *Acantholinqua*, *Hucho*, *Salvelinus*, and *Oncorhynchus*), the sub-family Salmonines, one of three making up the Salmonid family (Fig. 1) (Nelson, 1994). This genus comprises only two species, *Salmo trutta*, the brown trout and *Salmo salar* the Atlantic salmon, since the rainbow trout (or steelhead), *Salmo gairdneri* and the cut-throat, *Salmo clarkii* have recently been reclassified in the genus *Oncorhynchus* (Smith and Stearley, 1989). But the systematics of the genus *Salmo* is still not quite clear as some subspecies of the brown trout can be considered as species of the genus *Salmo* according to author (i.e. *Salmo* (*Trutta*) *carpio* and *Salmo* (*Trutta*) *marmoratus* (see Fig. 2) (Elliott, 1994; Berrebi, 1997).

The ancestors of the Salmonid family appeared at the start of the Cretaceous period (between 63 and 135 million years ago) (Legendre, 1980). According to Tchernavin (1939), they would originate in freshwater. The sub-family Salmoninae appeared after the tertiary period, more accurately the Miocene (between 13 and 25 million years ago)

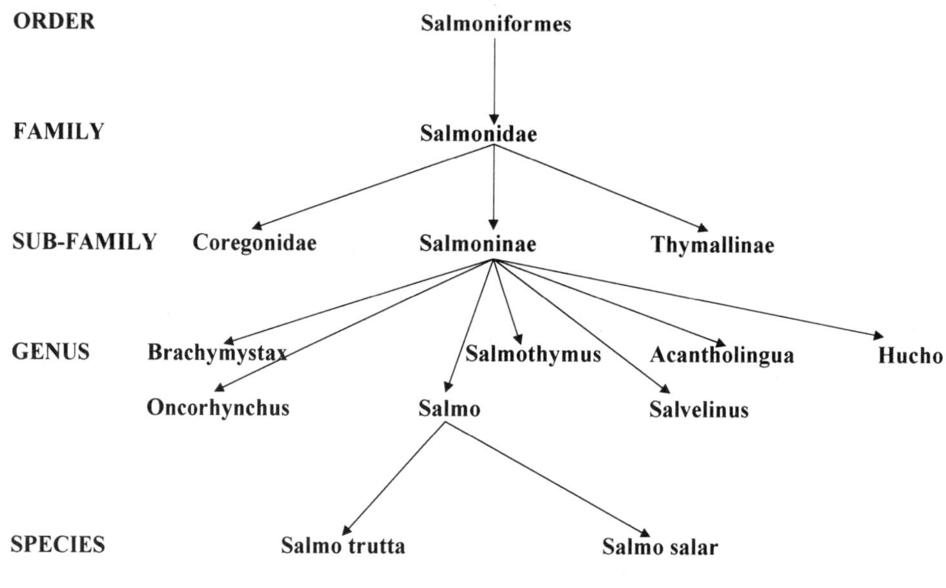

Fig. 1.

(Legendre, 1980). The three main genera (*Salmo*, *Oncorhynchus*, *Salvelinus*) known today appeared more recently (at the start of the Pleistocene) from this common ancestry (Jones, 1959). The separation of the American and Euro-Asiatic continents and the subsequent succession of glaciation ages of the Pleistocene and recent eras has not only resulted in the differentiation of these three genera but also the appearances of numerous lower level taxa (Jones, 1959; Hoar, 1976). However, according to Tchernavin (1939), anadromous behaviour appeared at the start of the Ice Age.

However, Balon (1980) considers that this behaviour existed before speciation in the Salmonid family and that the marine types, mainly of the genus *Salmo* would have formed the origin of the freshwater forms. From this well-argued theory of Balon (1980), Thorpe (1982) deduced that the salmonids were in fact primitive teleosts, probably of marine origin, and that certain species may have progressively lost their anadromous behaviour.

The formation of the sub-family Salmonines occurred at the end of the last glaciation period, about 10 000 years ago. It was linked to temporary geographical barriers caused by the advancing or retreating glaciers and to temperature changes in the oceans (Jones, 1959). In the case of the trout, *Salmo trutta*, the appearance of the anadromous type at the time of the glaciations provided the origin of a large variety of freshwater European forms (Lelek, 1980) (Fig. 2).

III. DISTRIBUTION

1. Original distribution (Fig. 2)

The brown trout is essentially present, in its many forms, across Europe. The northernmost limits of its distribution stretch from Iceland to the USSR (north of the Volga),

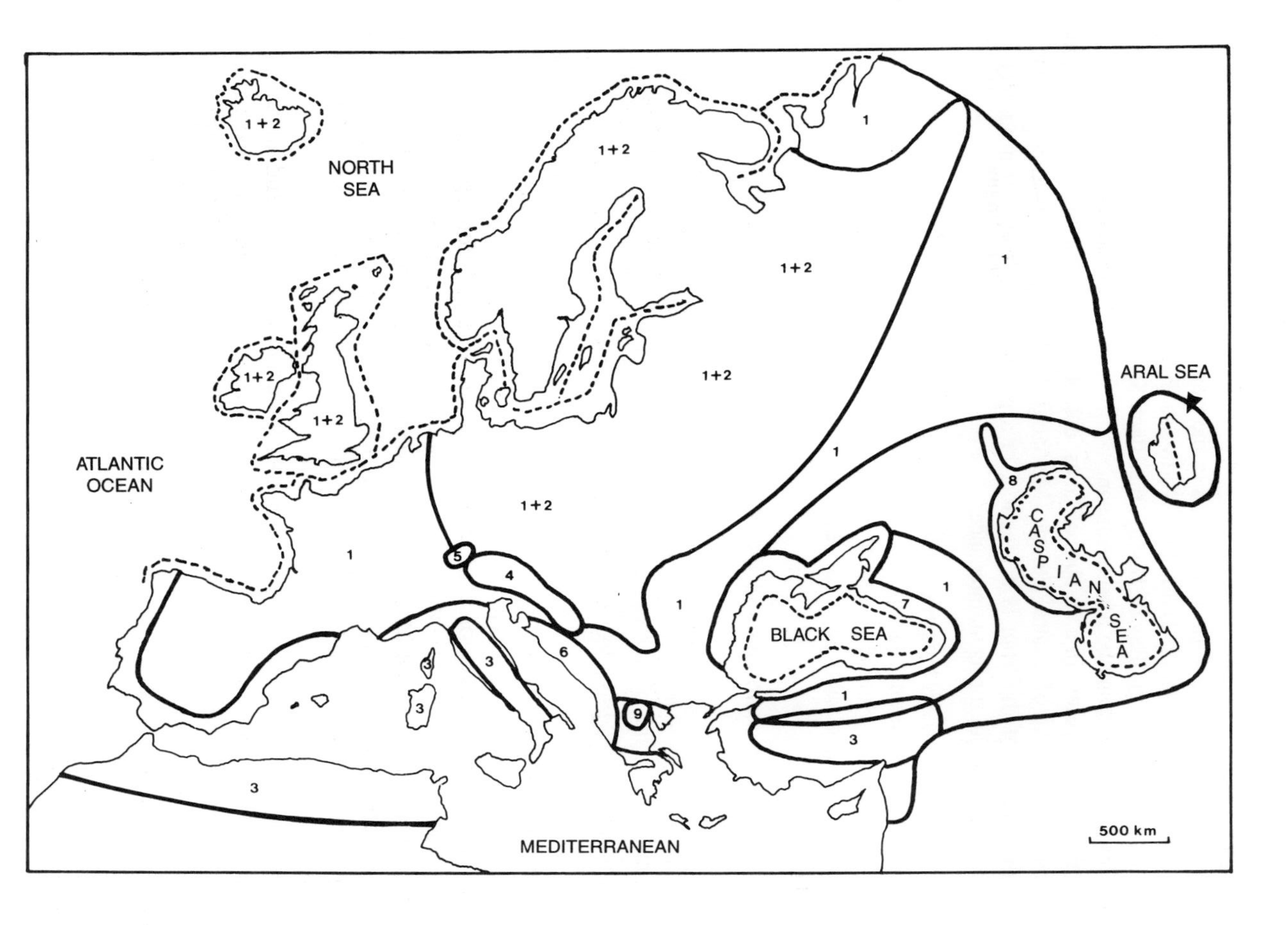
NORTH SEA
ATLANTIC OCEAN
ARAL SEA
BLACK SEA
CASPIAN SEA
MEDITERRANEAN
500 km
1 + 2
1
3
4
5
6
7
8
9

Fig. 2.

along the north of Scandinavia. Its southernmost limit is at the level of the Atlas mountains (Algeria and Morocco) including Sicily and Sardinia. From west to east, the trout is distributed from the European Atlantic front to the buttresses of the Himalayas, including the Caspian and Aral Seas.

The marine anadromous form is localized in the water courses feeding the White Sea and the Gulf of Cheshkaya, the Baltic Sea, North Sea, Irish Sea, the English Channel, the Atlantic Ocean as far as the Bay of Biscay, the Black Sea, Caspian Sea and the Aral Sea. The sea trout is absent from the Mediterranean.

The lake form is present in a number of lakes, notably in the Alps, Scandinavia, Great Britain and in the north of Central Europe (Melhaoui, 1985). Within this area, the longitudinal distribution of the trout is a function, in a given environment, of a number of characteristics essential for its maintenance:

- a narrow range of water temperatures (an average of less than 20°C in summer);
- current speeds, medium to strong;
- good water quality with pH values close to neutral;
- accessibility to favourable areas for reproduction (clean bottom with coarse gravel and pebbles).

This original area of distribution of the brown trout has been modified by man in two ways:

- restriction, mainly in the last two centuries, following industrial and agricultural development (dams, pollution, increased water levels etc.) (Thibault, 1983; Crisp, 1989);
- extension following introduction and restocking (Thibault, 1983).

2. Current distribution (Fig. 3)

The first introductions in France started in the middle of the nineteenth century, benefiting from the discovery of artificial fertilization or insemination (Thibault, 1983). They were followed by many introductory operations around the world, between 1952 and 1969 (MacCrimmon and Marshall, 1968; MacCrimmon *et al.*, 1970). The primary motivation for such operations was the angling interest for the species (MacCrimmon and Marshall, 1968; Hardy, 1972). Most of these introductions have been successful, since the brown trout is established in 24 countries and the number of failures has been very limited; Mexico, Jamaica, Malawi, Uganda, Colombia and Ecuador (MacCrimmon and Marshall, 1968; MacCrimmon *et al.*, 1970). In some cases, resident populations have developed migratory forms, either in a lake (New Zealand, Hardy (1972); Chile, Boeuf (1986); Kerguelen Islands, Davaine and Beall (1988)) or in the sea (Kerguelen Islands, Davaine and Beall (1988); Deep Valley of Columbia in the United States, Bisson *et al.* (1986)).

This introduction of the trout was likely to have certain repercussions on the presence and biology of indigenous fish species. Thus in North America, the trout sometimes supplanted the brook charr, *Salveninus fontinalis*, because of its better capacity for adapting to environmental changes (MacCrimmon and Marshall, 1968). In the same way in Australia, it provoked the eradication of certain species of Galaxiidae, because of heavy predation (Jackson and Williams, 1980). Furthermore, in New Zealand streams,

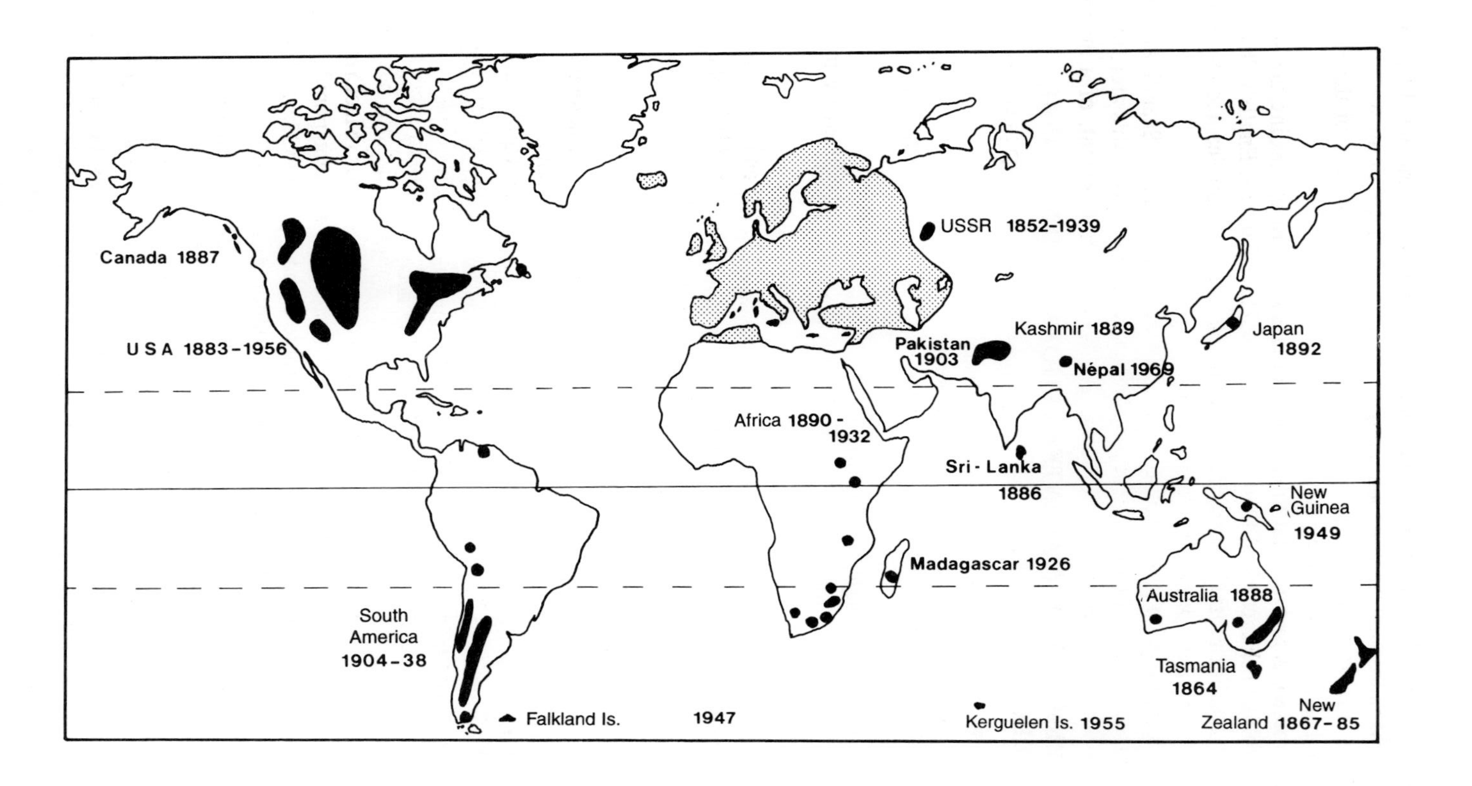

Canada 1887
U S A 1883-1956
South America 1904-38
Falkland Is.
1947
Africa 1890 - 1932
USSR 1852-1939
Pakistan 1903
Kashmir 1889
Népal 1969
Japan 1892
Sri - Lanka 1886
Madagascar 1926
New Guinea 1949
Australia 1888
Tasmania 1864
Kerguelen Is. 1955
New Zealand 1867-85

Fig. 3.

the introduction of brown trout provoked subtle but important changes in the fish predation regime (McIntosh and Townsend, 1995).

Of all the introduced salmonid species, the brown trout is the one which has established itself best outside its original distribution. Its great capacity for adaptation to diverse environments and its great tolerance in the face of habitat changes, have been the factors controlling the success of these introductions. However, the most important determining factor has been that of temperature (MacCrimmon and Marshall, 1968).

IV. THE BROWN TROUT: AN INTERNATIONAL AND NATIONAL RESOURCE

The brown trout is exploited throughout its distribution. Its national and international importance as a renewable resource is at three levels.

1. The passion for angling/sport fishing

It is difficult to determine the economic value of this fishery; it depends, on the one hand, on the country and, on the other hand, on the form of exploitation.

- *For trout in rivers*, there is a chronic absence of statistics on fisheries and methods of exploitation on a world level, except for in New Zealand (Graynoth, 1974a,b; Table 1) and in Quebec (Dumont and Mongeau, 1990). In France, only the following data are available:
  - — the length or area of water in the salmonid fishery network, i.e. corresponding to waters of the first category: 125 000 km of water course (45.4% of flowing network) and 66 000 hectares of still water (lakes etc.).
  - — the number of individuals destined for repopulation and produced in different forms by fish culture. For example, in 1985, 783 million eyed eggs, alevins and juvenile trout were sold to AAPP departmental federations (Benoît, 1986). The selling price of these products has varied from 1984 to 1989 according to the stage and establishment, between 40 and 80FF per thousand (Benoît, 1986; Charbonnel, 1989).
- *For sea and lake trout*, information is a little more precise, partly because capture statistics are drawn up, as in Great Britain (Table 1), and partly because permits are issued to individuals allowing numbers to be monitored (as for sea trout in France). Thus, in 1987 and 1988, the number of sea trout anglers was around 4000 (Tendron, 1989), corresponding to a mean rise in fishing rights of 0.55 million francs, a value very likely underestimated. But the number of anglers decreased for 10 years to attain 2030 in 1997 (Fournel, 1998). In France, the economic value of sport fishing for sea trout and lake trout can only be evaluated from the expenses involved (the selling of caught fish is forbidden). Since 1992, a voluntary declaration of sea trout catches has been instituted at a national level (Fournel, 1998), as has been done for some time for lake trout in Lakes Léman (Gerdeaux *et al.*, 1988) and Annecy (Gerdeaux, 1988).

2. The importance of commercial fishing

This concerns only lacustrine and marine forms. The activity is clearly demonstrated in the case of the sea trout in Great Britain (Table 1). However, it is difficult to compare the data for that country to local French examples. At a national level, the importance of the professional fishery must be looked at in relative terms. Thus, lake trout only represented 4% of the tonnage captured in Lake Léman in 1988 (Gerdeaux *et al.*, 1988). Similarly, sea trout contributed only 4% and 1.2% to the total revenue of the estuarine fishery in the Adour in 1988 (Prouzet *et al.*, 1988) and 1993 (Pronzet *et al.*, 1994) respectively.

3. The aquaculture interest

The brown trout has never been able to rival the rainbow trout in terms of cost of production in commercial hatcheries in fresh water, mainly due to low growth rates and low stocking levels. Until now, the brown trout has been raised mainly for the production of juveniles for restocking purposes (Chevassus and Fauré, 1988).

Recently, however, the economic pressures present in aquaculture and the necessity for diversification of products have led to a promising development of culture in the sea of brown trout (Chevassus and Fauré, 1988). In fact, three aspects justify the use of this species in aquaculture, in comparison with rainbow trout (Quillet *et al.*, 1986): very good summer survival, autumn transfer to the sea possible after 10–11 months culture and a growth in the sea more rapid than in fresh water, limited however by sexual maturation which could be avoided by the production of sterile, triploid individuals (Chevassus *et al.*, 1985).

V. THE BROWN TROUT: A COMPLEX ECOLOGICAL MODEL TO BE STUDIED IN THE NATURAL ENVIRONMENT

The reproductive cycle of the trout has been the subject of numerous studies and its mechanisms are now well understood (Breton and Billard, 1984). It is similar for the biological cycles of the different forms. In contrast, the demographic strategy of this species with the option to migrate, is not always easy to define and remains a function of the great ecological plasticity which expresses itself by the close relationship between population characteristics and potentialities and natural environment in relation to spatial scale (stream order). The complexity of this demographic strategy is clearly illustrated using two examples:

- In river trout, there are interactions between the populations living in the main river and in the tributaries, mainly determined by the way of their issue (juveniles) (Baglinière, *et al.*, 1989);
- Since the introduction of sedentary stocks in the Kerguelen Islands, the habitat has been colonized by the three forms of the species (river, lake and sea) and it is difficult to study the biology of the forms separately (Davaine and Beall, 1988).

This multiplicity of biological cycles results in the coexistence in the same watershed, of populations of trout resident in small brooks in which certain females mature at a length of 10 cm at 2 years old, and sea trout reproducing at a length of 70–80 cm after three seasons of growth in the sea. Added to this are not only the complexity of natural factors

Table 1. Some data on the economic value of the brown trout resource

Type of fishing	Country	Region	Period	Trout form	Annual catch		Market value or expenditures (millions FF)	Reference
					Number (× 10^3)	Tonnage (metric tons)		
Sport fishing	New Zealand		1947–1968	river[1]	35–40	—	1.2[2]	Graynoth (1974a)
	Great Britain	England Wales	1983–1986	sea	35	—	181.0[3]	Elliott (1989)
		Scotland	1983–1986	sea	45.5	8.0	236.7[3]	Anon (1989)
	France	North West rivers	1997	sea	1.8	3.6	—	Fournel (1998)
		Lake Léman	1987–1988	lake	—	4.0		Gerdeaux *et al.* (1988)
		Lake Annecy	1984–1986	lake	—	0.5		Gerdeaux (1988)
Commercial fishing	Great Britain	England Wales	1983–1986	sea	76		394.5[3]	Elliott (1989)
		Scotland		sea	82.7	88.3	430.4[3]	Anon (1989)
	France	Upper-Normandy Coasts	1986–1988	sea	3	6.9	0.3–0.6[4]	Euzenat *et al.* (1991)
		Gironde	1983–1984	sea		3.0	0.24	Boigontier (1987)
		Adour estuary	1988	sea	1.25	3.5–4.0	0.20	Prouzet *et al.* (1988)
			1994			0.8–1.2	0.06	Prouzet *et al.* (1994)
		Lake Léman	1987–1988	lake		14.25[5]	0.64	Gerdeaux *et al.* (1988)
		Lake Annecy	1983–1986	lake		0.25		Gerdeaux (1988)

(1) In this study, the species mainly caught was Rainbow trout.
(2) The value of expenditures did not take the rate of current French money into account, so it is underestimated.
(3) The economic value of the fishery is based by multiplying the number of annual catches by 500 pounds sterling (Elliott, 1989).
(4) The sale price of sea trout range from 38 to 90 FFper kg (Fegard, personal communication).
(5) The sale price of lake trout is about 45 FF per kg (Gerdeaux, personal communication).

acting on the survival and growth of the trout but also man's influence. While certain general rules apply, as far as natural regulation and effects of human activities are concerned (Elliott, 1989; Crisp, 1989), knowledge is far from complete.

VI. CONCLUSION: THE REQUIREMENT FOR BALANCED KNOWLEDGE AT A NATIONAL LEVEL

Before proposing new studies on brown trout, it was necessary to review the available knowledge. This objective was attained with the opening of an international colloquium on brown trout organized by the Institut National de la Recherche Agronomique and the Conseil Supérieur de la Pêche which took place in September 1988 in the Paraclet Centre. Most of the chapters presented here come from different themes approached during the congress (biology of wild populations, position in the ecosystem, management, study methods) and constitute the basis of the present work. Thus different aspects of the biology and ecology of the trout (*Salmo trutta* L.) in France are presented in this book and concern:

- the position of the trout in the ecosystem by characterization of its habitat, its feeding and social behaviour in the natural environment. While the last two topics mainly concern the juvenile stages (up to 2 years old), where the three forms cannot be separated, the habitat is analysed across the specific requirements, in relation to age, size and reproductive stage, essential for the river ecotype. This characterization of habitat also allows us to approach interspecific competition, since the pressure from populations of other species coexisting with the trout can be an important factor in the regulation of species numbers;
- the biological characteristics of different forms of trout present in France, using regional examples. These are: a river in Brittany for all the ecological aspects of the river trout, including some data on density and growth, extended to the whole French territory, Lake Léman for the lake trout and lastly several rivers from Upper and Lower Normandy for sea trout;
- the characterization of the genetic diversity of the French populations and in particular the river and sea-going forms, while touching on the role of population restocking with the evidence of gene flow between natural and farmed populations;
- the analysis, from a historic perspective, of the management of natural populations of brown trout in connection with the ecological potentials of the species and the actions of man on the fish habitat.

BIBLIOGRAPHY

Anon., 1989. Annual review 1986–1987. *Freshwat. Fish. Lab.* Pitlochry, DAFS, 35 pp.

Arrowsmith E., Pentelow F. T. K., 1965. The introduction of Trout and Salmon to the Falkland Islands. *Salmon Trout. Mag.*, **174**, 119–129.

Baglinière J. L., Maisse G., Lebail P. Y., Nihouarn A., 1989. Population dynamics of Brown Trout (*Salmo trutta* L.) in a tributary in Brittany (France): spawning and juveniles. *J. Fish. Biol.*, **34**, 97–110.

Balon E. K., 1980. Early ontogeny of the lake charr, *Salvelinus* (Cristivomer) *namaycush*. In *Charrs* E. K. Balon (Ed), Junk, The Hague, 485–562.

Behnke R. J., 1972. The systematics of Salmonid fishes of recently glacied lakes. *J. Fish. Res. Board Can.*, **29**, 639–671.

Benoit G., 1986. Estimation de la production de l'aquaculture continentale française. *Aqua Revue*, **9**, 7–11.

Berrebi P., 1997. Introduction d'espèces dans les milieux aquatiques d'eau douce: le impacts génétiques. *Bull. Fr. Pêche Piscic.*, **344/345**, 471–487.

Bisson P. A., Nielsen J. L., Chillote M. W., Crawford B., Leider S. A., 1986. Occurence of Anadromous Brown Trout in two Lower Columbia river tributaries. *North Am. J. Fish. Mngt.*, **6**, 290–292.

Bœuf G., 1986. La salmoniculture au Chili. *Piscic. Fr.*, **84**, 5–35.

Boigontier B., 1987. Présentation des données recueillies par le CEMAGREF sur les Salmonidés de l'estuaire de la Gironde. CEMAGREF, division ALA, Bordeaux, 9 pp.

Breton B., Billard R., 1984. The endocrinology of teleosts reproduction: an approach to the control of fish reproduction. *Arch. Fishereiwiss.*, **35**, 55–74.

Chevassus B., Faure A., 1988. Aspects techniques de l'aquaculture. Evolution technologique de la filière salmonicole. *Piscic. Fr.*, **92**, 5–14.

Chevassus B., Quillet E., Chourrout D., 1985. La production de truites stériles par voie génétique. *Piscic. Fr.*, **78**, 10–19.

Crisp D. T., 1989. Some impacts of human activities on trout, *Salmo trutta*, populations. *Fresh-water Biol.*, **21**, 21–23.

Davaine P., Beall E., 1988. Cycle vital et analyse démographique d'une population de truite commune (*Salmo trutta* L.) acclimatée dans les Iles Kerguelen (TAAF). Colloque sur la truite commune, le Paraclet, 6–8 September 1988.

Dumont D., Mongeau J. R., 1990. La truite brune (*Salmo trutta*(dans le Québec méridional. *Bull. Fr. Piscic.*, **319**, 153–166.

Elliott J. M., 1989. Wild brown trout *Salmo trutta*: an important national and international resources. *Freshwater Biol.*, **21**, 1–5.

Elliott J. M., 1994. *Quantitative Ecology and the Brown Trout.* Oxford University Press, Oxford, 286 pp.

Euzenat G., Fournel F., Richard A., 1991. La truite de mer (*Salmo trutta* L). en Normandie/Picardie. In J. L. Baglinière and G. Maisse (Eds), *La Truite: Biologie et Écologie*, INRA Paris, 183–214.

Fournel F., 1998. La truite de mer en France. Année 1997. Rap. CSP, Paris, 47 pp.

Frost W. E., Brown M. E., 1967. *The Trout.* Collins, London, 236 pp.

Gerdeaux D., 1988. Synthèse des connaissances actuelles sur le peuplement piscicole du lac d'Annecy, octobre 1988. Biln piscicole et halieutique. *St. Hydrobiol. Lac.*, INRA, Thonon les Bains, 43 pp.

Gerdeaux D., Buttiker B., Pattay D., 1989. La pêche et les recherches piscicoles en 1988 sur le Léman. *Rapport Annuel 1988,* CIPEL, 7 pp.

Graynoth E., 1974a. The Auckland Trout Fishery. *N.Z.M.A.F. Fish. techn. Rep.*, **89**, 22 pp.

Graynoth E., 1974b. New Zealand angling 1947–1968. An assessment of the national angling diary and postal questionnaire schemes. *N.Z.M.A.F. Fish. Techn,. Rep.*, **135**, 70 pp.

Hardy C. J., 1972. South Island Council of acclimatisation societies. Proceedings of the Quinnat Salmon fishery Symposium 2–3 October 1971—Ashburton. *N.Z.M.A.F. Fish. Techn. Rep.,* **83,** 298 pp.

Hoar W. S., 1976. Smolt transformation: Evolution, Behaviour and Physiology. *J. Fish. Res. Board Can.*, **33**, 1234–1252.

Jackson P. D., Williams W. D., 1980. Effects of brown trout, *Salmo trutta* L., on the distribution of some native fishes in three areas of Southern Victoria. *Aust. J. Mar. Freshwater Res.*, **31**, 61–67.

Jones J. W., 1959. *The Salmon.* Collins, London, 192 pp.

Krieg F., 1984. Recherche d'une différenciation génétique entre populations de *Salmo trutta.* Thèse 3e cycle Fac. Sci. Univ. Paris Sud Orsay, 92 pp.

Legendre V., 1980. Les âges géologiques et quelques uns de leurs vivants d'après les fossiles. *M.L.C.P., Service de l'Aménagement et de l'exploitation de la Faune, Montréal, Province du Québec*, 1 p.

Lelek A., 1980. Les poissons d'eau douce menacés en Europe. Conseil de l'Europe, Strasbourg. *Sauvegarde la Nature*, Vol. 18, 277 pp.

Lesel R., Therezien Y., Vibert R., 1971. Introduction des Salmonidés aux Iles Kerguelen. I—Premiers Résultats et observations préliminaires. *Ann. Hydrobiol.*, **2**, 275–304.

MacCrimmon H. R., Marshall T. L., 1968. World distribution of Brown Trout, *Salmo trutta. J. Fish. Res. Board Can.*, **25**, 2527–2548.

MacCrimmon H. R., Marshall T. L., Gots B. L., 1970. World distribution of Brown Trout, *Salmo trutta*: further observations. *J. Fish Res. Board Can.*, **27**, 811–818.

McIntosh A. R., Townsend C. R., 1995. Contrasting predation risks presented by introduced brown trout and native galaxias in New Zealand streams. *Can. J. Fish. Aquat. Sci.*, **52**, 1821–1833.

Melhaoui M., 1985. Eléments d'écologie de la truite de lac (*Salmo trutta* L.) du Léman dans le système lac-affluent. Thèse 3e cycle, Fac. Sci., Univ. Paris VI, 127 pp.

Nelson J. S., 1994. *Fishes of the World*, 3rd edn. John Wiley, New York, 600 pp.

Prouzet P., Martinet J. P., Casaubon J., 1988. Rapport sur la pêche des marins pêcheurs dans l'estuaire de l'Adour en 1988. Rap. IFREMER/DRV/RH/St Pée-sur-Nivelle, 15 pp.

Prouzet P., Martinet J., Cuende F. X., 1994. Rapport sur la pêche des marins pêcheurs dans l'estuaire de l'Adour en 1993. Rap. IFREMER/DRV/RH/St-PEE, Stat. Hydrobiol. INRA St-Pee, 19 pp.

Quillet E., Chevassus B., Krieg F., Burger G., 1986. Données actuelles sur l'élevage en mer de la truite commune (*Salmo trutta*). *Piscic. Fr.*, **86**, 48–56.

Smith G. R., Stearley R. F., The classification and scientific names of Rainbow and Cutthroat Trouts. *Fisheries*, **14**, 4–10.

Tchernavin V., 1939. The origin of Salmon. *Salmon Trout Mag.*, **95**, 120–140.

Tendron G., 1989. Rapport d'activité du Conseil Supérieur de la Pêche 1988. *Rap. Conseil Supérieur de la Pêche*, Paris, 98 pp.

Thibault M., 1983. Les transplantations de Salmonidés d'eau courante en France, saumon atlantique (*Salmo salar* L.) et truite commune (*Salmo trutta* L.). *C.R. Soc. Biogeogr.*, **59**, 405–420.

Thorpe J. E., 1982. Migration in salmonids, with special reference to juveniles movements in freshwater. In *Proceedings of Salmon and Trout migratory behaviour symposium*, E. L. Brannon and E. O. Salo (Eds), Univ. of Washington, School of Fisheries, Seattle, Washington, USA, 86–97.

Part I

River trout

1

Biology of the brown trout (*Salmo trutta* L.) in French rivers

G. Maisse and J. L. Baglinière

I. INTRODUCTION

The brown trout (*Salmo trutta*) is one of the best-known fish species in Europe. In France, its biology has been the subject of many individual studies which have only allowed an imperfect view of an extremely complex cycle.

In 1960, Vibert, based mainly on work led by M. Huet in the Belgian Ardennes, distinguished several types of biological cycle:

(1) The existence of semi-migratory trout, moving between a habitat which is good for their growth and one apparently only good for reproduction and early growth.
(2) The existence of trout occupying a habitat favourable to both reproduction and growth which, as a result, are sedentary. Are there in certain sectors, migratory trout which are superimposed on the sedentary trout populations? This certainly seems possible.

This complexity of the cycle is also described by Arrignon (1968) at the conclusion of a study of the migratory behaviour of the trout in the Seine basin: 'It appears to us to be very difficult to say definitely, from the meristic characteristics of the subjects examined, whether the migratory character is an attribute of the species, the race or the individual'.

In other words, the plasticity of the brown trout is important and it is an oversimplification to try to present a synthesis of knowledge about river trout without reference to sea or lake trout.

In this presentation, we apply ourselves principally to the knowledge acquired on the Scorff, a river in the Massif Armoricain, during 11 years of study, from 1973 to 1984 (Maisse and Baglinière, 1990).

The Scorff is a coastal river in South Brittany, 75 km long, with a catchment area of 480 km^2. The gradient varies from 1.5 per thousand on schists to 7 per thousand on granites and is modified locally by the presence of millstreams and dams. The geological nature of the basin and the oceanic climate to which it is subjected, give the Scorff a regime of high winter water levels and low summer and autumn water levels. The water quality is good despite the presence of industrial hatcheries and several fish farms; it is slightly acidic (pH = 6.5) and is low in calcium (5 mg/l) (Euzenat and Fournel, 1976; Champigneulle, 1978; Bourget-Rivoallan, 1982). The Scorff basin and the sampling sites are shown in Fig. 1.

Fifteen species of fish are present apart from the brown trout, including the Atlantic salmon (*Salmo salar*), the eel (*Anguilla anguilla*), the pike (*Esox lucius*), the bullhead (*Cottus gobio*) and the stone loach (*Nemacheilus barbatulus*) (Baglinière, 1979). In addition, some selected studies are presented, from which we will attempt to demonstrate the status of the brown trout in French rivers (Fig. 2). While on this subject we should note that the work, on growth, of Cuinat (1971) is the only one to provide an overall, national dimension.

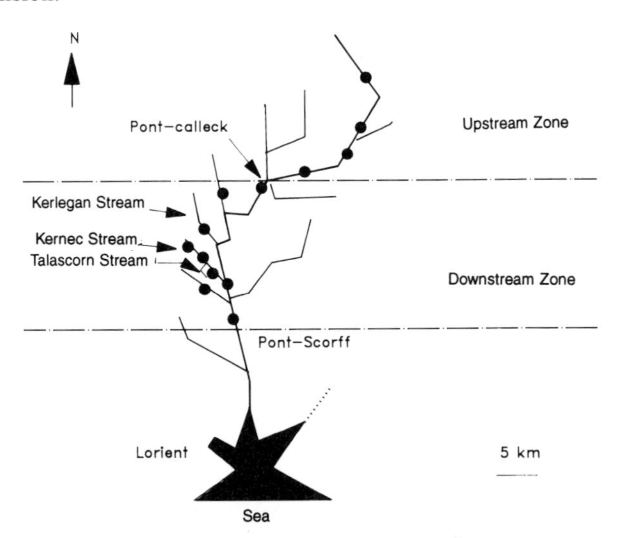

Fig. 1. Location of sampling sites on the Scorff catchment area, 1973–1984.

II. POPULATIONS IN RELATION TO THEIR LOCATION

1. The Scorff

Autumn electrofishing, carried out at various locations in the basin, shows variable dispersal of the population, depending on the age classes in relation to the environment (Table 1, Fig. 3). The highest densities of 0^+ juveniles are found in the tributaries, for which the head areas of the basin play the role of nursery. Within the Scorff, the most favourable habitats are the riffles and the rapids (flowing waters, $v > 40$ cm/s, shallow, 10–40 cm deep and with large-sized substrate particles; Champigneulle, 1978), with a predominance of 1^+ individuals. In the river, the 0^+ age class is better represented upstream than downstream. In deep water (more than 60 cm deep), less suitable for sampling by electrofishing, older, larger individuals are present, but have not been counted (Baglinière *et al.*, 1979a).

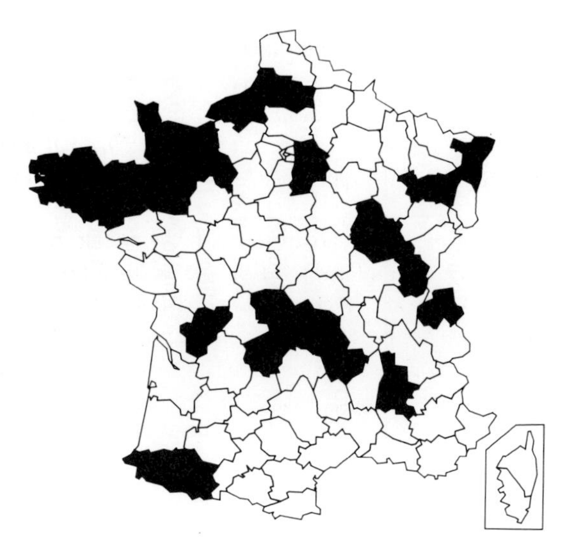

Fig. 2. Regions in which cited studies were carried out.

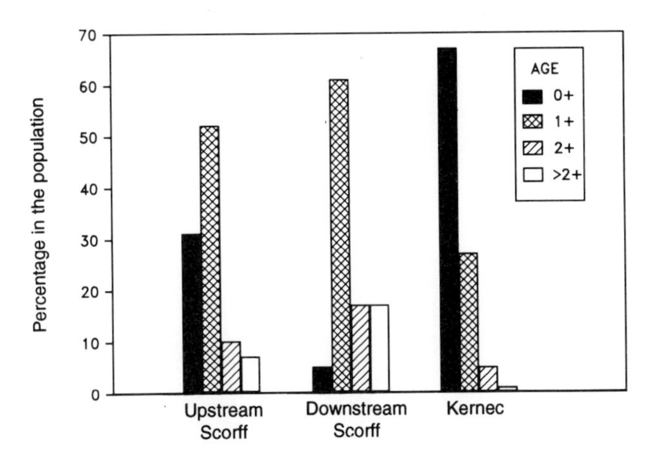

Fig. 3. Demographic structure of the brown trout in the Scorff catchment, according to zone (after Nihouarn, 1983).

An important bank effect was shown by Baglinière and Arribe-Moutonet (1985) in the main river: 80% of the population are found near the banks, boulders, or islands of emerged vegetation, no matter what age class or habitat (Table 1), which should be considered in connection with success of spawning, climatic conditions and intra- and inter-specific competition. As regards the latter, the existence in the Scorff basin of a population of Atlantic salmon leads to the coexistence of the two species in a number of areas (Baglinière and Champigneulle, 1982).

In the main water course, the distribution of juveniles of the two species appears, however, to be different, with the absence of the trout in the zones with high current

Table 1. Densities of brown trout in the Scorff water course, according to age and zone. Habitat: H1 = riffles and rapids; H2 = deep with pools

Location	Zone	0^+ ind/100 m^2	1^+ ind/100 m^2	$> 1^+$ ind/100 m^2	Study period	References
Upstream	Scorff, H1	0.7–2.9	0.5–4.5	0.3–1.4	1976–1980	Baglinière and Champigneulle, 1982
	Scorff, H2	0.0–0.2	0.0–3.3	0.0–2.4	1976–1980	Baglinière and Champigneulle, 1982
	Tributary	72.9	3.0	0.0	1977	unpublished data
Downstream	Scorff, H1	0.0–1.1	0.4–9.2	0.3–2.8	1976–1980	Baglinière and Champigneulle, 1982
	Scorff, H2	0.0–0.3	0.0–4.0	0.4–2.3	1976–1980	Baglinière and Champigneulle, 1982
	Tributary	5.1–23.3	4.7–13.2	1.0–5.0	1975–1983	Nihouarn, 1983a; Baglinière *et al.*, 1989
	Headstream	38.0–64.0	0.5–1.0	0.0–0.00	1982–1983	Baglinière *et al.*, 1989
	Sub-tributary	6.9–50.0	4.0–14.0	2.0–6.0	1975–1983	Nihouarn, 1983; Baglinière *et al.*, 1989

speed, frequented by the salmon (Baglinière and Arribe-Moutonet, 1985). In the streams and in particular those of the Kernec, an inverse correlation exists between the densities of 0^+ juveniles of the two species (Fig. 4; Baglinière and Maisse, 1989), and the indications are that such small tributaries are of great value for recruitment in trout but not very productive for juvenile salmon (Baglinière *et al.*, 1994).

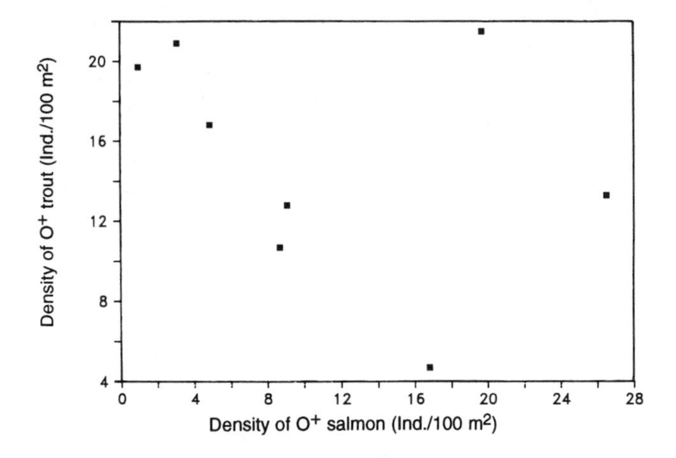

Fig. 4. Coexistence of juvenile brown trout and salmon in the Scorff catchment; relationship between the two species in the same sectors of the Kernec Stream (after Baglinière & Maisse, 1989).

A more precise study of the seasonal change of the population of trout in the middle section of the Kerlégan stream was led by Euzenat and Fournel (1976), from April 1974 to October 1975. This study shows that in this stream the population is mainly composed, in March and April, of fish aged 2 years and older, while in summer we witness a rejuvenation of the population with the appearance of 0^+ fish and an increase in the density of 1^+ fish; this age structure is maintained until October (Fig. 5).

These seasonal fluctuations can be related to two events: the emergence of alevins at the beginning of spring and the autumnal and, more importantly, spring migration of fish of one or two summers' growth, into the main river (Euzenat and Fournel, 1976; Nihouarn, 1983a). The numbers of these migrants, constituting about 80% of the 1^+ fish, produce very important annual fluctuations (Fig. 6).

2. French rivers

The situation described for the Scorff is generally representative of the rivers frequented by the trout. The heads of the basins are populated almost exclusively by 0^+ fish, the streams having more varied populations, with notably 1^+ fish, whose relative importance grows in the main river (Table 2). Locally, this scheme can be reapplied, with a lower proportion of 0^+ and 1^+ in the Pyrenean streams (where old individuals (6^+) are found) and higher in Normandy rivers frequented by the sea trout. In the latter, the impossibility of distinguishing sea trout from sedentary brown trout, brings an element of doubt to the measurement, as sea trout generally reproduce in the main river.

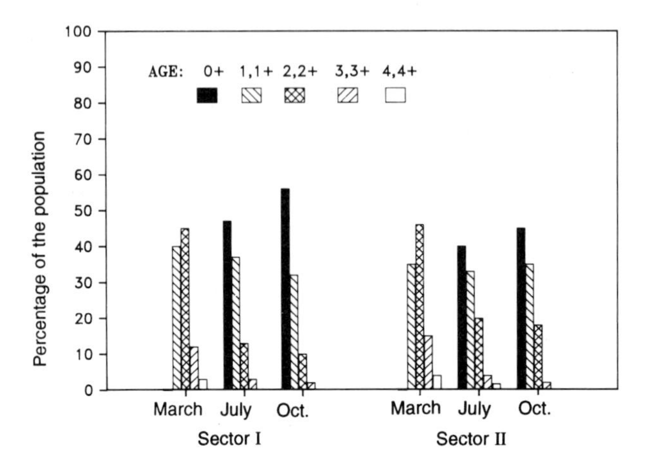

Fig. 5. Seasonal change in the demographic structure of the brown trout population of two sectors of the Kerlégan Stream, a tributary of the Scorff (after Euzenat and Fournel, 1976).

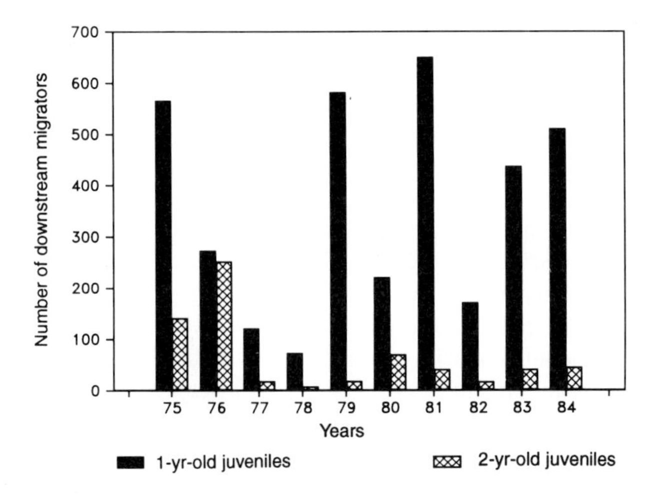

Fig. 6. Annual change in the number of juvenile brown trout caught in a downstream trap in the Kernec Stream, a tributary of the Scorff (after Euzenat & Fournel, 1976, Nihouarn, 1983, Baglinière *et al.*, 1989).

III. GROWTH OF INDIVIDUALS

1. The Scorff

In general terms, the size of brown trout decreases from downstream to upstream in the Scorff, and from the tributaries to the sub-tributaries (Fig. 7). Study of annual and environmental factors shows that the differences in growth of cohorts is established very early, at 0^+ (Table 3); however, the juveniles from the streams migrating into the main river, with its superior feeding capacity, will benefit from growth conditions such that

Table 2. Relative importance of 0^+ and 1^+ age classes in brown trout populations in water courses in continental France, with reference to: (1) Angelier, 1976; (2) Anon., 1983; (3) Anon. 1984; (4) Anon. 1986; (5) Baglinière, 1981; (6) Baglinière & Champigneulle, 1982; (7) Baglinière *et al.*, 1989; (8) Barré, 1972; (9) Benard, 1984; (10) Champigneulle *et al.*, 1988; (11) Chancerel, 1971; (12) Changeux, 1988; (13) Euzenat & Fournel, 1976; (14) Fournel & Euzenat, 1979; (15) Fragnoud, 1987; (16) Gayou & Simonet, 1978; (17) Neveu & Echaubard, 1982; (18) Neveu & Echaubard, 1983; (19) Neveu & Echaubard, 1984; (20) Neveu, unpublished; (21) Nihouarn, 1983a; (22) Nihouarn, 1983b; (23) Nihouarn, 1983c; (24) Ombredane, 1989; (25) Ombredane *et al.*, 1988

	Headstream		Stream		River		
Region	0^+	1^+	0+	1^+	0+	1+	References
Haute-Normandy			85		10–80	15–60	8,14
Maine/B.-Normandy			60–80	10–40	10–40	30–90	2,3,5,9,22,23,
Brittany	90–100	00–10	60–70	20–30	00–30	50–60	4,6,7,13,21,24,25
Basque country			30–40	20–40			11,20
Pyrenees			10–40	20–30	00–20	20–40	1,16
Massif Central	90–100	00–10	40–70	20–40	20–50	30–50	12,17,18,19
Préalpes			70–90	05–10	00–90	05–50	15
Alps (Léman catchment)			10–40	10–20			10
Jura			15	20	10–20	10–40	15

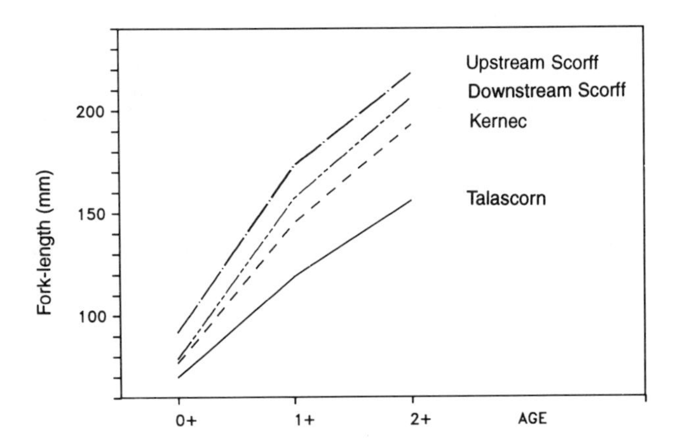

Fig. 7. Mean growth of brown trout in the Scorff catchment area (after Baglinière & Maisse, 1990).

they compensate for their delay in growth compared with the individuals born in the main river (Baglinière and Maisse, 1990).

Table 3. Linear mean monthly growth rate ($G = 2.8$ (log L2 – log L1)/(t2 – t1)) of brown trout in the Scorff catchment area (cohorts'77–'81) according to age and environment (after Baglinière & Maisse, 1980)

	0^+	1^+	2^+
Talascorn	0.55	0.12	0.06
Kernec	0.60	0.15	0.07
Upstream Scorff	0.62	0.16	0.06
Downstream Scorff	0.69	0.15	0.06
Interannual effects	***	NS	NS
Environmental effects	***	***	NS

In the Kernec stream, Baglinière and Maisse (1990) propose a Von Bertalanffy model to describe the growth of 0^+ individuals as a function of the sum of mean daily temperatures (Fig. 8). This interpretation implies an asymptotic size at one year, which can be related to the biogenic capacity of the environment, which becomes progressively limiting; the growth rate can only, in this case, increase as a consequence of a change in environment, such as in the stream, the river or even the sea.

Lastly, it must be stated that certain human activities can modify the growth locally: Bourget-Rivoallan (1982) thus showed that growth is greater downstream of a salmonid farm at Pont-Calleck, in comparison with upstream.

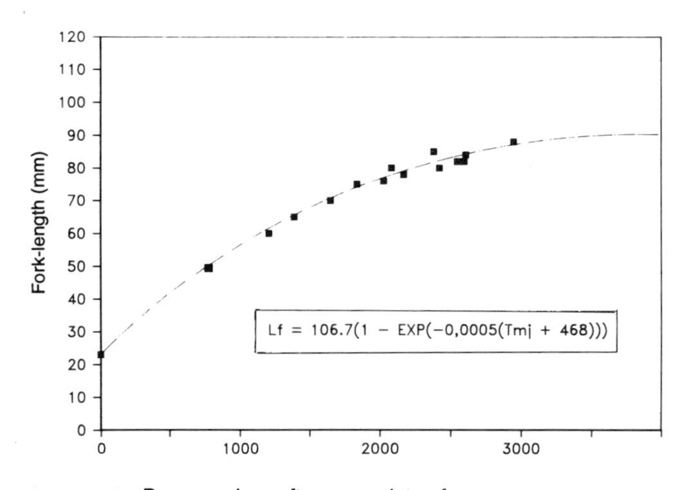

Fig. 8. Model of juvenile brown trout growth in relation to water temperature in the Kernec Stream, a tributary of the Scorff (after Baglinière & Maisse, 1990).

2. French rivers

At a national level (Table 4), growth is variable according to region and basin. According to the data at our disposal, the regions can be divided into three categories:

- high growth: Haute Normandy, Basse-Normandy-Maine, Poitou-Charentes, Léman Alpes-Bassin, East Parisian basin, Alsace;
- medium growth: Brittany, Basque country, Massif central, Alps;
- low growth: Pyrenees, Jura.

This classification should, however, be confirmed by complementary studies, especially in the undersampled regions.

While, as shown by Cuinat (1971), the 'calcium' and 'slope-width' index are good indicators of growth, other factors must be taken into consideration, such as temperature, which penalizes the mountain areas, with the effect of the thawing of snow. An exception is the good growth noted in an alpine tributary of Lake Léman where the lake trout come to spawn. This leads to the consideration of a genetic component in certain basins frequented by migratory stocks.

IV. SEXUAL MATURATION

1. General features

In the parents, the sex ratio is always in favour of the males (Euzenat and Fournel, 1976; Nihouarn, 1983; Maisse *et al.*, 1987; Baglinière *et al.*, 1987). The age structure of the parents varies according to their sex and origin (Table 5): in the females, the majority are in the 2^+ age class, whereas most of the males are 1^+; the 1^+ age class is better represented in the females in streams than in the main river; the converse is true for the males. The maximum age is 5^+ years (Baglinière *et al.*, 1987).

2. Age and size at first maturity

(a) Males

Some individuals mature at 0^+, but the great majority of males mature for the first time at 1^+. The rate of maturation at this age depends on the growth at 0^+ and the mean size of mature males 1^+ is greater than of those which are immature (Maisse *et al.*, 1987).

(b) Females

Euzenat and Fournel (1976) consider the sexual maturation of 1^+ females to be very rare. However, more recent studies, notably those using serodiagnostic methods of sexing (Le Bail *et al.*, 1981) showed that this phenomenon is not insignificant (Baglinière *et al.*, 1981; Nihouarn, 1983a; Maisse *et al.*, 1987; Baglinière *et al.*, 1987). Nevertheless, the great majority of females mature for the first time at 2^+. Of those maturing at 1^+, the maturing females are larger than the males and immature females of the same origin (Fig. 9) (Baglinière *et al.*, 1981; Maisse *et al.*, 1987), but their size varies according to location (Table 6). As for the males, it appears that growth at 0^+ is one of the key factors affecting precocious maturation.

Table 4. Growth of brown trout in continental France, with reference to: (1) Angelier, 1976; (2) Anon., 1983; (3) Anon., 1984; (4) Anon., 1986; (5) Baglinière, 1981; (6) Baglinière & Champigneulle, 1982; (7) Baglinière *et al.*, 1989; (8) Barré, 1972; (9) Benard, 1985; (10) Champigneulle *et al.*, 1988; (11) Chancerel, 1971; (12) Changeux, 1988; (13) Cuinat, 1971; (14) Cuinat & Dumas, 1973; (15) Euzenat & Fournel, 1976; (16) Fragnoud, 1987; (17) Gayou & Simonet, 1978; (18) Neveu, unpubl. (19) Neveu & Echaubard, 1982; (20) Neveu & Echaubard, 1983; (21) Neveu & Echaubard, 1984; (22) Nihouarn, 1983a; (23) Nihouarn, 1983b; (24) Nihouarn, 1983c; Ombredane, 1989; (26) Ombredane *et al.*, 1988; (27) Porcher-Dechar & Porcher, 1974; (28) Prouzet *et al.*, 1977; (29) Steinbach, unpubl.

Region	Temperature	Ca^{2+} index	Slope/width index	Length $0^+/1$	Length $1^+/2$	Length $2^+/3$	Referencess
Haute-Normandy	05–22	9	1–3	90–140	170–210	250–320	8,13,27
Maine/Basse-Normandy	05–20	4–10	2–8	70–110	150–220	200–330	2,3,5,9,23,24
Brittany	05–22	1–5	3–5	70–110	120–200	140–270	4,6,7,15,22 26,28
Poitou-Charentes	07–20	9	5–6	110–120	210–260	300–350	29
Basque country	05–20	3–8	1–8	70–80	120–130	160–300	11,13,18
Pyrenees	00–17	5–7	6–9	60–70	90–130	120–180	1,14,17
Massif Central	03–17	1–6	2–8	70–80	130–150	150–260	12,13,19,20,21
Pré-Alpes	03–24	7–9	5–6	70–120	120–160	180–250-	16
Alps, Léman catchment	01–17	7–8	7	90–110	160–210	270	10
Jura	< 15	7–9	5–6	80–110	110–140	150–180	16
Paris catchment East		9	4–7			250–300	13
Alsace		5–9	3–5			220–310	13

Table 5. Distribution (%) of each sex among the trout spawners according to the age for each origin (from Maisse & Baglinière (1990))

Origin	Catching method	Females			Males			References
		1+	2+	⩾ 3+	1+	2+	⩾ 3+	
Lower Scorff	Census	19	62	19				Baglinière *et al*. (1981)
	Upstream trap	21	66	13	80	18	2	Baglinière *et al*. (1987)
Kernec	Census	33	51	16	44	33	23	Maisse *et al*. (1987)
Talascorn	Census	50	42	8	41	41	18	Unpublished data

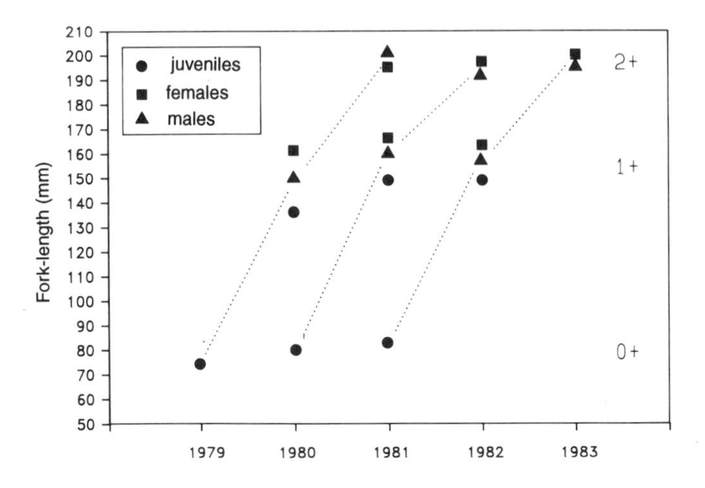

Fig. 9. Relationship between growth and maturation in the population of brown trout in the Kernec Stream, a tributary of the Scorff (after Maisse *et al.*, 1987).

Table 6. Comparison of autumnal sizes of 0+ and 1+ individuals in three different locations: Scorff river, Kernec and Talascorn Streams (from Maisse & Baglinière (1990))

Geograpahical location	0+		Mature 1+ females		Others 1+	
	L	range	L	range	L	range
Lower Scorff	95.3	73–114	186.5	180–235	169.4	142–235
Kernec Stream	80.0	51–118	160.7	140–182	139.9	100–225
Talascorn Stream	68.1	52–83	129.3	106–148	119.6	101–148

L = average fork length (mm)

3. Female fecundity

Euzenat and Fournel (1976) described the development of the gonado-somatic index in trout from the Scorff: the gonads enter a phase of active vitellogenesis from the month of May, which accelerates greatly in September, to reach a GSI of 20% at spawning.

The relationship between fecundity and fork length was given by Euzenat and Fournel (1976) for fish in the Scorff ($\log F = 3.31771 \log L - 5.010$) and by Maisse *et al.* (1987) for fish in the Kernec stream ($\log F = 1.64 \log L - 1.385$). It is difficult to compare these two equations as they were not established over the same range of size, however, where there is an overlap, the river females have, on average, a greater fecundity than those in the stream. Despite this lower individual fecundity, the total reproductive potential of the breeding fish in the Kernec tributary is greater than or equal to that of the migratory spawners, of which there are generally fewer (Baglinière *et al.*, 1987).

V. SPAWNING

1. Duration and location

In general terms, the spawning period of trout starts in November and finishes at the end of January, exceptionally at the end of February (Euzenat and Fournel, 1976; Baglinière *et al.*, 1979b; Nihouarn, 1983a; Baglinière *et al.*, 1989). In the downstream part of the Scorff, the trout spawn mainly in the tributaries, while in the upstream part of the basin, reproduction occurs in the main river as well as in the tributaries (Baglinière *et al.*, 1979b).

2. Spawning-linked activities

The activity of reproductive migration (Fig. 10) is one of the characteristics of the behaviour of breeding fish in the downstream Scorff, which are going to spawn in the tributaries. This migration concerns mainly the spawners present near the confluence of the river and the tributary. Euzenat and Fournel (1976) distinguished several phases of migration:

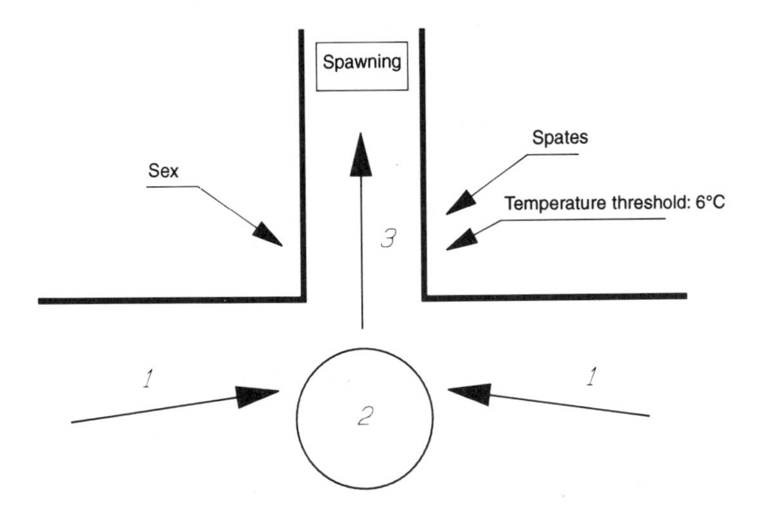

Fig. 10. Diagrammatic representation of spawning behaviour of breeding brown trout in the downstream sector of the Scorff. 1, approach of the confluence between river and spawning stream. 2, station holding at the mouth of the stream, waiting for favourable conditions. 3, upstream migration in the tributary according to sex, affected by certain environmental factors. (After Euzenat & Fournel, 1976, and Baglinière *et al.*, 1987.)

- a first phase of approaching the confluence of the river and the spawning stream (ascending or descending movement);
- stationing at mouth of stream waiting for favourable conditions;
- ascent into the tributary, modulated by a certain number of environmental factors.

The stimuli for the latter phase have been closely studied by Baglinière *et al.* (1987) in the Kernec stream: ascending the stream is conditioned by the appearance of spates; in mean hydrological conditions, a minimum temperature threshold of 6°C exists.

Elsewhere, these same authors have shown that the receptivity to stimuli varies according to sex; the males are receptive earlier than the females which migrate mainly after strong spates; the receptive stage of females occurs just before ovulation.

Spawning activity is mainly diurnal and does not appear to depend on temperature conditions (Baglinière *et al.*, 1979b). The role of flow has not been clearly established; in the tributaries, it does not appear to have a marked influence, while spates hinder redd-building activity in the main stream (Baglinière *et al.*, 1979b). However, the observation conditions vary, especially the turbidity of the water, and it is difficult to determine the real effects of external factors.

3. Characteristics of spawning sites and spawning success

Delacoste (1995) has shown that the physical habitat is the key factor in the abundance of redds. The majority of spawning sites are situated in an area where the current is accelerating; 84% of spawning sites observed in the river are hollowed out in gravel between 2 cm and 5 cm in diameter; in tributaries the granulometry is finer; 62% of redds are made of gravel 2 mm–2 cm in diameter (Nihouarn, 1983a).

The *in situ* study of eggs laid in the redds indicate that the rate of fertilization is high (93%, Nihouarn, 1983a). The rate of hatching is not known, but can be estimated at 80% according to the works of Witzel and MacCrimmon (1983) on the relationship between the granulometry of the redds and embryonic survival.

VI. POPULATION DYNAMICS

1. Survival rates

Table 7 gives an estimation of annual survival rates in the Kernec and Talascorn streams, in the absence of angling. These results confirm high mortalities, typically recorded during the first year. The survival rate during the second year is variable and depends partly, for the Kernec stream, on the number of migrating juveniles (20–35% of 0^+); for the whole basin, the survival rate from 0^+ to 1^+ is probably greater than 50%. After spawning, the annual survival rate of males is less (30%) than that of the females (50%); however, in the streams, the catches by angling appear to affect the females in particular, of which the 4^+ age class may totally disappear from the Kernec stream (Maisse *et al.*, 1987). This does not appear to be the case in the main river, where the sex ratio of caught fish is in equilibrium (Baglinière *et al.*, 1979b).

2. Biological cycle

Baglinière *et al.* (1989) proposed a biological cycle describing the relationship between populations in the downstream part of the Scorff and the Kernec stream (Fig. 11). This interpretation differs from that proposed by Euzenat and Fournel (1976), in the importance given to the spawning population in the stream. These authors consider that the majority of migrating juveniles result directly from the spawning of migratory parents, which has been proved in the case of small upstream tributaries, whereas Baglinière *et al.* (1989) consider that the spawners in the larger streams are the progeny of migratory spawners and produce the majority of migratory juveniles. If the latter interpretation is verified, the recruitment of juveniles in the main river becomes more difficult to model.

Table 7. Survival rate of brown trout in the Scorff catchment area, at each stage, estimated for the Kernec and Talascorn Streams (from Maisse & Baglinière (1990))

	Kernec	Talascorn	References
Fertilization rate	93	—	Nihouarn 1983
Emergence rate	(80)	(80)	*
Survival rate 1 year after emergence	5 [3–7]	7 [3–12]	
Survival rate 2nd year	40[1] [30–50]	40 [30–50]	[1] after Baglinière *et al.*, 1989
Survival rate 3rd year	40[2] [30–50]	40 [30–50]	[2] Maisse *et al.*, 1987
Survival rate 4th year	30[2] 50[2]	25 30	[2] Maisse *et al.*, 1987
Survival rate 5th year	30[2] 50[2]	15 0	[2] Maisse *et al.*, 1987
Survival rate 6th year	0	0	

* Estimation after Witzel & MacCrimmon (1982), taking into account the substrate particle size of the spawning grounds (Nihouarn, 1983).

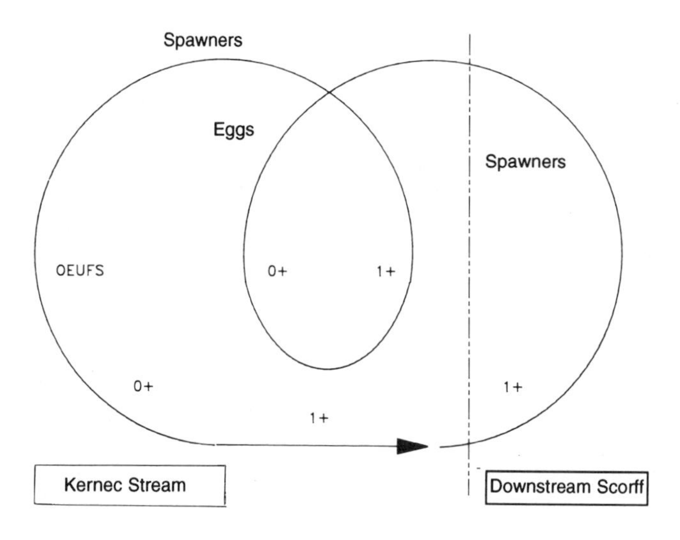

Fig. 11. Simplified biological cycle of the brown trout in the lower part of the Scorff (after Baglinière *et al.*, 1989).

In Fig. 12, we present a diagram of the dynamics of the trout in the entire Scorff basin. This diagram does not take into account the stocking carried out by the angling clubs, the impact of which is little understood; a study of genetic characterization by Krieg and Guyomard (1985) showed the original character of trout in Brittany and in particular in the Scorff, indicating a minimal effect due to restocking from fish culture.

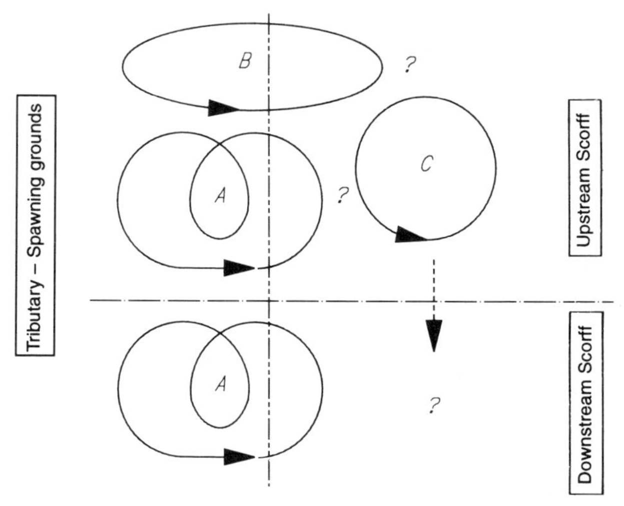

Fig. 12. Different biological cycles of the brown trout observed in the Scorff catchment area. A, cycle with intermediate 'stream' generation between main river spawners and juveniles produced from the ocean. B, cycle where the stream plays a limited role in growth of juveniles up to 1^+. C, cycle occurring entirely in the main river.

VII. CONCLUSION

The population of brown trout in the Scorff has been the subject of numerous complementary studies which have allowed the knowledge of its biology to be rendered almost complete.

Studies carried out on other rivers of the Massif Armoricain (Prouzet *et al.*, 1977; Prouzet, 1981) give the impression that the biological cycle proposed for trout in the Scorff could be adopted without fundamental change. At a national level, similar situations may be found; however, temperature and hydraulic conditions are very varied and there are many feeding areas (streams, rivers, lakes, sea); the biological cycle of the brown trout presents numerous variants. The only major constraints are linked to water quality (summer temperature <25°C; $6 < pH < 9$) and to conditions for reproduction (non-clogged substrate, minimum temperature <12°C). In the absence of aggravated pollution, modifications of the environment with respect to these vital minima would lead to changes in the biological cycle which would put the survival of the species at risk (e.g. populations in large reservoirs). This capacity for adaptation in the brown trout makes it one of the most interesting species to study.

Research into brown trout populations can be divided into two main themes: the biogenic capacity of the environment and the adaptive strategies of the species.

The updating of the concept of 'biogenic capacity' which, according to Léger (1910) 'is the expression of the nutritional level of a water course from the point of view of food for the fish' appears to apply. This idea of Léger's appears, at first glance, to be very promising: 'It is a hugely important factor for the estimation of the economic value of a river' but the author tempers the enthusiasm of the reader: 'the appreciation of the biogenic capacity of a water course depends on so many different factors that one can and must fervently confess that it is not amenable to precise mathematical evaluation. It is more a case of judgement than measurement'. It is better to admit that the concept of 'scale of biogenic capacity', although fictitious, corresponds to the expectations of all fisheries managers. Léger presented this as one of the fundamental elements of the calculation of the annual yield of a water course, a determinant element in management: 'while this yield is attained by the natural population, it is superfluous or even harmful to practise artificial restocking, when the biogenic capacity increases. If it is attained, the difference would serve as a basis for evaluating the number of restocking individuals to add in order to attain the normal yield'.

Biogenic capacity occupies a prime place in the 'hydrological monograph' proposed by Léger for 'proceeding with all the rigour of a scientific method to try to evaluate the waters'. But, for Léger, one had to possess 'a certain clinical habit ..., and a lot of judgement, to co-ordinate all these factors and deduce from them on a scale of 0 to 10, the synthetic figure which would express this capacity'. This appeal to experts was not necessarily a bad method for short-term management, as shown by Chaveroche and Sabaton (1988) in their main description of habitat quality.

However, Baglinière and Thibault (1982) 'concentrating on the two essential elements which emerge from the years of study on the Scorff (complexity of ecological phenomena and importance of natural fluctuations)' insist on 'the necessity of long-term research on the ecology of natural populations'; this leads to the definition of objective indices, the developments of which can be measured identically over the decades.

Cuinat (1971) had already had this aim in formulating growth (size to 3 years) and density as functions of a 'calcium index' and a 'slope-width index'. Today these techniques of multidimensional analysis allows us to appreciate in an objective way, the collection of 'essential factors of biogenic capacity' as described by Léger in 1910:

- the wealth of fish population in the water course at the moment considered;
- the aquatic animals which can serve as food for the trout;
- the terrestrial animals, mainly insects, which can fall into the water;
- plants, including algae, mosses and higher plants which live in the water and nourish the animals in the first group;
- vegetation on the banks and orientation of the water course in relation to prevailing winds;
- the nature of the bottom;
- physical qualities (mainly temperature) and chemical characteristics of the water;
- current speed and general status of waters'.

In the light of the results obtained from the Scorff from 1973 to 1984, it appears that the analysis of growth and autumn populations of small 0^+ trout as a function of various environmental, trophic and behavioural parameters, would be particularly promising for developing and confirming the ideas of Léger.

The other topic of research which appears to require development, concerns the adaptive strategies of the species. The study of the phenomena of regulation and compensation within a population, linking the constituent sedentary and migratory groups (sea trout or lake trout) is complementary to that demonstrated in the previous topic. In view of the management of different ecotypes, it is fundamental to determine the influence of heredity and environment on their phenotypic differentiation.

In the framework of this global approach to the species, it will be necessary to investigate the biology of individuals if we are to understand the resources of the population in terms of adaptation. The plasticity of the species depends in particular on the heterogeneity of their physiology and behaviour. In terms of management, it is important to know if a population ('local stock') possesses, in its individuals, the same evolutionary potential as the species.

BIBLIOGRAPHY

Angelier M. L., 1976. Le peuplement piscicole du ruisseau de la Mousquere (Hautes-Pyrénées). *Ann. Limnol.*, **12**, 299–321.

Anon., 1983. Etude du peuplement piscicole du bassin de la Touques. Rapport Fédération des APP du Calvados-Conseil Supérieur de la Pêche, Délégation de l'Ouest. 34 pp.

Anon., 1984. Etude piscicole du bassin de la Sée (Manche). Rapport Conseil Supérieur de la Pêche. Cesson Sévigné. 27 pp.

Anon., 1986. Situation halieutique du Trieux (1982–1985). Rapport Lab. Ecol. Hydrobiol. INRA, Rennes. 15 pp.

Arrignon J., 1968. Comportement de l'espèce *Salmo trutta* dans le bassin de la Seine. *Bull., Fr. Piscic.*, **229**, 117–122.

Baglinière J. L., 1979. Les principales populations de poissons sur une rivière à Salmonidés de Bretagne-Sud, le Scorff. Cybium, 3rd series, **7**, 53–74.

Baglinière J. L., 1981. Etude de la structure d'une population de truite commune (*Salmo trutta* L.) dans une zone à barbeau. *Bull. Fr. Piscic.*, **283**, 125–139.

Baglinière J. L., Arribe-Moutounet D., 1985. Microrépartition des populations de truite commune (*Salmo trutta* L.), de juvéniles de saumon atlantique (*Salmo salar* L.) et des autres espèces présentes dans la partie haute du Scorff (Bretagne). *Hydrobiologia*, **120**, 229–239.

Baglinière J. L., Champigneulle A., 1982. Densité des populations de truite commune (*Salmo trutta* L.) et de juvéniles de saumon atlantique (*Salmo salar* L.) sur le cours principal du Scorff (Bretagne): préférendums physiques et variations annuelles (1976–1980). *Acta œcol., œol. Appl.*, **3**, 241–256.

Baglinière J. L., Champigneulle A., Nihouarn A., 1979b. La fraie du saumon atlantique (*Salmo salar* L.) et de la truite commune (*Salmo trutta* L.) sur le bassin du Scorff. *Cybium*, 3rd series, **7**, 75–96.

Baglinière J. L., Le Bail P. Y., Maisse G., 1981. Détection des femelles de Salmonidés en vitellogénèse. 2. Un exemple d'application: recensement dans la population de truite commune (*Salmo trutta*) d'une rivière de Bretagne-Sud (Le Scorff). *Bull. Fr. Piscic.*, **283**, 89–95.

Baglinière J. L., Maisse G., 1989. Dynamique de la population de juvéniles de saumon atlantique (*Salmo salar*) sur le ruisseau de Kernec. *Acta œcol., œcol. Appl.*, **10**, 3–17.

Baglinière J. L., Maisse G., 1990. La croissance de la truite commune (*Salmo trutta* L.) sur le bassin du Scorff. *Bull. Fr. Pêche Piscic.*, **318**, 89–101.

Baglinière J. L., Maisse G., Le Bail P. Y., Nihouarn A., 1989. Population dynamics of brown trout (*Salmo trutta* L.) in a tributary in Brittany (France): spawning and juveniles. *J. Fish. Biol.*, **34**, 97–110.

Baglinière J. L., Maisse G., Le Bail P. Y., Prevost E., 1987. Dynamique de la population de truite commune (*Salmo trutta* L.) d'un ruisseau breton (France)—Les géniteurs migrants. *Acta œcol., œcol. Appl.*, **8**, 201–215.

Baglinière J. L., Nihouarn A., Champigneulle A., 1979a. L'exploitation des Salmonidés à la ligne sur le Scorff, rivière de Bretagne-Sud. *Bull. Fr. Piscic.*, **272**, 94–115.

Baglinière J. L., Thibault M., 1982. Les difficultés d'une gestion rationnelle: l'exemple des populations naturelles de Salmonidés, saumon atlantique et truite commune, sur le Scorff. Assoc. Internat. Entretiens Ecol. Colloque sur *la production et la commercialisation du poisson d'eau douce*, 30 March–1st April 1982, 26 pp.

Baglinière J. L., Le Louarn H., Hamonet J. M., 1991. Du Brochet en première catégorie; étude de la Boutonne. *Eaux Libres*, **5**, 8–11.

Barre N., 1972. Dynamique d'une population de truites sur en secteur d'une rivière normande, l'Andelle. Thèse doctorat vétérinaire. Faculté de Médecine Créteil. 117 pp.

Benard A., 1985. Monographie des bassin versants de l'Orne et de l'Huisne (département de l'Orne): milieu naturel et peuplements piscicoles. DAA '*Protection et aménagement du milieu naturel*', ENSA Rennes. 34 pp.

Bourget-Rivoallen S., 1982. L'impact des piscicultures sur les rivières à Salmonidés. DAA '*Protection et aménagement du milieu naturel*', ENSA Rennes. 29 pp.

Champigneulle A., 1978. Caractéristiques des juvéniles de saumon atlantique (*Salmo salar* L.) en relation avec l'habitat sur le cours principal du Scorff. Thèse doctorat 3e cycle. Biol. Anim. Fac. Sci. Univ. Rennes, 92 pp.

Champigneulle A., Melhaoui M., Maisse G., Baglinière J. L., Gillet C., Gerdeaux D., 1988. Premières observations sur la truite (*Salmo trutta* L.) dans le Redon, un petit affluent frayère du lac Léman. *Bull. Fr. Pêche Piscic.*, **310**, 59–76.

Chancerel F., 1971. Dynamique d'une population de truites dans un ruisseau des Pyrénées atlantiques, le Lizunia. DAA '*Halieutique*'. ENSA Rennes. 62 pp.

Changeux T., 1988. Situation halieutique d'une rivière à truite *Salmo trutta* L.: le Haut-Torion. DAA '*Halieutique*'. ENSA Rennes. 45 pp.

Chaveroche P., Sabaton C., 1988. Quinze experts analysent l'habitat de la truite fario (*Salmo trutta* fario). Colloque sur la truite, centre du Paraclet, 6-7-8/09/88. résumé de la communication, 2 pp.

Cuinat R., 1971. Principaux caractères démographiques observés sur 50 rivières à truites françaises. Influence de la pente et du calcium. *Ann. Hydrobiol.*, **2**, 187–207.

Cuinat R., Dumas J., 1973. Diagnose écologique en cours d'eau à Salmonidés, méthode et exemple. Rapport Organisation Nations Unies Alimentation Agriculture, Rome, 1973, 124 pp.

Delacoste M., 1995. Analyse de la variabilité spatiale de la reproduction de la truite commune (*Salmo trutta* L.). Etude à l'échelle du micro et du macro habitat dans 6 rivières des Pyrénées Centrales. Thèse Doct. Inst. Nat. Polytechn. Toulouse, 133 pp.

Euzenat G., Fournel F., 1976. Recherches sur la truite commune (*Salmo trutta* L.) dans une rivière de Bretagne, le Scorff I. Caractéristiques démographiques des populations de truite commune de la rivière Scorff et des affluents. 2 Premiers éléments d'une étude de dynamique de population de truite commune. Thèse doctorat 3e cycle Biol. Anim. Fac. Sci. Univ. Rennes, 213 pp.

Fournel F., Euzenat G., 1979. Etude sur les Salmonidés migrateurs du bassin de l'Arques, Part 1. *Bull. Inf. C.S.P.*, **114**, 25–49.

Fragnoud E., 1987. Préférences d'habitat de la truite fario (*Salmo trutta fario* L., 1758) en rivière, (quelques cours d'eau du Sud-Est de la France). Thèse doctorat 3e cycle 'Ecologie Fondamentale et Appliquée des Eux Continentales'. Univ. Claude Bernard-Lyon I. 435 pp.

Gayou F., Simonet F., 1978. Dynamique des populations de truites (*Salmo trutta fario*, L.). Aménagements piscicoles en haute vallée d'Aure. Thése doctorat 3e cycle 'Sciences et Techniques en Production Animale'. Inst. Nat. Polytech. Toulouse. 244 pp.

Krieg F., Guyomard R., 1985. Population genetics of French brown trout (*Salmo trutta* L.): large geographical differentiation of wild populations and high similarity of domesticated stocks. *Genet. Sel. Evol.*, **17**, 225–242.

Le Bail P. Y., Maisse G., Breton B., 1987. Détection des femelles de Salmonidés en vitellogénèse. 1. Description de la méthode et mise en œuvre pratique. *Bull. Fr. Piscic.*, **283**, 79–88.

Leger L., 1910. Principes de la méthode rationnelle du peuplement des cours d'eau à Salmonidés. *Ann. Univ. Grenoble*, **22**, 533–568.

Liebig H., 1998. Etude du recrutement de la truite commune (*Salmo trutta*) d'une rivière de moyenne montagne (Pyrénées Ariégeoises). Effets de la gestion par éclusées d'une centrale hydroélectrique. Thése Doct. Inst. Nat. Polytechn. Toulouse, 200 pp.

Maisse G., Baglinière J. L., Le Bail P. Y., 1987. Dynamique de la population de truite commune (*Salmo trutta*) d'un ruissseau breton (France): les géniteurs sédentaires. *Hydrobiologia*, **148**, 123–130.

Maisse G., Baglinière J. L., 1990. The biology of brown trout, *Salmo trutta* L., in the river Scorff, Brittany: a synthesis of studies from 1973 to 1984. *Aquac. and Fish. Mgmt.*, **21**, 95–106.

Neveu A., Echaubard M., 1982. Inventaire piscicole: résultats obtenus au cours des pêche électriques de septembre 1982 dans la région de Besse-en-Chandesse (Puy-de-Dôme). Rapport 1982. Lab. Ecol. Hydrobiol. INRA, Lab. Zool. INA Paris. 20 pp.

Neveu A., Echaubard M., 1983. Inventaire piscicole: résultats obtenus au cours des pêches électriques de septembre 1983 dans la région de Besse-en-Chandesse (Puy-de-Dôme). Rapport 1983. Lab. Ecol. Hydrobiol. INRA, Lab. Zool. INA Paris. 19 pp.

Neveu A., Echaubard M., 1984. Inventaire piscicole: résultats obtenus au cours des pêches électriques de septembre 1984 dans la région de Besse-en-Chandesse (Puy-de-Dôme). Rapport 1984. Lab. Ecol. Hydrobiol. INRA, Lab. Zool. INA Paris. 14 pp.

Nihouarn A., 1983a. Etude de la truite commune (*Salmo trutta* L.) dans le bassin du Scorff (Morbihan): démographie, reproduction, migration. Thése doctorat 3e cycle, Biol. Anim. Fac. Univ. Rennes, 73 pp.

Nihouarn A., 1983b. Les cours amont et moyen de la Sienne. Etude des populations piscicoles. Rapport Conseil Supérieur de la Pêche DR no. 2, 14 pp.

Nihouarn A., 1983c. Situation piscicole du bassin due Couesnon, état naturel et influences humaines: pollutions, aménagements hydro-agricoles, repeuplements. Rapport Conseil Supérieur de la Pêche DR no. 2. 36 pp.

Ombredane D., 1989. Les peuplements et habitats piscicoles de l'Elorn en 1988. Rapport Lab. Ecol. Hydrobiol.—Dept. Halieutique ENSAR. 14 pp.

Ombredane D., Haury J., Thibault M., 1988. Etude des peuplements piscicoles de l'Ehorn en relation avec les habitats aquatiques en Octobre 1987. Rapport Lab. Ecol. Hydrobiol. INRA—Dept. Halieutique ENSA. 24 pp.

Porcher-Dechar C., Porcher J. P., 1974. Etude hydrobiologique et écologique du bassin de la Bresle. DAA 'Protection et aménagement du milieu naturel'. ENSA Rennes. 101 pp.

Prouzet P., 1981. Caractéristiques d'une population de Salmonidés (*Salmo salar* et *Salmo trutta*) remontant sur un affluent de l'Ehorn (rivière de Bretagne-Nord) pendant la période de reproduction 1979–1980. *Bull. Fr. Piscic.*, **283**, 140–154.

Prouzet P., Harache Y., Danel P., Branellec J., 1977. Etude de la croissance de la truite commune *Salmo trutta fario* (L.) dans deux rivières du Finistère. *Bull. Fr. Piscic.*, **267**, 62–84.

Vibert R., 1960. Bases rationnelles de la gestion piscicole. Les truites sont-elles migratrices ou sédentaires? *Bull. Fr. Piscic.*, **196**, 107–110.

Witzel L. D., MacCrimmon H. R., 1983. Embryo survival and alevin emergence of brook char, *Salvelinus fontinalis*, and brown trout, *Salmo trutta*, relative to redd gravel composition. *Can. J., Zool.*, **61**, 1783–1792.

2

The habitat of the brown trout (*Salmo trutta* L.) in water courses

J. Haury, Dominique Ombredane and J. L. Baglinière

I. INTRODUCTION

Within its overall area of distribution, the brown trout (*Salmo trutta* L.) is not always present, nor is it evenly distributed in the water courses. Among the factors regulating natural populations of trout, environmental conditions are of great importance and allow the definition of the habitat of this species.

From a theoretical point of view, the 'habitat ... is defined in relation to a species, as a collection of elements of the biotope, which serve its needs satisfactorily and, by extension, the assembly of biotopes in which the species is found' (Blondel, 1979). Trout habitat can therefore be defined by general parameters characterizing not only the range of distribution of this species, but also governing its longitudinal distribution within a hydrological network. Lastly, at the level of a section of river, other descriptive elements of the habitat determine the population structure of the brown trout, a territorial species, with quite precise environmental requirements (Kalleberg, 1958; Timmermans, 1960; Allen, 1969; Taube, 1974).

Thus, the definition and characterization of the trout's habitat require different factors depending on the spatial scale chosen. It is therefore necessary to start by establishing the terminology employed for the different scales of habitat study and to identify which are the determining factors in the trout's ecology which are likely to be good descriptors of its living environment (Chapter 1).

On this basis, it is possible to characterize the trout zone in a water course, and its longitudinal variations (Chapter 2).

In addition, the habitat requirements of the brown trout vary in relation to the development cycle (eggs, alevins, juveniles, adults) and their activities at a particular time (reproduction, rest, feeding). Some stages are therefore more distinct as a result of their special needs: the juveniles and their growing areas, the adults for their holding and reproductive areas (Chapter 3). The absence of morphological differences in the juveniles of the three ecological forms of the species *Salmo trutta* L. (lake, sea and resident) which can co-inhabit one river, does not allow the final specific habitats to be distinguished. On the contrary, for the adult stage, the feeding areas are distinct and only the resident form is considered there; the spawning zones of the three forms overlap each other, but the characteristics of the redds differ.

Lastly, from a management perspective of trout populations, a dynamic study of the habitat is presented; natural modifications of the habitat and their consequences for the trout are analysed and the influence of human activities detailed (Chapter 4). Models of stock estimation in relation to habitat characteristics (Chapter 5) are now complementary tools for managers.

The subject, which is presented essentially with French examples and mainly from the Armoricain region, is discussed together with overseas literature. The observations in the natural environment are complemented by results of experiments under controlled conditions. Some original results on Armoricain rivers show certain aspects of the utilization of the habitat by different age classes of fish, spatial relations with other species of fish, etc.

II. DESCRIPTION OF TROUT HABITAT

Two complementary approaches show the relationship between the trout and its living environment.

- the knowledge of ecological factors and quantification of their scale, allowing determination of the ecological range of the species;
- the description of natural, colonized biotopes, which are determined by the same ecological factors.

1. Characteristic habitat features

Whatever the scale of the study, the habitat is defined by physical and biotic factors. The latter are difficult to quantify, especially food resources which result from the condition and functioning of the ecosystem; vegetation is studied because of its role in shelter and in space structuration. For the former, all the physical descriptors, affecting the ecology of the species, are analysed one by one, as has been done by a number of authors (Huet, 1962; Brown, 1975; Crisp, 1989).

(a) Flow (current)

Flow is an essential parameter in the trout's habitat. It has a major role in the functioning and structuring of the ecosystem (Butcher, 1933; Huet, 1962; Bournand, 1963). It operates in two ways:

- directly, by supporting individual trout, which are particularly well morphologically adapted to the current (Huet, 1962; Brown, 1975) and show a net positive rheotactic effect (Baldes and Vincent, 1969), at the same time limiting its energy cost (Bachman, 1984; Fausch, 1984), which leads to studies on fish morphometry;
- indirectly, through many effects on the ecosystems: invertebrate drift, oxygenation of the redds (Crisp, 1989), selection and modification of substrates as sedimentation (Butcher, 1933).

(b) Morphology of the stream bed

Depth plays an important role in the positioning of trout (Egglishaw and Shackley, 1982; Heggenes *et al.*, 1995). As it is inversely correlated to the current speed, for given width and current speed, depth itself can play a role as a shelter, especially in winter (Chapman and Bjornn, 1969).

The local morphology of the bed also affects the positioning of individuals. Submerged banks provide potential hiding places (Nihouarn, 1983). The width of the bed also has to be taken into account: in major water courses, the trout rest close to the banks (Lindroth, 1955), especially juveniles (Roussel and Bardonner, 1999). There is a real 'bank-effect' as was shown by Baglinière and Arribe-Moutounet (1985).

The slope of the stream bed directly determines the current speed and is in fact an important factor in the localization of spawning areas (Champigneulle, 1978).

(c) Substrate particle size (granulometry)

The granulometry of the bed is a result of the current speed, depth and geological nature of the basin slopes. As long as the substrate particle size is sufficiently large, it plays a role as shelter from fast-flowing water and protection from predators (Jenkins, 1969; Heggenes, 1988a; Baran *et al.*, 1995a). The substrate particle size determines the number of refuges, whose utilization depends on the size of the fish. Indeed, it has to be able to fit in the spaces between the substrate particles (Heggenes, 1988b).

In addition, the substrate particle size is a determining factor in the choice of redd sites for spawning fish (Ottaway *et al.*, 1981). An excess of fine sediment can clog up the redds, causing a lack of oxygen for the eggs (Peters, 1967; Reiser and White, 1988; Grant *et al.*, 1986) and too large a particle size can limit the possibilities for hollowing out a redd.

(d) Light

Light determines positioning and orientation by vision and phototactism, controls the development cycle and modifies other parameters of the fishes' habitat, such as temperature and dissolved oxygen (via photosynthesis). Phototactism changes during the development cycle; the alevins are negatively phototropic until yolk sac resorption is complete, while the free-swimming stage (fry) shows positive phototropism (Ottaway and Clarke, 1981). Lastly, in adult trout, progressively stronger negative phototropism has been noted by many authors and is believed to be shelter-seeking behaviour (Butler and Hawthorne, 1968; Haury and Baglinière, 1990).

(e) Temperature

Temperature is one of the major controlling factors of salmonid ecosystems. The trout is considered as a stenotherm of cold waters (Mills, 1971; Brown, 1975). Temperature has an effect at two levels:

- direct action as a behaviour regulator (migration, reproduction) and above all, controlling the ecophysiology of the trout. In the natural environment, optimum temperatures are between 7 and 19°C (Frost and Brown, 1967) or 7 and 17°C (Mills, 1971). Under experimental conditions, respiratory metabolism and activity increase with temperature; if other parameters are not limiting, the lethal temperature is 25°C

(Charlan, 1962). Growth of juveniles in relation to water temperature and diverse biological parameters was modelled by Elliott (1984a) and Baglinière and Maisse (1990).

- indirect action consists of modifying other habitat characteristics, especially dissolved oxygen concentration, growth of vegetation and development of benthic invertebrates.

(f) Dissolved oxygen

Dissolved oxygen is vital for trout survival; it is considered to be a very demanding species (Schindler, 1953, in Huet, 1962). The minimum concentrations required are between 5.0 and 5.5 mg l^{-1} and the minimum level of oxygen saturation should normally be 80% (Mills, 1971).

(g) Other water quality parameters

The water quality required by the trout should conform to the following:

- the pH must be between 5 and 9.5 (Mills, 1971). Values below 4.5 lead to death of the alevins (Crisp, 1989). A pH of less than 7 is harmful to the spermatozoa and therefore unfavourable to reproduction (Gillet & Roubaud, 1986);
- excessive suspended matter can lead to the clogging of the gills in the most serious cases (Barton, 1977, in Grant *et al.*, 1986); it could be the main reason for low survival rates of eggs and alevins before emerging (Massa *et al.*, 1998);
- ions causing acute toxicity episodes are mainly nitrites (Lewis and Morris, 1986), aluminium (Ramade, 1982) and heavy metals (Alabaster and Lloyd, 1980);
- variations in hardness, concentrations of dissolved oxygen and pH, can lead to indirect toxicity as a result of dissolving and/or reduction, of aluminium, manganese or iron, and by reduction of nitrogenous nitrogen to ammoniacal nitrogen (Bremond and Vuichard, 1973; Brooker, 1981; Vander Borght *et al.*, 1982);
- the trophic levels of water favour the potential biological production of rivers; for the trout this is particularly true for the levels of calcium (Timmermans, 1960; Crisp, 1963; Cuinat, 1971), but also for nitrates (Binns and Eiserman, 1979) and phosphates.

(h) Macrophytes and bank-side vegetation

The submerged macrophytes and the river vegetation, either overhanging or submerged, play a direct role as shelter (Boussu, 1954; Egglishaw & Shackley, 1977) and also in the structure of the habitat (Fig. 1a) by offering visual boundaries (Jenkins, 1969; Haury and Baglinière, 1996). The nature of 'hiding places' offered and their structure are still not well understood and probably depend on the morphology of the species concerned. Stumps and roots constitute specific shelter (Milner *et al.*, 1978; Lewis, 1969; Neveu, 1981).

The indirect roles of vegetation are very varied;

- firstly, the macrophytes modify the flow, slowing the current inside the clumps (Karlström, 1977; Dawson, 1978) and accelerating it between them. This hydraulic aspect is accompanied by action in the erosion and sedimentation cycle (Dawson *et al.*, 1978, Haury, 1985), the trapping of fine sediments (Roussel *et al.*, 1998), limiting the turbidity of the water and, for the water buttercups which regrow in the autumn, the blocking of the redds during the spawning period;

- by photosynthesis, the macrophytes contribute to the oxygenation of the water and modify the pH (Westlake, 1975). When they proliferate, the macrophytes can cause major increases in pH (up to 9), leading to ammoniacal toxicity which results in fish deaths; this was observed following proliferations of buttercups in the Semois, in Belgium (Vander Borght *et al.*, 1982; Peltre *et al.*, 1998), and create anoxic conditions at the end of the night;
- submerged macrophytes constitute a support for a varied invertebrate fauna (Meriaux and Verdevoye, 1983; Tiberghien, 1985). Lastly, the vegetation on the banks supports a terrestrial fauna, some of which falls into the water and is utilized by the trout (Neveu, 1981).

2. Categories used in the study of fish-breeding habitats

The term habitat covers different spatial scales (Fig. 2) which each require precise description:

- the **area of distribution**, a geographical zone where natural or introduced populations of trout are found in the water course, is presented by others (Baglinière, 1990); it will no longer be considered here;
- the **trout zone** is the part of the water course colonized by the trout (preferred living area); the zone used corresponds to a collection of sites actually colonized within the trout zone and represents the value of the habitat in this water course (Fragnoud, 1987);
- the **ecological sector** corresponds to large hydrodynamic units in the 'trout zone', essentially characterized by the same flow and homogeneity of slope (Malavoi, 1988, 1989; Paris, 1989; Haury, 1990), and sometimes by macrohabitat linear density (Ombredane *et al.*, 1995);
- the **segment** (Malavoi, 1989) is an intermediate step corresponding to the portion of the water curse which has one or more sequences; in Breton rivers, for example, these are stretches of river between mill dams (Champigneulle, 1978; Haury, 1988a);
- the **sequence** (Fig. 1b) corresponds to the succession of types of flow, often marked by the alternation of shallow/deep (Ilies and Botosaneanu, 1963, in Welcomme, 1985; Cuinat, 1980; Wasson *et al.*, 1981; Malavoi, 1988) and it is the heterogeneous study area often used as representative of a segment or even an ecological sector;
- the **macrohabitat** (or even morphodynamic unit) is a portion of the water course presenting homogeneity for the principal ecological factors (Malavoi, 1989). In most studies of salmonids, many different typologies have been made (Table 1):
  - — for general cases such as those of Bisson and Sedell (1982, in Welcomme, 1985) for the USA or of Malavoi (1988 and 1989) for high-energy water courses in Eastern France (Fig. 3);
  - — for particular rivers such as those of Champigneulle (1978) for the Scorff in Brittany. These classifications show quantitative limits for the differentiation of macrohabitats.

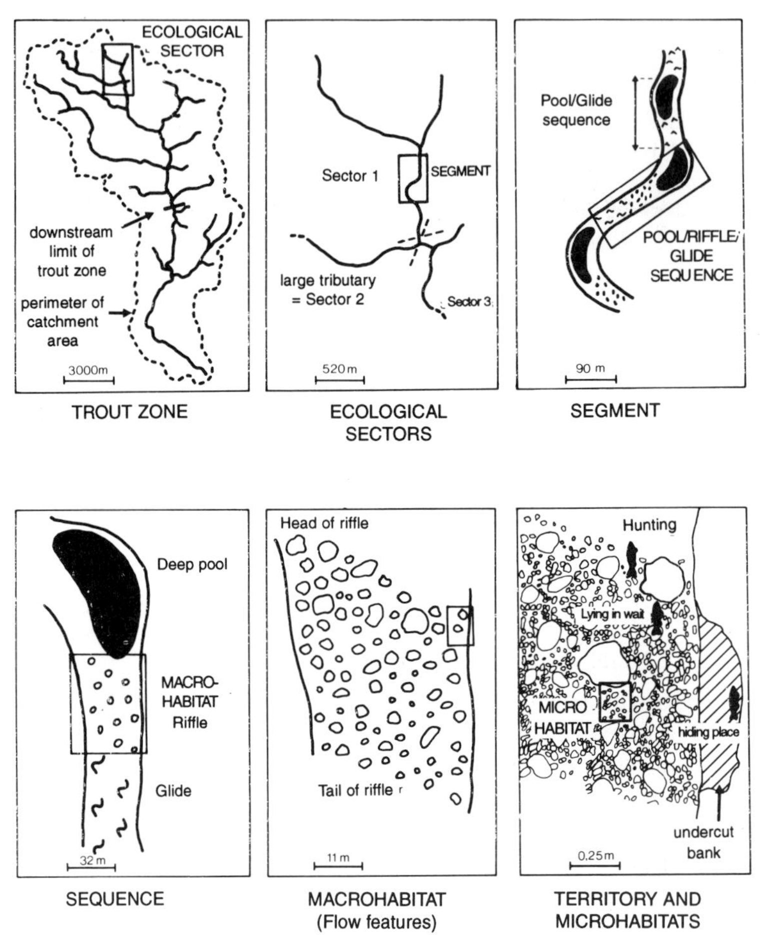

Fig. 2. Scales of study and morphoecological subdivisions in a trout water course (modified from Malavoi, 1989).

Using multivariate and cluster analysis, attempts at microhabitat typologies were made for each ecological zone of a river (Ombredane *et al.*, 1995) or for a geographical area with a hierarchy of suitable variables (Delacoste *et al.*, 1995a).

While the denomination of macrohabitats (riffle, rapids (glide, flat), deep) (Fig. 1c–f) diverges a little, the combination of descriptive factors is effectively the same and essentially takes into consideration, the current speed and type of flow, the depth and the granulometry. Other parameters have also been considered: the macrophytes (Champigneulle, 1978), the morphology of the banks (Malavoi, 1989), the slope (Delacoste *et al.*, 1995a), and so forth.

- the **microhabitat**, the basic physical environment, is homogeneous for granulometry, depth and current speed;

Fig. 1. Views of various scales of study habitat for brown trout fisheries and characteristic macrohabitats. a, microhabitats and vegetation; b, series of macrohabitats; c, riffle; d, rapids; e, pools, slow-moving; f, deep water.

Table 1. Characteristics of macrohabitats and correspondence between the various descriptions. 1: Malavoi, 1989; 2: Bisson and Sedell, 1982 in Welcomme, 1985; 3: Champigneulle, 1978

Name(s)	General characteristics	Authors	Other characteristics
Riffle	significant slope shallow water depth moderate current speed and turbulence	1, 2, 3	3: $h < 30–40$ cm $v > 40$ cm.s^{-1} often in wide section of river
Rapid	slope > 4% water depth low to medium current speed and turbulence very significant	1, 2, 3	1: substrate particle size fine in counter-currents 3: $h > 40$ cm when stream bed is narrowed $v > 40$ cm.s^{-1}
Cascade	slope very steep but irregular alternation of chutes (average depth low, very fast flowing) and pools (depth medium to high, current speed slow to nil) substrate particle size very large (rocks)	1, 2	
Step pool	intermediate between rapid and cascade alternation of transverse ridges of boulders (very shallow, current medium to strong) and small pools or glides (medium depth, current slow to medium)	1	
Chute	due to an outcrop of underlying rock or a geological fault alternation of vertical flow (shallow depth, very fast current) and pools (very deep, current speed slow to nil, fine substrate particle size)	1	

Table 1. (*continued*)

Name(s)	General characteristics	Authors	Other characteristics
Flats	Very uniform habits		
Run, shoal	medium slope shallow water depth medium current speed with little turbulence medium substrate particle size	1, 3	3: h < 40 cm 20 < v < 40 $cm.s^{-1}$
Lotic channel	apart from being deeper, same characteristics as the run or shoal	1, 3	3: h > 40 cm 20 < v < 40 $cm.s^{-1}$
Glide, flat, slick	distinguished from the two types above by a slower current speed and no surface turbulence	2, 3	2: straight, with no obstacles 3: h < 60 cm and v < 20 $cm.s^{-1}$ substrate often sandy
Pools	significant depth current speed low to nil variable substrate particle size	1, 2, 3	3: h > 60 cm v < 20 $cm.s^{-1}$
Pool	at bends in water courses depth decreases towards the inside of the bend	1	
Plunged pool	occurs when a river passes over an obstacle and hollows out the river bed; substrate and depth variable	1, 2	
Backwater pool, scour pool	hollowing out of the stream bed or bank by the eddies caused by the partial obstruction of the bed; significantly deep; low to no current speed, sometimes with a counter current	1, 2	

Table 1. (*continued*)

Name(s)	General characteristics	Authors	Other characteristics
Alcove, lateral scour pool	occurs when water is diverted by a large obstacle in the bank same characteristics as the backwater pool	1	
Dammed pool	medium to little slope water depth increasing from upstream to downstream slow current speed fine substrate	1, 2	1: transverse profile horizontal
Lentic channel	upstream of certain features of the bed causing an obstruction such as a narrowing or bridge etc. same characteristics as dammed pool	1	
Off channel pond	temporary habitat liable to flooding, associated with a gravel dam depth medium to high current speed very slow to nil substrate particle size fine (sand, silt)	1, 2	
Trench pool	long, deep, arising downstream of a large obstacle coarse, stable substrate	2	

- the **home range**, the area used by the trout, consists of a juxtaposition of microhabitats (Allee *et al.*, 1949, in Shirvell and Dungey, 1983; Shirvell and Dungey, 1983) whose roles in the life of the individual are different (Baldes and Vincent, 1969): refuge, rest and feeding. The territory is the 'protected' and defended part of the home range (Gautier *et al.*, 1978).

In fact, the two principal levels taken into account in the study of the relationships between trout and habitat remain the ecological sector and the macrohabitat. They will be

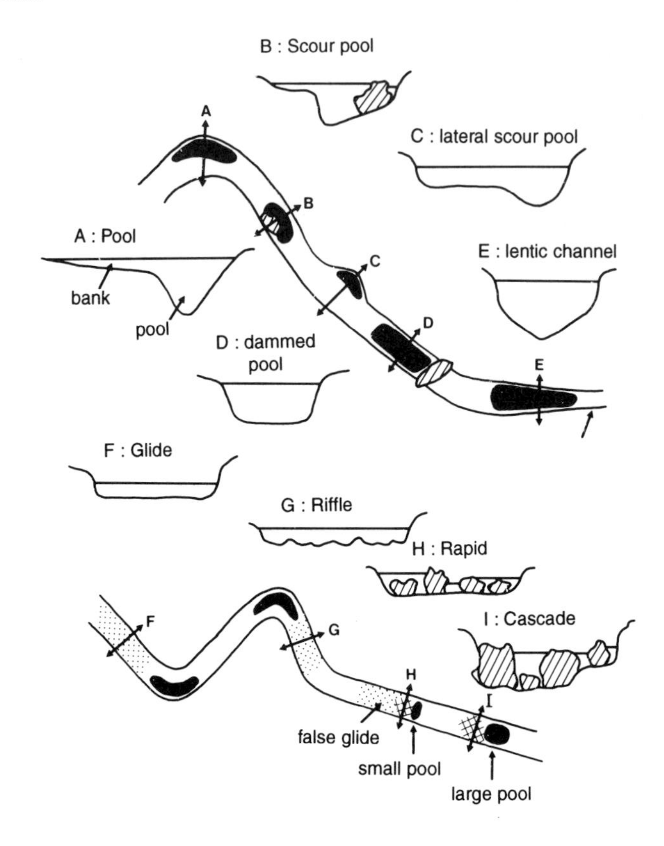

Fig. 3. Main features of habitats in high energy water courses (after Malavoi, 1989).

examined more precisely later. However, ethological studies refer to the home range and its use (Heland *et al.*, 1995; Gaudan and Heland, 1995).

III. TROUT RIVERS

Rivers are generally characterized and compared to each other by geographical, geomorphological, physiographical, thermal, hydrological (mean specific flow and drainage order) or chemical (conductivity, calcium concentration etc.) criteria.

In general, from its source to its estuary, a hydrographic network shows a continuous gradient of physical and chemical conditions as it is supported by the river continuum concept, which is relatively recent (Vannote *et al.*, 1980). Now the concept of a 'fluvial hydrosystem' which considers bidirectional flux, becomes the predominate theory (Amoros and Petts, 1993). Nevertheless, while these concepts are acceptable in theory, in practice it is easier to subdivide the whole into portions (or zones), presenting quasi-homogeneous characteristics from a structural and functional point of view.

More precise analytical approaches allow a zone (e.g. trout zone) to be divided into ecological sectors characterized by the predominance of an age class or by a precise function in the biological cycle of a species.

1. The place of salmonid waters in the longitudinal zonation of water courses

Numerous subdivisions of water courses into longitudinal zones are proposed in the literature. Some of the ecological zones are defined by their salmonid population (as opposed to cyprinid or estuarine populations), see, for example, Dumas and Haury, 1995. The relative importance of the salmonid zone determines whether a river network can be considered as a trout river or not.

Some typologies are based on the hydrographic ramifications of the network and take into account the relative importance of water courses which can be calculated in two ways (Fig. 4):

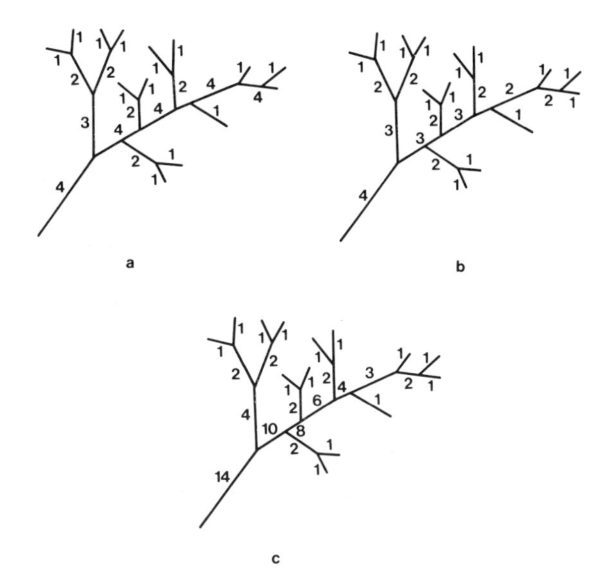

Fig. 4. Numbering systems for drainage orders in a water course. a: Horton, 1945, in Welcomme, 1985; b: Strahler, 1957 in Welcomme, 1985; c: Shreve, 1966 in Beaumont, 1975.

- the drainage order as in Horton (1941, in Welcomme, 1985) and Strahler (1957, in Welcomme, 1981) follows the rule: a little stream with no tributary is order 1; the confluence of two order 1 streams given an order 2 stream, etc. For Horton (system in Welcomme, 1985), the drainage order highest up the hydrographic network affected the assembly of the main stream, whereas the Strahler system did not distinguish this last point;
- the accumulation of the drainage orders after each confluence of the Shreve system (1966, in Beaumont, 1975) leads to a better representation of the relative importance of the flows in the branches.

The utilization of the **stream order** allows a more natural regrouping of the zones of a water course in order to compare the biological communities. Indeed, the factors of slope, mean and maximum depth are positively linked to the stream (Barila *et al.*, 1981). These changes of characteristics accompany an increase in the diversity of habitats (Gorman and Kerr, 1978), which results in an increase in the diversity of the breeding ground (Barila *et al.*, 1981). In practice, Vannote *et al.* (1980) propose a structuring of water courses into three zones defined not only by drainage order but also by the ratio of energy/respiration and by the specific diversity of biological communities.

Another 'universal' zonation of rivers, applicable to the entire world, is that of Illies and Botosaneanu (1963, in Hawkes, 1975). From upstream to downstream, they distinguished three super-zones: the 'crenon' where there are springs, then the 'rhitron' and lastly the 'potamon'; the last two differ essentially in temperature, as in the rhitron, the mean monthly temperature does not exceed 20°C (25°C in the tropics). Lastly, within each super-zone, the criteria of width and drainage order lead to subdivisions and they propose eight zones in total (Table 2).

In 1928, Carpenter (in Hawkes, 1975) defined longitudinal zonations of European rivers, based on the presence or absence of some fish species. The five zones thus determined were later characterized by using physical and biological parameters (fish and benthic invertebrates). The two upstream zones, named 'head stream' and 'trout becks' were mainly defined by the presence of *Salmo trutta.*

Ricker (1934, in Huet, 1962, and in Hawkes, 1975) distinguished two main categories of rivers using the criteria of current speed and width. Within the latter, he defined the salmonid waters as those where the summer temperature did not exceed 24°C; these waters can be divided into two categories according to their $CaCO_3$ concentration.

The zonation described by Huet (1949, 1954) defined four longitudinal zones, for western Europe, each with the name of a fish (from upstream to downstream: trout, grayling, barbel and bream) based as much on the fish associations and the dominance of some species and previously described by two physical parameters: slope and size. A diagram demonstrates this method of classification, known as the 'slope rule' (Arrignon, 1972). Huet (1962) considers as salmonid waters, mixed or cyprinid waters, those in which the temperature does not exceed 20°C, 22°C or 25°C respectively. Further measurements, always as a function of slope and size, allow the distinction to be made between lower and upper trout zones (Arrignon, 1970). In order to extend this sectorization to all of Europe, Arrignon (1972) proposed three diagrams which take into account the climatic and geological differences of the biogeographical regions.

Taking into account the physical and biological heterogeneity of streams, French legislation has defined two categories based on fish fauna: the first category having mainly a salmonid population, the second category with a dominant cyprinid population (Arrignon, 1976). This classification of streams is very different from the biological reality. Since 1978, the Directives of the EC have set guidelines and rules for the quality of salmonid (Table 3) and cyprinid waters. For the present, there is under discussion a European directive on the ecological quality of waters that leads us to distinguish five levels of quality comparing the observed biocoenoses to those expected depending on natural features without any human influence.

The use of computing and multivariate analysis have allowed a finer approach of the

Table 2. Comparison of longitudinal zonations demonstrating the correspondence of zones and the factors on which they are based (after Hawkes, 1975)

Carpenter (1928)*	Ricker (1934)*	Huet (1954)*	Illies & Botosaneanu (1963)**	Verneaux (1973–1974)***
Great Britain	Ontario rivers	Western Europe		Europe
— zoological communities	— flow — width — temperature — $CaCO_3$ concentration	— fish communities — slope — width	— temperature — width — drainage order	— zoological communities (1)
Head stream	/	/	Eucrenon	B0
	Spring creeks	/	Hypocrenon	B1
Trout beck	Swift trout stream	Trout zone	Epirhitron	B2
	Slow trout stream		Metarhitron	B3
				B4
Minnow reach		Brook charr zone	Hyporhitron	B5
				B6
	Warm rivers		Epipotamon	B7
Upper reach		Barbel zone		
Lower reach	/	Bream zone	Metapotamon	B8
Brackish estuary	/	/	Hypopotamon	B9

(1) Habitat types representative of the theoretical typological levels.
* in Hawkes, 1975.
** in Hynes, 1970.
*** in Verneaux, 1980.

Table 3. Water quality standards for salmonid fisheries, from the EC journal 14/08/78 directives adopted by the council of Ministers, 30/05/78 (summarized after Martin, 1979)

Parameters	Guide	Essential
Temperature (°C)		21.5°C (•) 10.0°C (•) during reproduction
	Increase downstream of an outflow:	1.5°C (•)
Dissolved oxygen ($mg \cdot l^{-1}$ O_2)	50% > 9 100% > 7	50% > 9 never less than 6 $mg \cdot l^{-1}$ (•)
pH		6–9 (•)
Suspended matter ($mg \cdot l^{-1}$)	< 25 (.)	
$DB0_5$ ($mg \cdot l^{-1}$ O_2)	< 3	
Nitrites ($mg \cdot l^{-1}$ $N0_2$)	< 0.01	
Ammonia ($mg \cdot l^{-1}$ NH_3)	< 0.005	< 0.025
Ammonium ($mg \cdot l^{-1}$ NH_4)	< 0.04	< 1 (•)
Residual chlorine ($mg \cdot l^{-1}$ HCl)		< 0.005
Total zinc ($mg \cdot l^{-1}$ Zn)		< 0.3 (*)
Soluble copper ($mg \cdot l^{-1}$ Cu)	< 0.04 (*)	

(•) Dispensations possible.
(*) For a water hardness of 100 $mg \cdot l^{-1}$ $CaCO_3$. Standards are established for different levels of hardness.

zonation of rivers. The best-known example is the **biotypological** system, which Verneaux (1973) described, firstly in the Doubs (Jura), afterwards extended to all French, and then European streams (Verneaux, 1980). Thus, ten typological levels, characterized by a community of fish and benthic macroinvertebrates, occur from the source to the estuary of the river course. Statistical analyses allow one to define the typological level of a station:

- the theoretical typological level T (Fig. 5) is calculated from the physical factors corresponding to the morphology of the station and the area of study (Verneaux, 1977a);

- the biological typological level B (Fig. 5) is determined from the observed fish community (Verneaux, 1977b); the trout is present in typological levels B1 to B6;
- the comparison between the theoretical (T) and biological (B) values, allow the development of a diagnosis: a disagreement can be attributable to particular historical conditions of a biogeographical or palaeogeological nature, or to 'abnormal' ecological conditions (pollution of water, habitat deterioration and so on).

Longitudinal zonation of rivers which distinguish the trout zone refer essentially to four physical factors: width, stream order, slope and temperature. Taking into account the common characteristics shared by the zones defined by different authors (Table 2), the biotopes colonized by the trout (*Salmo trutta*) in western Europe are characterized by:

- a width between less than 1 m and 100 m
- low stream order, possibly equal to 1
- slope >4.5 per 1000 (1.4 per 1000 in calcareous regions)
- mean monthly summer temperature less than 20–22°C

Such criteria, which are not absolute, show the wide ecological range of the brown trout.

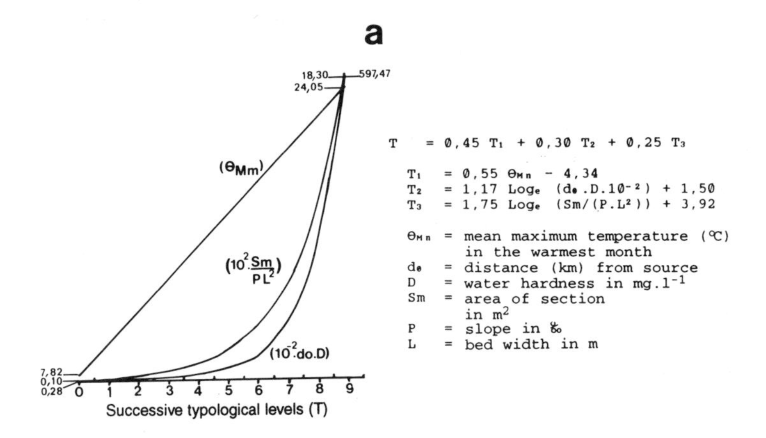

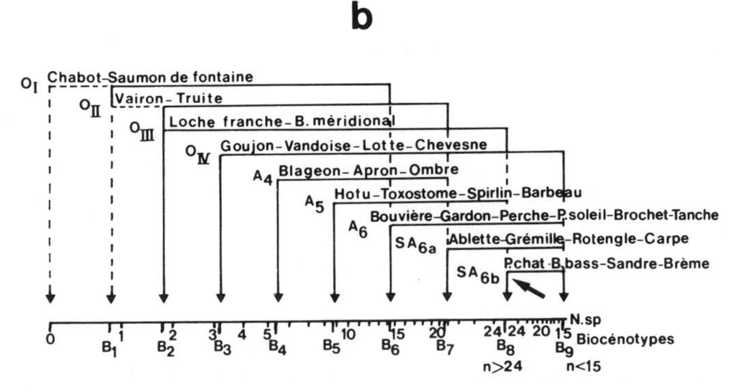

Fig. 5.

2. Longitudinal structure of the trout zone

Numerous studies show a longitudinal variability in trout densities, in the age spectra of populations and in the growth of individuals, under the influence of habitat parameters.

(a) Spatial variability of numbers and cohorts

Studies on the River Scorff (Morbihan, Brittany) since 1972 (Euzenat and Fournel, 1976; Nihouarn, 1983; Maisse and Baglinière, 1990) show an increase in the density of trout from downstream to upstream in the main flow, but equally from the river to the tributaries and sub-tributaries. Work on other water courses—River Bresle in Picardy (Arrignon, 1972), River Redon in the Alps (Melhaoui, 1985), River Elorn in Finistère (Ombredane, 1989)—corroborates these observations.

In addition, there is segregation of the different age classes within the trout zone. Thus, Nihouarn (1983) showed that the 0^+, 1^+ and $\geqslant 2^+$ cohorts represent 21%, 51% and 28% respectively of the total numbers in the main water course of the River Scorff, 48%, 38% and 14% in a tributary (Kernec brook) and 67%, 26% and 7% in a sub-tributary (Talascorn brook). This distribution can be explained by the differences in width of the three systems. In the River Elorn, there is a gradual increase in density of 0^+ and 1^+ fish from downstream to upstream in the main water course, which continues into the tributaries (Ombredane *et al.*, 1988; Ombredane, 1989). In the Kernec brook (flowing into Scorff basin), the same distribution of age classes from downstream to upstream was found (Baglinière *et al.*, 1989). This fish zonation is superimposed on that of the macrophytes which thus combine to provide an integrated descriptor of trout habitat (Haury and Baglinière, 1990). Nevertheless, in some cases, segregation does not exist (Baran *et al.*, 1993).

In order to compare all these observations, total densities and those of each age class of trout are compared in relation with the cumulative order of Shreve (1966, in Beamont, 1975). Thus, in two Breton rivers, Scorff and Elorn, the total densities, more particularly that of the 0^+ age class, increase as cumulative stream order decreases (Fig. 6). The change in the densities in 1^+ parr is similar up to order 2 on the Elorn and order 3 in the Scorff. Individuals older than 2 years are present in maximum densities in sections of the order 3 to 20. These results illustrate the intraspecific spatial segregation and above all the important role of the head waters (cumulative orders 1, 2 and 3) for the production of the juvenile 0^+ age class.

The spatial segregation of 0^+ trout and salmon is linked to the ability of the spawning fish to return to the different parts of the hydrogeographical network, then to hydroclimatic conditions during spawning and the emergence of the alevins and lastly to possible interspecific competition which is difficult to detect (Baglinière and Maisse, 1989; Heggenes *et al.*, 1998). On the River Scorff, the salmon is absent from the headwaters (Baglinière, 1979) and, in the Gaula in Sweden, juvenile trout represent 19% of young salmonids in the main river and 94% in the tributaries (L'Abbelund *et al.*, 1987 in Heggberget *et al.*, 1988). However, sometimes, interspecific competition between 0^+ trout and salmon may explain the reduction in densities of trout when salmon is present. Thus, on the Quillivaron brook, a cumulative order 12 tributary of the Elorn, not colonized by salmon, there are medium densities of trout (mostly 0^+ and 1^+), comparable to those of stations of order 1 or 2 where salmon is present in low densities (Fig. 6).

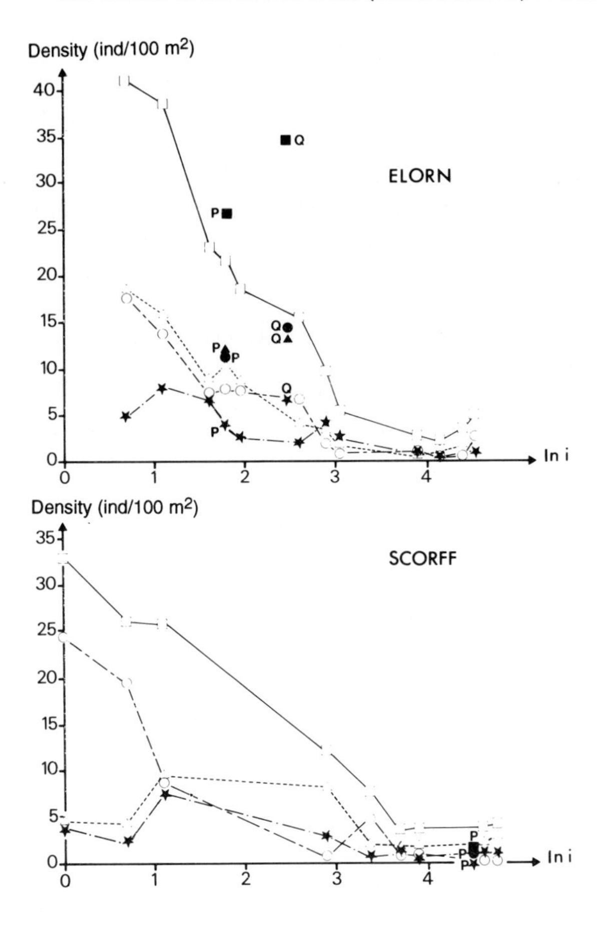

Fig. 6. Longitudinal variation in total density of trout and density by age class (ind/100 m^2), in relation to the Naperian logarithm of the cumulative drainage order of Shreve (lni) for the Elorn (Finistère) and the Scorff (Morbihan).

(b) Growth

Habitat factors are responsible likewise for variations in biological characteristics of trout populations (growth, survival) between biogeographical areas, but also within the same hydrogeographic network.

Crisp (1963) emphasized the low growth of trout in mountainous, oligotrophic and acid environments. Cuinat (1971) positively linked growth to two parameters, Ca concentration and slope/width index, survival (from 0^+ to 4^+ years) increasing with the slope-width index. In a Breton brook, Baglinière and Maisse (1990) showed that shade (coming from a forest) caused a local decrease in growth of juvenile trout.

Numerous examples show the longitudinal zonation of growth patterns of trout (0^+ and 1^+) under the influence of environmental conditions and not population density (Euzenat and Fournel, 1976; Egglishaw and Shackley, 1977). The mean size of individuals decreases not only from downstream to upstream in a river (Euzenat and Fournel, 1976;

Melhaoui, 1985; Ombredane, 1989) but also from the main river to tributaries and sub-tributaries (Baglinière and Champigneulle, 1982; Baglinière *et al.*, 1989).

Thus, trout streams, the trout zone and its different ecological sectors are defined and characterized by physical, chemical and biotic factors, including the presence of the species *Salmo trutta*. But the absence of the species in a river, seen as a 'favourable' zone, can be circumstantial (pollution, climatic parameters, exhaustion of stocks) or structural (isolation of colonized zones).

IV. HABITAT UTILIZATION BY THE TROUT

After the macrodistribution of trout in a stream has been analysed, the particular characteristics of the habitats of different stages of this species are defined, especially with regard to their territoriality.

1. Breeding areas

(a) Spawning zones

Suitable spawning sites for the brown trout are above all determined by the flow and size of the substrate particles and particle aggregation (Baglinière *et al.*, 1979; Witzel and MacCrimmon, 1983; Delacote *et al.*, 1995b). However, the accessibility of potentially favourable areas is a function of the hydroclimatic conditions: increase in flow and water temperature greater than a threshold of 6–7°C (Euzenat and Fournel, 1976; Campbell, 1977; Nihouarn, 1983; Baglinière *et al.*, 1987).

Thus, on the River Scorff, the spawning areas are situated mainly in the downstream tributaries, as well as in the upstream and middle reaches of the main river (Euzenat and Fournel, 1976; Champigneulle, 1978); the downstream limit of the spawning in the main river depends on the flow conditions at the beginning of the spawning period (Baglinière *et al.*, 1979; Nihouarn, 1983); the mean winter flow plays a role in the relative importance of the different spawning zones in this tributary/sub-tributary system. In these brooks, the main spawning zones are characterized by a slope of 1–3%, a mean water depth of 13 cm, a relatively great current speed and a substrate mainly of gravel (< 2 cm) and pebbles (2–5 cm) (Euzenat and Fournel, 1976).

There is no real spatial segregation of spawning zones in the three forms of *Salmo trutta*: resident, lake and sea. Melhaoui (1985) shows, notably for the River Redon, tributary of Lake Léman and for two tributaries of Lake Annecy, that while lake trout reproduce exclusively in the downstream zones, spawnings of resident trout are found throughout the river system. Similarly, Campbell (1977) noted an overlap between the spawning zones of resident and sea trout. More generally, the spatial segregation for spawning of the different species or ecological forms of salmonids is only a reality at the macrohabitat level (Heggberget *et al.*, 1988). In most cases, different-sized spawners occupy different spawning grounds. This results in a relative spatial segregation of breeding salmon and trout on the River Scorff during the spawning period; the downstream zone of the main river is mainly frequented by salmon while the upper part and the tributaries are used by the trout (Baglinière *et al.*, 1979).

Trout spawn in winter at temperatures between 4 and 10°C inclusive, according to Huet (1962) and between 2 and 10.5°C, according to Baglinière *et al.*, (1979). This reproductive activity can apparently be reduced by strong floods (Nihouarn, 1983).

The optimum temperature for the eggs is between 2 and 6°C (Vernidub, 1963, and Kokurewicz, 1971, in Alabaster and Lloyd, 1980) and extreme lethal temperatures are less than 0°C and 15–16°C respectively (Junwirth and Winkler, 1984, and Humpesch, 1985, in Crisp, 1989).

(b) Location of redds within macrohabitats

Spawning macrohabitats are mainly shallow areas—riffles and glides (Baglinière *et al.*, 1979), with spawning beds (redds) generally localized at the head of riffles or at the end of the glide (Witzel and MacCrimmon, 1983), or sometimes at mid-riffle (Euzenat and Fournel, 1976). Favourable habitats are characterized by medium flows (lower limit 10–20 cm s^{-1} (Crisp and Carling, 1989)). Huet (1962) stated that spawning is not possible in areas with zero current speed. These spawning areas are also characterized by medium-sized substrate particles (pebbles, gravel) (0.6–5.4 cm in diameter, according to Jones and Ball (1954), and 2–3 cm mean diameter according to Crisp and Carling (1989)).

Variation in slope, resulting in increased current, is favourable to spawning (Vaux, 1962, in Euzenat and Fournel, 1976; Melhaoui, 1985; Fragnoud, 1987). These hydrodynamic factors prevent clogging by fine sediments and ensure circulation of water within the redd; Crisp (1989) summarizes their beneficial effects on oxygenation of the eggs and the alevins and on the elimination of toxins.

However, in some places, the redds are concentrated in convex banks, protected from violent currents during spate (Baglinière *et al.*, 1979). In addition, redds are often situated close to shelter as shown by Euzenat and Fournel (1976) on a tributary of the River Scorff: 90% of recorded redds were close to hollowed out banks, submerged vegetation, or still and deep areas. Similarly, Witzel and MacCrimmon (1983) observed that 84% of redds were located as close as 1.5 m to the bank (mainly, close to trunks and branches of trees).

Shirvell and Dungey (1983) observe a constancy in the sites of spawning from one year to the next, with a slight extension (resulting in an increase in the variance of mean characteristics of habitats) when the densities of spawners are too great. This results in significant flexibility in the species in its selection of microhabitats for reproduction (Heggberget *et al.*, 1988). Ottaway *et al.* (1981) concluded that the trout appears to spawn everywhere that the substrate is sufficiently fine to be moved. [In an Alpine mountain boulder brook, spawning occurs in a great diversity of microhabitats particularly with more than 50% in places with low water speed. If spawning occurs firstly in riffles, it develops in other microhabitats before salmation of the preferred one (Champigneulle and Lagia, unpublished data).] However, the phenomenon of redd overcutting in favourable areas is common, leading to mortality of the first-laid eggs (Crisp, 1989). Nihouarn (1983) observed an accumulation of redds in one habitat, while others, also favourable, were ignored. This kind of spatial competition at reproduction can become a limiting factor in the overall recruitment in a water course (Chapman, 1966).

Heggberget *et al.* (1988), in a study of spatial segregation of spawning in salmonids, showed that trout spawn closer to the banks than salmon in wide rivers; at a mean distance between 5 and 10.5 m for *S. trutta* versus 6.8–16.9 m for *S. salar*. For both species, they noted the likelihood of redds being located at less than a metre from the bank and in depths of less than 10 cm.

(c) Characteristics of redds

Redds are usually elliptical in shape. They are made up (Fig. 7) of an elongated dome, under which the eggs are found, and of a circular trough immediately upstream (Fragnoud, 1987). Parameters used to describe the location and form of redds are water depth, current speed, granulometry, surface and distance from the bank. Also, one must consider the depth of the egg pocket, which is linked to their vulnerability; a greater depth is protective against spates but also corresponds to a risk of asphyxiation, if sedimentation is significant (Crisp and Carling, 1989). For these descriptors, values found in the literature are not truly comparable, taking into account the different methodologies employed (Table 4). Apart from the methodological disparities, the variability in numbers observed can be explained not only by the diversity of the water courses examined, but also by the size differences in the spawners concerned. Indeed, the area of the redds and the depth of the egg pocket are proportional to the size of the spawners (Ottaway *et al.*, 1981; Crisp and Carling, 1989). Moreover, the largest individuals spawn in the deepest zones with highest current speeds (Ottaway *et al.*, 1981) which was confirmed by Nihouarn (1983) for trout and appears to be generally true for all salmonids. Thus Elliott (1984b) showed that the mean burial depth of eggs is 17 cm for sea trout (length: 25–45 cm) and 4 cm for resident trout (length: 17–30 cm). Similarly, redds of lake trout are larger, deeper and made in larger-sized substrate than those of stream-resident trout (Melhaoui, 1985).

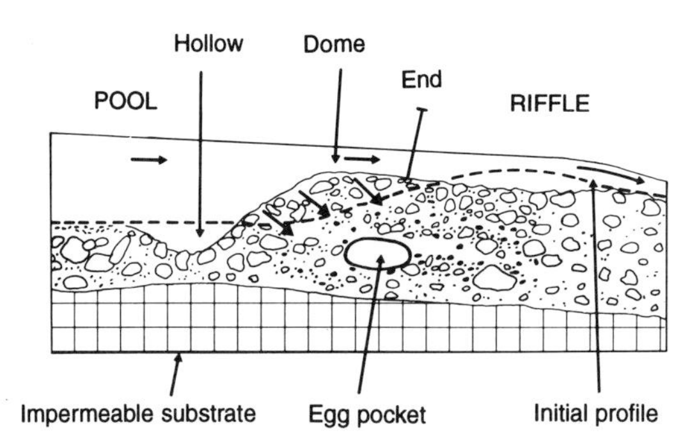

Fig. 7. Diagram of the position of a brown trout (*S. trutta*) redd in a pool–riffle sequence and of the water circulation (modified from Ottaway *et al.*, 1981, and Reiser *et al.*, 1985 *in* Fragnoud, 1987).

2. Preferred habitats of juveniles and adults

The characteristics of preferred habitats of the free stages, studied in most cases at low water, are summarized in Table 5; the values correspond to density optima (e.g. for Baglinière and Champigneulle, 1982) or to the maximum frequency of limited observations of individuals (peaks in Bouee curves of preferenda, 1978 in Fragnoud, 1987, and

Table 4. Main characteristics of spawning sites (redds) of resident brown trount (*Salmo trutta*) as measured by different authors

	Heggberget *et al.*, 1988 (Sweden)		Shirvall and Dungey, 1983 (New Zealand) $l = 0$ to 40 m	Fragnoud, 1987 (Eastern France) $\bar{l} = 11.6$ m	Nihouarn, 1983 (Brittany–France)		Witzel and MacCrimmon, 1983 Ontario (USA) $\bar{l} = 5.3$ m
	Large rivers	Small rivers			$l = 7$ to 21 m	$l = 1.5$ to 3 m	
Depth (cm)							
— mean at centre of egg pocket	43.1 (± 17.9)	50.0 (± 15.5)					
— at snout of fish			31.0 (± 14.4) [6–82]				
— mean depth of redd				23.9 [5–51]	76% between 30 and 60	98% < 30	25.5 (± 8.0) [7–58]
Current speed (cm·s^{-1})							
— at mid-depth above dome	38.3 (± 17.7)	27.4 (± 13.6)					46.5 (± 14.9) [10.8–80.2]
— at 2 cm above dome			39.4 (± 11.0) [5–75]				
— average at the redd				20.5 [0–70]	58% between 48 and 75	75 between 30 and 75	

Table 4. *(continued)*

	Heggberget *et al.*, 1988 (Sweden)		Shirvall and Dungey, 1983 (New Zealand)	Fragnoud, 1987 (Eastern France)	Nihouarn, 1983 (Brittany–France)		Witzel and MacCrimmon, 1983 Ontario (USA)
	Large rivers	Small rivers	$\bar{l}$ = 0 to 40 m	l = 11.6 m	l = 7 to 21 m	l = 1.5 to 3 m	$\bar{l}$ = 5.3 m
Substrate particle size							
— mean diameter	11.5 (± 5.5)	8.1 (± 2.7)	14.0 (± 6.0)				6.9 (± 2.8)
— preferred maximum size				2–64	20–50	2–20	
Area (m^2)	4.7 (± 2.9)	2.3 (± 3.3)		9,32 max.: 1.57			
Trout characteristics	1.8 kg	0.8 kg	42 cm	30 cm			[18–54.5 cm]
No. of observations	36	125	140	620	38	85	112

(± X) = standard deviation of the mean
[a–b] = range of values found
l and $\bar{l}$ = width and mean width

Table 5a. Preferred habitat and microhabitat of 0^+ trout (opt, optimum; pres, presence; dmax, maximal density)

Trout length (cm)	Environment	Depth (cm)	Dominant substrate particle size	Current speed ($cm.s^{-1}$)	Observations	Source
	Swedish rivers	20–30	Gravel		Trout and salmon (spring) Shelter close to banks	10
11–12	River Ontario	43–46 42		2.2 to 4.7 13.6	Winter—cover compulsory Summer—cover not compulsory	12
4.3	Experimental channel		opt: 50–70 mm excl: < 30 mm			9
	Experimental channel	15–90	Sand and pebbles	20 to 30	Aquatic vegetation used as shelter	
	Scorff (Brittany)	dmax: 10–40	Shingly	dmax: > 40	Other salmonids summer low water, close to banks	3
	Scotland	pres: 15–35 excl: < 10–15			Little utilization of undercut banks	11
	Wales	20–30			Riffles and rapids	13
	Sweden	10–12			Sea trout	14
5.2–7	Ireland	opt: 0–15 excl: > 30			Variations between years and sites. Intraspecific competition	6

Table 5b. Preferred habitat and microhabitat of 1^+ trout (for key to abbreviations and references, see Table 5a)

Trout length (cm)	Environment	Depth (cm)	Dominant substrate particle size	Current speed ($cm.s^{-1}$)	Observations	Source
4 to 12	Oignin (Jura)	opt: 20 excl: < 5 > 70	Gravel–small pebbles		71 fished sites	1
4–12	Sémine (Jura)	opt: 30 excl: < 15 > 60	Boulders			1
	Scorff			dmax: ≤ 28		4
	Wales	20–40			Rapids and pools	13
	Experimental channel			25	1 cohort only	5
	Sweden	15–25	Rocks		Sea trout	14
10.6–11.6	Ireland	opt: > 15 excl: < 15				6
	Scotland	pres: (20) 25–35 exl: < 15(19)			Marked utilization of undercut banks	11

Table 5c. Preferred habitat and microhabitat of ⩾ 2-year-old trout (see Table 5a for key to abbreviations and sources of information)

Trout length (cm)	Environment	Depth (cm)	Dominant substrate particle size	Current speed ($cm.s^{-1}$)	Observations	Source
	Scorff	>40	Coarse > 2 mm	fast	Importance of shelter and safety; depth and/or cover	3
21.3	Experimental stream	excl: < 5.1		excl: < 12.2 excl: > 21.3	No cover, small space Directly lit, turbulent	8
> 12	Oignin (Jura)	opt: > 50 excl: > 15 >70	Small and large pebbles	opt: 5–15 excl: > 30	111 fished sites	1
> 12	Sémine (Jura)	opt: > 50 excl: < 5	Boulders	opt: 30 excl: > 100		1
15–20	River Ontario	53–76 43–59		5.7–16.0 5.4–17.2	Winter—cover not compulsory Summer—shelter not compulsory Presence of Arctic charr	12
	Wales	40			Pools	13
	Ireland	> 30 < 15			Slope: unfavourable factor	6
42	6 rivers (New Zealand)	65.0		26.7	140 sea trout	7

of Fragnoud, 1987). However, difficulties exist in the distinction between alevins (0^+), juveniles (1^+ to 2^+) and adults ($\geqslant 2^+$) as many authors mix the age classes and/or do not state the length of the trout studied.

Elsewhere in these macrohabitats, many functions of the life of the trout occur, associated with different microhabitats that make up a territory: study of the latter is, thus, essential for explaining intraspecific competition.

(a) Preferred habitats of alevins and juveniles ($\leqslant 1^+$) (Tables 5a and 5b)

The young stages (0^+ and 1^+) are found, according to most authors, in shallow water; the optimum is between 10 and 30–40 cm (Lindroth, 1955; Baglinière and Champigneulle, 1982; Fragnoud, 1987; Heggenes, 1988b; Heggenes *et al.*, 1998). However, small trout can be found in deeper water (Gaudin, 1981, in Fragnoud, 1987; Bovee, 1978, in Fragnoud, 1987; Raleigh *et al.*, 1984, in Fragnoud, 1987); analysis of the exact position of individuals in deep water allows some estimate of the use of such parts of the river system suggested by these authors. In addition, in winter, Cunjak and Power (1986) have observed small 0^+ trout at an optimum depth of 43–46 cm.

The current speed at the fish's snout is not generally high (< 20 cm s^{-1}) with very slow speeds for 0^+ trout in winter: 4.7 and 2.2 cm s^{-1} for little trout of 11.1 and 11.6 cm (Heggenes and Saltveit, 1989, in Heggenes, 1988b). Within the habitat, optimum mean current speeds are certainly much higher (most often between 20 and 50 cm s^{-1} (Heggenes, 1988b).

Preferred substrate for juveniles is essentially gravel and pebbles (Baglinière and Champigneulle, 1982; Glova and Duncan, 1985; Fragnoud, 1987) and most authors except Raleigh *et al.* (1984, in Fragnoud, 1987) indicate the absence of young trout from very fine substrates.

Lastly, the microhabitats for resting at night, possibly close to or within the hunting territories, where the young tend to hide, appear to be an important component of the juvenile habitat (Kalleberg, 1958). This puts emphasis on the role of shelter played by the substrate and depth (Heggenes, 1988b, 1988c).

These combinations of factors correspond to some morphodynamic units: thus on the Scorff the 0^+ fry are found in riffles and rapids where the current speed is characteristically above 40 cm s^{-1} and the substrate is pebbles (Baglinière and Champigneulle, 1982). However, day and night observations by diving gave information on trout activities and habitat used for these (Roussel & Bardonnet, 1995). The 1^+ trout use the pool–riffle sequence with a diurnal pattern: the pool is used at least twice as much during the night as during the day and corresponds to a resting area (Roussel, 1998).

(b) Preferred habitats of adults (aged $\geqslant 2^+$) (Table 5c)

The habitat of adult trout is mainly characterized by a relatively great depth (20–50 cm) and all the studies showed a link between the size of the trout and the depth of the habitats used (Bohlin, 1977; Egglishaw and Shackley, 1977; Karlström, 1977; Baglinière and Champigneulle, 1982; Heggenes, 1988b): while Kennedy and Strange (1982) found a positive correlation between the density of 1^+ and 2^+ individuals and the depth of the water ($r = 0.92$ between density of 2^+ trout and depth greater than 30 cm).

According to most authors (Bohlin, 1977; Baglinière and Champigneulle, 1982; Egglishaw and Shackley, 1982; Heggenes, 1988c), adult trout prefer relatively slow current speeds. However, Karlström (1977), like Baglinière and Champigneulle (1982) showed their presence in turbulent areas. Similarly, mature individuals in rivers in Ontario occupy deeper and faster flowing parts of the stream than the juveniles (Cunjak and Power, 1986). However, large variations in optimal speeds exist, depending on the water course; e.g. optima found in two adjacent rivers for trout of mean sizes 20.3 and 21.9 cm respectively, were 5.7 cm 3^{-1} and 16.0 cm s^{-1} (Cunjak and Power, 1986). This phenomenon has also been observed in two rivers in the east of France (Fragnoud, 1987).

Optimal substrate type for adults is larger than for juveniles, stones and boulders being preferred (Baglinière and Champigneulle, 1982, Fragnoud, 1987). Big trout hide in deeper waters where they can stay in areas of high sedimentation.

Many authors emphasize the requirement for shelter, possibly with a lot of shade, for adult trout (Lewis, 1969; Cuinat, 1980; Mesick, 1988; Haury and Baglinière, 1990), which is interpreted by Baglinière and Champigneulle (1982) as the result of the search for security by trout aged 2^+ or more. These shelters allow the trout to stay in areas where current speeds are high (Jenkins, 1969; Baglinière and Arribe-Moutounet, 1985) or can increase suddenly and violently (Heggenes, 1988c).

It is possible for one individual to use the broad scale of different currents and depths, according to its activities; calm and deeper areas for rest (holding station or shelter) or lying in wait, medium currents for feeding and hunting, as noted by Cuinat (1980); fish living in a pool frequently go to search for food at its upstream limit (the head of the pool) or downstream, where feeding activity is frequently seen. Shirvell and Dungey (1983) found for adult trout of 42 cm length, the optima for depth and speed respectively to be 65 cm and 26.7 cm s^{-1} for feeding and 32 cm and 36.9 cm s^{-1} for spawning. Areas for resting and/or refuge are characterized by lower flow speed (but sufficient to maintain direction) than those in full current; mean measured currents in shelter (Baldes and Vincent, 1969) for trout of mean length 21.3 cm, vary from 12.2 cm s^{-1} to 31.3 cm s^{-1}, maximum speeds in the habitat being respectively higher than 30.5 cm s^{-1} and 137.3 cm s^{-1}. These microhabitats for resting are found in contact with submerged structures, often roots or stumps (Milner *et al.*, 1978), larger substrates (Heggenes, 1988a, 1988b) and close to banks, submerged banks or overhanging banks (Egglishaw and Shackley, 1982; Baran *et al.*, 1995a) or branches close to the surface (Neveu, 1981). The trout is generally found close to the bottom, but it can leave it for a current speed less than 15 cm s^{-1} and find itself in open water, all while remaining within its territory (Kalleberg, 1958).

(c) Intraspecific competition

Intraspecific competition among cohorts is limited above all by a degree of spatial segregation of ages in the different macrohabitats of a sequence (Bohlin, 1978; Baglinière and Champigneulle, 1982, Nihouarn, 1983; Heggenes, 1988a, b and c; Ombredane *et al.*, 1988; Haury and Baglinière, 1990). But it is also limited by an unequal distribution across the river, more particularly in the large rivers where the juveniles mainly stay close to the banks (Lindroth, 1955; Bohlin , 1977; Karlström, 1977; Baglinière and Arribe-Moutounet, 1985; Roussel and Bardonnet, 1999).

The existence of a marked individual territory for the juveniles (Héland, 1971a, 1971b, 1980), then of a hierarchy among individuals of the same age for occupation of the best positions (Jenkins, 1969; Fausch and White, 1981; Bachman, 1984 Roussel, 1998) leads to intraspecific competition for individuals of the same age.

The size of the territory depends on the size of the fish and is more or less proportional to the cube of the length in the salmonids (Allen, 1969): several dm^2 for alevins (Héland, 1971a), to 58 g m^{-2} for fish greater than 18 cm (Souchon *et al.*, 1989). However, this proportionality is contested by Bachman (1984) who found a mean individual territory size of adult trout in Pennsylvania of 15.6 m^2, this area being acquired during the second, or even the first year and remaining stable thereafter. The size of the territory is also linked to behaviour as it depends on visual isolation, the latter being assured mainly by the substrate type, topography and vegetation. Lastly, it is determined by the availability of food from the invertebrate drift (Chapman, 1966; Grant and Noakes, 1987). However, Kalleberg (1958) observed that territorial limits are fluid, as confirmed by Bachman (1984).

In sheltering, the dominant trout is found in the upstream part of the hiding place, the best situation for profiting from drift; it can tolerate other individuals downstream and off to the side, as long as visual isolation is maintained (Butler and Hawthorne, 1968; Fausch and White, 1986). Elsewhere, in very strong flows, with current speeds outside the shelter of 91 cm s^{-1} to 183 cm s^{-1}, Baldes and Vincent (1969) observed that more trout can remain in the same deep zone. In the absence of shelter, intraspecific competition and aggressive behaviour are more pronounced, which affects the growth of both dominated and dominant individuals (Fausch and White, 1986).

For a given area, there is therefore a maximum number of territories, which can be slightly increased when population densities are high, according to Héland (1971b). The carrying capacity of the environment corresponds to the potential number of these territories. The density of trout will depend, partly on the importance of shelter and 'feeding grounds' (Kalleberg 1958; Allen, 1969; Nihouarn, 1983) and partly on the presence of appropriate areas for night-time rest (Chapman and Bjornn, 1969). Moreover, regulatory mechanisms of density can vary according to the season: Chapman (1966) considers that, in summer, it is most often food/space, or possibly shelter/space, mechanisms which regulate the salmonid populations, while in winter (if there is regulation), it is the quantity of available refuges from strong currents.

3. Interspecific competition

For an allopatric population, intraspecific competition allows the expression of the full ecological extent of the species, whereas in the case of sympatry, the optima for each species are displaced and their ecological extent reduced (Svärdson, 1949, and Nilsson, 1955, 1956, in Kalleberg, 1958). For example, the fact that two species have similar food requirements does not necessarily indicate that there is competition, as some plasticity exists in feeding habits (Kalleberg, 1958).

Interspecific competition occurs in the occupation of space and taking of food; it can also occur through predation. However, spatial segregation among different species in sympatry, demonstrated in the natural environment, can result in just as much choice for individuals (adaptation of the species to the physical environment) as real interspecific competition (Jones, 1975; Hearn, 1987).

(a) Competition with other salmonids

In a water course, there is competition between individuals for microhabitats favourable for feeding, sheltering or access to shelter (Fausch and White, 1981). The spatial segregation among different salmonids in sympatry is a result not only of habitat preference, but also the date of emergence of the alevins and their morphology (Hearn, 1987).

Competition with Atlantic salmon (Salmo salar)

Alevins and 0^+ juveniles of the two species together occupy shallow water such as riffles, where the salmon is usually predominant (Jones, 1975), whereas only trout of this age class are found in deeper waters (Jones, 1975; Baglinière and Champigneulle, 1982; Egglishaw and Shackley, 1982; Kennedy and Strange, 1982). Similarly, if 1^+ individuals of the two species are together in medium depth (slow-moving) waters, the trout again colonize deeper waters (Jones, 1975; Baglinière and Champigneulle, 1982; Egglishaw and Shackley, 1982; Kennedy and Strange, 1982), together with the older individuals (Jones, 1975).

In Swedish waters, Karlström (1977) showed that, whatever age class, salmon juveniles, compared to those of trout, colonize environments where the current speed is greater. Trout are present where the current is slow (< 50 cm s^{-1}) and where salmon are rare or absent. In a Breton river, Baglinière and Champigneulle (1982) showed that 1^+ salmon use rapids (with associated high current and medium depth), while trout of the same age colonize different kinds of habitat. Current speeds in areas where salmon are found are generally greater than 40 cm s^{-1} (Karlström, 1977; Neveu, 1981; Baglinière and Champigneulle, 1982; Baglinière and Arribe-Moutonet, 1985) whereas the trout has less marked preferences (20–70 cm s^{-1}, Karlström, 1977) and 28–61 cm s^{-1} for 0^+ (Baglinière and Arribe-Moutounet, 1985).

Lastly, juvenile salmon are found in habitats with finer substrate composition (pebbles, stones < 30 cm) than the trout (boulders/blocks) (Karlström, 1977).

In allopatry, the trout colonizes zones of higher current than when in sympatry with the salmon (Heggenes *et al.*, 1995), but it is never found where strong current and fine substrate particle size are combined (Karlström, 1977), as confirmed by Baglinière and Champigneulle (1982). In addition, in allopatry, the salmon can establish its territory in environments with weak or low currents or in those with large stones if the current is strong (Karlström, 1977). The spatial segregation observed when the two species are sympatric, can be explained by:

- a better morphological adaptation of salmon to strong currents; their pectoral fins are approximately 50% larger than those of the trout and are capable of deflecting the current (Jones, 1975);

- more aggressive behaviour and thus dominance of trout over salmon (Lindroth, 1955; Kalleberg, 1958; Jones, 1975) allows the trout to establish territories in environments where current speed is low (Karlström, 1977);
- faster growth of trout (Lindroth, 1955; Karlström, 1977).

In wide rivers, densities of trout are lower than those of salmon, because they colonize the entire width, whereas the trout are stationed by the banks (Lindroth, 1955; Karlström, 1977) where the current is weaker. Thus, the predominance of salmon in streams appears to result from their better morphological adaptation and their inter-action with trout (Jones, 1975). In small water courses, the trout is in the majority, because the areas which are more favourable to it (close to banks) are proportionally greater and because the trout chase salmon from zones which are favourable to both species (Karlström, 1977).

In brooks, competition between the two species can be limited by certain degree of visual isolation of individuals as a result of surface turbulence (Chapman, 1966). Elsewhere, the drift which constitutes a large part of the diet of the two species, is important. This large amount of available food reduces the space required by individuals to satisfy their needs (Egglishaw and Shackley, 1982).

In summary, as Heggenes *et al.* (1999) conclude in their synthesis the two species colonize the same kinds of habitat, but at equal age/length, salmon juveniles are found in shallower waters (Neveu, 1981; Egglishaw and Shackley, 1982; Kennedy and Strange, 1982), with stronger current speeds and finer granulometry (Karlström, 1977; Neveu, 1981; Baglinière and Champigneulle, 1982) and more illumination (Karlström, 1977) than for trout. Even though there is predation by large trout on salmon juveniles (Neveu, 1981), competition between cohorts is less pronounced than that between similar-sized individuals of the two species; this allows optimal utilization of the habitats in a river (Karlström, 1977). Lastly, competition affects survival and growth of both species (Kennedy and Strange, 1986).

*Competition with rainbow trout (*Oncorhynchus mykiss*), brook charr (*Salvelinus fontinalis*) and coho salmon (*Oncorhynchus kisutch*)*

These three species colonize the same environments as the brown trout (Karlström, 1977; Neveu, 1981) and all of them show a marked preference for shaded areas with overhanging shelter (Butler and Hawthorne, 1968; Fausch and White, 1981). The rainbow trout is less attached to shelter than the brook charr and especially the coho salmon (Butler and Hawthorne, 1968) for which this is the determining factor, whereas current speed is the most important factor for the rainbow trout (Lewis, 1969). Studies in a Pyrenean mountain stream show that the occupation rate of habitat by rainbow trout is greater for an allopatric population than for a sympatric population with brown trout. The competition is favourable to the brown trout for which the occupation rate of habitat is only dependent on shelter abundance (Baran *et al.*, 1995b).

The brook charr and brown trout appear to be mutually exclusive in Canadian rivers (Karlström, 1977) and those of the north-east United States (Fausch and White, 1981). In the natural environment, the brown trout dominates the brook charr in occupation of the best resting microhabitats, situated close to 'hunting grounds' that both species use alternately (Fausch and White, 1981). However, Fausch and White (1986) showed, in an experimental raceway, that competition was favourable to the brook charr and concluded that dominance is expressed differently according to the size and age of individuals and to the river environment.

Coho salmon alevins in sympatry with those of brown trout in a river are dominant because they emerge 2–3 weeks earlier and because they are larger at emergence (Fausch and White, 1986). In an experimental raceway, for equal sizes, competition for the best feeding microhabitats was also favourable to the coho (Fausch and White, 1986).

(b) Competition with other, non-salmonid, species

*Competition with the eel (*Anguilla anguilla*)*
The eel colonizes very diverse habitats, from flowing and stony areas (small individuals) to still and silted up areas (large individuals) (Jones, 1975, Baglinière, 1979; Neveu, 1981). Like the trout, the eel searches out shelter: it is found close to the banks where aquatic or bank cover is significant (Baglinière and Arribe-Moutounet, 1985), but it prefers the woody debris on the bottom, unlike the trout (Neveu, 1981). Eels can have a marked effect on salmonids, more in terms of competition than predation (Baglinière, 1979).

*Competition with the minnow (*Phoxinus phoxinus*)*
Minnows are generally attracted to deep habitats with low current speed where they cohabit with $\geqslant 1^+$ trout (Baglinière, 1979) and $\leqslant 2^+$ trout (Jones, 1975). Nevertheless, they are also found in more moderate habitats (Jones, 1975; Neveu, 1981), even in shallow areas with strong current speeds (Baglinière and Arribe-Moutounet, 1985). Neveu (1981) suggests that their rarity in deep habitats in the bottom of the Nivelle is a result of predation by large trout, a phenomenon mentioned by Héland (1973).

In an experimental stream, Héland (1973) showed predation by minnows on trout alevins at emergence, which actively avoided their predators while rushing downstream. In addition, he indicated that the microhabitats frequented by these two groups of individuals are not the same and that feeding competition would be limited as their selected prey are different for the two species.

*Competition with the bullhead (*Cottus gobio*)*
Relations between the adults of this species and trout alevins, studied in an experimental channel, are similar (predation, active evasion) to those seen previously for the minnow (Gaudin & Héland, 1984). Brown trout fry avoided bullheads and reduced their feeding activity during daylight (Bardonnet and Héland, 1994).

However, no predation of this kind was found by Crisp (1963) in the natural environment although he stated that trout weighing more than 75 g could feed on bullheads. The bullhead essentially colonizes stony and high current habitats, more specifically the upstream reaches of rivers and tributaries (Jones, 1975; Baglinière, 1979; Ombredane, unpublished data). However, they are also found in slow-moving habitats (Jones, 1975; Baglinière and Arribe-Moutounet, 1985). Because of its habitat preferences, the bullhead is only in partial competition with the trout for food (Crisp, 1963), even though the latter feeds mainly at the surface and the former is benthic, as is the Atlantic salmon, with which it is entirely in competition (Baglinière, 1979). So, within a single riffle, bullheads and brown trout coexist because of a patchy distribution (Roussel and Bardonnet, 1997).

Some other species colonize the same environments as the trout. The stone loach (*Nemacheilus barbatulus*) is found alongside 0^+ trout juveniles in medium depth habitats (Jones, 1975; Baglinière, 1979; Baglinière and Arribe-Moutounet, 1985). Gudgeon (*Gobio gobio*) are found with older trout in deeper habitats where flows are moderate (Jones, 1975; Baglinière, 1979; Neveu, 1981).

These interspecific competitions are thus a significant parameter of habitat utilization by the trout and so in the carrying capacity of an environment, but they are still poorly understood because these 'accompanying species', that often constitute the majority of the biomass in some habitats in salmonid rivers (Baglinière, 1979), are little studied. Lastly, Hearn (1987) emphasized that when sympatric populations are reduced by natural conditions (drought or flood), interspecific competition becomes very weak, almost non-existent.

V. TEMPORAL VARIATIONS IN HABITAT AND EFFECTS OF HUMAN ACTIVITY

The physical, chemical and biotic characteristics of habitats vary according to daily and seasonal cycles, but they can also be changed gradually by evolutionary processes, or more radically under the actions of man.

1. Natural variations

(a) Daily variations

The daily cycle is marked by cyclical variation in both light and temperature and has consequences for the activity of the trout.

Feeding activity is maximal at dawn and dusk and null at night (Lindroth, 1955; Roussel and Bardonnet, 1995, 1996); trout, hunting 'by sight', stop all feeding at around 0.02 lux (Kalleberg, 1958) probably due to the lack of efficiency in catching prey (Rader, 1997, in Roussel, 1998). During the night, salmonids detect the current both by sight and feel, whence the importance of substrate particles size and current (Arnold, 1974, in Ottaway and Clarke, 1981), while many displacements, such as the downstream movement of alevins, occur at night (Héland, 1980; Ottaway and Clarke, 1981).

The alteration between day and night also results in modifications of other biological activities (plant photosynthesis, invertebrate drift) and certain physical factors (dissolved oxygen tension, temperature).

The thermal amplitude varies according to the local conditions: it is noticeably reduced in shaded or forested waters (Edington, 1966) and in the mountains; conversely, in well-lit or south-facing environments, there can be major daily fluctuations (Crisp *et al.*, 1982). Excessive temperature and daylight reinforce the sheltering behaviour of the trout; when the water is too warm, they shelter in the deeper areas downstream of the currents.

(b) Seasonal variations

Seasonal variations in temperature, which are different depending on the kind of spring (Westlake *et al.*, 1972; Crisp *et al.*, 1982) have an influence on the migration and spatial distribution of trout.

When the water temperature drops below a minimum value, the trout find a wintering shelter, in slow-flowing, deep environments, with cover (Chapman and Bjornn, 1969; Bjornn 1971; Cunjak and Power, 1986); when the water temperature remains above 7°C, the same phenomenon does not appear to exist (Allen, 1969). Excessively low temperatures (< 4.5°C) inhibit the displacement of alevins (Raleigh, 1971, in Ottaway and Clarke, 1981) and reduce their aggressive behaviour. Similarly, they inhibit spawning migrations (Euzenat and Fournel, 1976).

From the end of winter and throughout spring, movements of the population occur in order to satisfy nutritional requirements and to find new feeding territories (Kalleberg, 1958; Chapman and Bjornn, 1969; Roussel and Bardonnet, 1997; Ombredane *et al.*, 1998).

During low water, both current speed and surface area of water decrease, resulting in deposition of suspended matter, mainly silt (Haury, 1985; Haury and Baglinière, 1996), less oxygenation and increased temperature. The hiding places are restricted therefore to pools which are often the only areas of water remaining. Biological low waters occur when slow currents represent more than 70% of the water surface (Dumont and Rivier, 1981). Exceptional droughts have a strong negative effect on recruitment, growth and population dynamics (Elliott, 1984a, 1985; Gourand *et al.*, 1998; Haury and Richard, unpublished data).

The effect of spates depends on the season when they occur. The repercussions of spring spates have been studied experimentally by Ottaway and Clarke (1981): strong ones, reshaping the bottom, can destroy the redds and alevins emerging from the redd are very vulnerable to the current, especially when the variations in current are sudden and violent. For older fish, sheltering behaviour limits the negative effect of spates (movement of individuals downstream) (Baldes and Vincent, 1969; Heggenes, 1988c).

Seasonal variations in macrophyte cover also lead to seasonal diversification of the habitat and modification of trophic resources. Jenkins (1969) observed changes in the positioning of trout in relation to macrophyte cover: he concluded that the social structure and delineation of territories are more stable in rocky streams than in water courses with much vegetation. Some studies dealing with relationships between macrophyte cover and fish populations have shown that great seasonal changes in habitat structure and fish micro-distribution occur in a sequence, where there is much spring vegetation. While in riffles few salmonids were found during spring, they were much more numerous during autumn, during low water level when macrophytes had almost disappeared (Baglinière and Arribe-Moutounet, 1985; Haury and Baglinière, 1996).

2. Impact of human activity on the habitat

The impact of human activity on the habitat can be expressed in terms of deterioration of water quality and modifications of physical characteristics of the habitat.

The effects of 'pollution' are extremely variable and cannot be generalized, as their impact depends not only on the nature of the effluent, but also on the receiving water course, and on the frequency and duration of the pollution. Fig. 6 well illustrates this phenomenon with an effluent from a fish farm: on the River Elorn, it resulted in an increase in trout densities (made up of 0^+) and an increase in growth (Ombredane *et al.*, 1988); on the River Scorff, such an effluent also led to increased growth, but conversely a decrease in trout numbers (Baglinière, 1983). Also all the changes in water quality are not detailed in these studies. However, it is important to note some effects of suspended matter which come from different human activities and particularly from agricultural practices. The effects of fine particles on embryo-larval stages of trout were studied within brooks localized in agricultural basins. The sedimentation of these particles reduces oxygen inside the redds and more egg mortality occurs between the eye stage and hatching (Massa, unpublished data).

(a) Physical changes to bed and banks

Straightening and widening results in the disappearance of submerged shelters, reduction in bank vegetation (80% of water courses studied by Elser (1968)) and a very large reduction in the heterogeneity of the bottom. The largest substrate particles (a few stones and gravel remain) are clogged by sediment resulting mainly from the works and then from the erosion of the banks. One of the main dangers for the trout is the reduction in summer flow and increase in water temperature. Strong light on the bed of the stream results in algal proliferation at first and often rapid plant colonization by helophytes (Dutartre and Gross, 1982). In managed sectors, Elser (1968) observed a 74% decrease in trout numbers as well as a change in proportion of 'accompanying species'. In fact, widening and straightening has the same impact as organic pollution on the river (Huet and Timmermans, 1976).

Restoration which involves only localized modification of the bed of the stream and removal of of excess vegetation, limits these unwanted effects (Cuinat, 1980; Gross and Dutartre, 1980) but leads to regular management of bank trees.

Cleaning of rivers constitutes the initial phase of the establishment of a maintenance programme. On the Scorff, it results in an increase in areas of flowing water (from 20% to 50%), a freeing of coarse substrates favourable for spawning and clearing out (–35 to 65% of fine sediments, resulting in better oxygenation of the bottom: +1 to 2 mg l^{-1} of O_2), an influx of light and increase in macrophytes cover (increased by two to five times), a slight modification of the temperature regime (+2°C at extreme low water), and an enhancement of invertebrate diversity and production (Champigneulle, 1978).

The consequences observed, in salmonid populations are (Champigneulle, 1978):

- a decrease in trout 1^+ or older (–70%) that is attributed to removal of hiding places, but that equally could be due to decreased import of invertebrates from the banks;
- an increase in 0^+ salmon (+300%).

Regular maintenance of the banks must be carried out after such a cleaning, if the creation of new imbalances is to be avoided.

Cutting macrophyte vegetation has moderately negative effects on fish populations (reduction in potential food items and shelter, notably). Also, in some English regions, weed-cutting must occur after the emergence of a mayfly (*Ephemerella ignita*) which is a particularly important component of the food of trout (Armitage, FBA, Wareham, UK, personal communication). In experimentally managed riffles where water crowfoot had been cut away from half the area, densities of brown trout were much lower in the bare parts, probably owing to lack of shelter and increased competition with salmon (Roussel *et al.*, 1998). Elsewhere, if the number of trout territories increases with the macrophyte cover by offering visual boundaries, then the available area for salmonids decreases. So, too much vegetation could be inauspicious for salmonids (Haury and Baglinière, 1996).

Extraction of gravel from the bed of water courses has as its consequences (Cuinat, 1980):

- the resuspension of fine sediments leading to a reduction of photosynthesis and a clogging of the bottom;
- deepening and uniformization of the river bottom;
- the destabilization of banks and stream beds and, notably, the disappearance of the alternating pool/riffle pattern;
- the reduction or disappearance of rheophilic species such as trout.

(b) Modifications of flow and dams

The historic management of rivers for different uses, such as hydraulic energy (millstreams) or eel fisheries, has resulted in a diversification of flow and water height conditions within inter-dam areas, thus increasing the number of morphodynamic units (Champigneulle, 1978; Haury, 1988a, 1988b).

The impact of impassable **dams** is certainly known to thwart the migration of anadromous species such as the sea trout (Elliott, 1989) and to render certain potential spawning zones in the tributaries completely inaccessible to trout (Brooker, 1981), isolating populations (Elliott, 1989) or competing species (Kaeding, 1980). However, Brooker (1981) indicated that the regulation of flow and the prevention of large spates can have a positive effect on spawning and recruitment of young trout.

Water abstraction reduces the flow, notably in summer, and can lead to a great increase in temperatures (Brooker, 1981). This effect is the more important the narrower the bed is in relation to the remaining flow. Quite often, this reduction in flow results in proliferation of macrophytes (Decamps and Capblancq, 1980; Khalanski *et al.*, 1985; Haury *et al.*, 1996). Yet in areas of least flow, the arrival of better-quality water from the tributary can lead to an improvement in certain habitat parameters (Haury, 1987).

In addition, **discharge of water** has two effects:

- rapid increase in flow resulting in mechanical shocks which can lead to the washing downstream and/or mortality of juvenile trout (Elliott, 1987). Hydropeaking management leads to a forced drift of post-emergent trout fry (Liebig *et al.*, 1998). Elsewhere, significant quantities of sediments can be resuspended. However, the restitution of water when the water levels are low attenuates the harmful effects of summer droughts and can contribute favourably to fish production;
- the discharge of water, whose temperature or physico-chemical quality is modified, leading to a change in habitat conditions, often detectable for several kilometres downstream, with possible fish mortalities (Brooker, 1981).

One solution for ensuring that discharged water is of adequate quality is to calculate the flow required to maintain fish life. The idea of a 'biological reference flow' introduced by Dumont and Rivier (1981) was based on the hypothesis that biological equilibrium is optimal when maximum diversity of the flow patterns is attained. The authors consider that there must be equal distribution of slow ($v < 30$ cm s^{-1}), medium ($30 \leqslant v \leqslant 80$ cm s^{-1}) and fast ($v \geqslant 80$ cm s^{-1}) flows. Models simulating the habitat conditions as a function of flows (PHASBIM—Bovee and Milhous, 1978; Souchon *et al.*, 1989) aim at quantifying the effects of hydraulic modifications.

(c) Habitat management by positive arrangements

Some arrangements can ameliorate human intervention within an environment, or even improve the 'natural state of rivers'.

In some uniform watercourses, with muddy, silted or sandy substrates, the creation of irregularities in flow with deflectors can possibly allow the build-up of gravels and thus increase the productivity of salmonids (Hunt, 1969; Arrignon, 1976; Milner *et al.*, 1985). Similarly, the effect of sand removal is favourable to trout populations (Hansen *et al.*, 1983). The installation of deflectors, accompanied by vegetation growth and then management of the trees and shrubs along the banks, results in narrowing the stream and deepening the bed, which is favourable to larger individuals, and is particularly beneficial in drought years (White, 1975). In the case of deforestation, creation of shelter, mainly by immersion of tree branches, is recommended, and too drastic cleaning of the banks should be avoided (Champigneulle, 1978). Particular attention has already been paid to riparian forest and woody debris, especially in mountainous and hilly watercourses (Piégay and Maridet, 1994; Thévenet, 1998). The woody debris changes flow, creates dams, gives shelter to fish etc. Thus, comparing stretches with and without riparian forest, Maridet and Souchon (1995) found fish densities (mainly brown trout) of 65 g/m^2 and 13 g/m^2 respectively in weighted usable areas measured using the PHASBIM method.

However, habitat management, for example by inserting small dams, deflectors and artificial hiding places, can modify the equilibrium of populations, as Saunders and Smith (1962) showed with the brook charr. By following such developments, they noted a decrease in the number of young individuals and a doubling in the number of older brook charr, which has since been confirmed for various species of salmonid (Hunt, 1969; White, 1975).

The necessity of chronological series and of long-term monitoring in order to judge the efficiency of management of a river is therefore essential and has already been emphasized by White (1975) and Hunt (1976).

VI. ESTIMATION OF TROUT POPULATIONS FROM HABITAT CHARACTERISTICS

Some authors (Cuinat, 1971; Philippart, 1978, in Welcomme, 1985; Lanka *et al.*, 1987) have attempted to model fish production (kg/ha/yr) or biomass (kg/ha), more precisely the trout for sections of the trout zone, using the physical factors which characterize them. Others (Binns and Eiserman, 1979; Raleigh *et al.*, 1986, in Heggenes, 1988b) have established models for the estimation of trout stock (mainly biomass) *in situ* in a macrohabitat, mainly from its characteristics.

1. Models for the trout zone

The best-known models in Europe are those of Léger–Huet (Léger, 1910, and Huet, 1964, in Arrignon, 1970), for the theoretical estimation of fish production, of Phillipart (1978 and 1978a, in Welcomme, 1985) for biomass of fish, and of Cuinat (1971) for biomass and production of trout (*Salmo trutta*) in French rivers.

Lanka *et al.*, (1987) established predictive models for biomass of 'trout' (genera *Salmo* and *Salvelinus*) for rivers in Colorado, Missouri and Wyoming. These authors use, for similar accuracy of estimations, either physical parameters characterizing a water course (including a pool-riffle sequence), or geomorphological variables of the river and its basin. Moreover, they distinguish between forested streams and 'uncovered' ones.

Table 6, which summarizes the many parameters used in these models, shows the importance of the role that Ca content of the water plays in the fish production of a river. The calcium effect was shown by Timmermans (1960) and Crisp (1963). Timmermans (1960) compared acidic water courses where the mean biomass of trout is 74 kg ha^{-1} with alkaline waters where the mean trout biomass reaches 166 kg ha^{-1}. Other parameters such as riverside and aquatic vegetation in 'uncovered' environments favour production (Mills, 1967).

2. Models for macrohabitats

Grant *et al.* (1986) emphasized the importance of a precise description of habitats before a predictive model can be made. These authors have shown that for similar units, characterized by 20 or so factors and in the same watercourse, no significant differences were found in the salmonid populations. They could thus show actual changes in the populations due to a change in the habitat.

A hydrological simulation model (PHASBIM) was developed by Bovee and Milhous (1978) and Bovee (1982 in Souchon *et al.*, 1989). The major habitat parameters (depth, current speed, granulometry) were assessed at the same time as the trout population at a given point in time. From a change in flow, which could be caused by hydraulic management of the watercourse, the model deduced the modifications to depth and current speed parameters. This led to the evaluation of the impact of such a perturbation of the habitat on trout, and therefore on its populations. Following a similar process, new developments and applications (repopulation) have been proposed by Souchon *et al.* (1989) under the name 'microhabitat method'.

Binns and Eiserman (1979) established the 'Habitat Quality Index' (HQI) which predicts, for Wyoming rivers, the stock of all mixed salmonid species (*Salvelinus fontinalis*, *Salmo trutta*, *Oncorhynchus clarkii*, *Oncorhynchus gairdneri*). This model explains 97% of the variation in autumnal biomass of salmonids between the macrohabitats, mainly from their specific characteristics: current speed, percentage of bank with shelter, width, proportion of banks eroded, nature of substrate, nitrate concentration. Added to these variables are parameters affecting spatio-temporal factors which limit the salmonid biomass present in autumn: mean daily flow at low water, annual variations in flow,

Table 6. Various prediction models for fishery stocks, salmonid or trout stocks, in relation to the characteristics of the section of a water course

Author	Parameter estimated	Formula for model	Variables used
Léger-Huet *	Fishery production P ($kg \cdot ha^{-1} \cdot yr^{-1}$)	P = B. W. K.	W = width of water course B = biogenic capacity K = production coefficient, a function of temperature acidity, types of fish and their age
Cuinat (1971)	Biomass of trout T ($kg \cdot ha^{-1}$) in Massif central (France)	$T = 41.26 + 16.4\ S/W$ or $T = 18.07\ S/W + 7.28\ Y - 3.37$	S = slope W = width Y = calcium concentration index
Phillipart (1978) **	Biomass of fishes for Belgian rivers ($kg \cdot ha^{-1}$)	$I = 6.9569\ W^{1.8456}$ or $I = -295 + 0.19\ A + 5.72\theta + 17.585\ S + 16.8\ W$	W = mean width θ = mean temperature S = mean slope A = alkalinity
Lanka *et al.* (1987)	Biomss of salmonids M ($kg \cdot ha^{-1}$) for Colorado, Missouri and Wyoming (USA)	M = f(W, W/H, S) or M = f(RT, RC, P, R, R/O, GS, ID) or M = function of all variables	W = mean width H = mean depth S = slope R = altitude of river section RC = central altitude of catchment area P = perimeter of catchment area R = Relief = source altitude/river section altitude O = distance from source GS = mean slope of water course ID = drainage density

* Léger, 1910 and Huet, 1964 (in Arrignon, 1968)
** Phillipart 1978 and 1978a (in Welcomme, 1985)

maximum temperature at low water. This model, applied to watercourses of southern Ontario, could not explain more than 9.2% of the variation in biomass between the different habitats despite the fact that certain factors were taken into account, whose influence had been demonstrated to be statistically significant (Bowlby and Roff, 1986). In addition, the relationship (regression and discriminant analysis) between biomass of salmonids and that of the micro-community (micro-organisms, invertebrates) or that of piscivorous fish was demonstrated (Bowlby and Roff, 1986).

Another model was developed by Raleigh *et al.* (1986, in Heggenes, 1988b), the 'Habitat Suitability Index' (HSI), for evaluating the potential carrying capacity of a habitat. This model involves 18 variables to describe the environment and is particularly relevant for the prediction of stock size for brown trout.

It is therefore possible to model the trout production for a given ecogeographical region, using characteristics, of which the most important seem to be depth, current speed, substrate and percentage shelter (Heggenes, 1988b). However, the routine use of such models does not seem to be very realistic because of the complicated nature of the protocols which must be worked through in order to acquire the data.

Lastly, the 'synthetic' curves of preferenda are too general and imprecise to be realistically usable, taking into account the movements of adults for reproduction, differences in positioning as a result of activities, the strong or weak effects of competition and variations between hydrographic reservoirs (Fragnoud, 1987; Heggenes, 1988b). This ecogeographical variability and eco-ethological plasticity of the species explains the limited value of such models and already renders pointless the exercise of putting together a general model for predicting a stock of salmonids, even the trout. Nevertheless, the setting up of models constitutes the first step for putting in place more simple tools for use in management.

New modelling developments

Many new developments in trout modelling using habitat features have been made to measure the effects of hydropeaking (Ginot, 1995; Liebig *et al.*, 1998) and of water abstraction (Johnson *et al.*, 1995) on suitable habitats for trout reproduction (Delacoste *et al.*, 1995b) together with critical remarks on these models (Fausch *et al.*, 1988; Pouilly and Souchon, 1995; Barnard and Wyatt, 1995; Barnard *et al.*, 1995).

A better understanding between fluvial hydraulics (studied at the reach scale) and ecology (needing local scale and microhabitat studies) should be attained, using statistical models of variables, such as depth, current velocities (Lamouroux, 1995). Now, some models can be easily used by operators to determine the carrying capacity of any stretch, such as E.V.H.A. (Ginot, 1995), but the field collection of data remains very long and tedious (Le Coarer and Dumont, 1995). In any case, the pertinency of preference curves should be reestablished in each ecoregion, as Delacoste *et al.* (1995b) showed for the reproductive habitat, which is proposed by Ginot.

A new concept was presented by Capra *et al.* (1995) to explain the differences of Weighted Suitable Area depending on flow: the Continuous Under Threshold Habitat Duration which is the real limiting factor during the reproduction stage as well as emergence of fry. Thus an improvement of model accuracy is given by using habitat chronicles together with detailed biological surveys. So this tool includes time-effect and gives a dynamic view of carrying capacity.

Presented as a tool for the manager, HABSCORE (Barnard and Wyatt, 1995; Barnard *et al.*, 1995) uses empirical statistical models of regression to estimate fish population, especially for brown trout (0^+, $>0^+$ and <20 cm and >20 cm), involving 130 independent variables.

A model on population dynamics regulated by habitat features was used to evaluate the impacts of various types of water use on the aquatic ecosystem, involving a LESLIE matrix and a PHASBIM method (Gouraud *et al.*, 1998).

VII. CONCLUSION–DISCUSSION

The description of the trout's habitat (from the individual to the population) in a river depends therefore on the scales of space and time considered. The acquired results point out some questions as to their representational value. Some research projects must be developed to arrive at an understanding allowing true management of populations in relation to the habitat.

1. Results acquired and their limitations

Several major points appear:

(a) A complementarity of studies in natural and experimental environments

Experimental studies on effects of abiotic factors are interesting because of the fundamental understanding, precise and reproducible, that they bring. They cannot elucidate the exact relationships between the environment and the population which colonizes it, but only show which limiting factor could be compensated by another parameter in uncontrolled conditions. Also it is necessary to conduct studies in a complementary way in the 'natural environment' on 'wild' populations and in controlled or artificial environments.

(b) Partial agreement of scattered results from preferred habitats

As a whole, the authors agree on the main characteristics of preferred habitats for the different stages of the trout. However, important divergences appear, from which arise quantified characteristics for the determination of the statistical relationship between the trout and its environment. This can be explained by the lack of information about ecogeographical differences between rivers and populations studied, but equally by the great ecological plasticity of the trout (Heggenes *et al.*, 1999). In fact, it is difficult to compare results obtained from a few diverse watercourses, with the wide range of study methods and protocols for different seasons (and flow conditions), as emphasized by Fragnoud (1987) and Heggenes (1988b).

(c) The problem of spatial integration of local results

It is imperative to ensure that all spatial scales of the trout's habitat are taken into account. A rigorous sampling plan and great prudence in the extension of local results at higher study levels are required because of the complementarity of:

- ecological sectors for completing the biological cycle;
- macrohabitats for holding of different cohorts;
- microhabitats to satisfy all the needs of the individual.

(d) The importance of temporal variations in the habitat–trout relationship

The dynamic study of habitat requires long-term research to determine the variations in function of the ecosystem and of the use of the environment by the trout. Notably, 'threshold' or 'catastrophic' effects of exceptional droughts in the natural environment (Elliott, 1984a, 1985) or major spates which destroy the redds (Crisp, 1989) are taken into consideration. Lastly, the effects of severe accidental pollution or management which destroys, at least temporarily, the population in an entire water course, must be considered.

2. Perspectives

The objectives of studies on the habitat of the brown trout are summarized by Elliott (1989) as 'to arrive at a mathematical model for predicting the optimal densities of trout of different populations, the maximum growth rate in different habitats and the effects on wild populations of natural events (such as drought or spate) and human activities'. One way of modelling requires the estimation of habitat value (potential usage) of a river used by trout.

Consequently, diverse avenues of research, requiring certain methodological definitions, can be identified:

(a) It is necessary to achieve a standardization of methods of habitat description for the different scales of study, in order to obtain comparable results (nature of the factors examined, mode and frequency of measurements). At this level, parameters related to macrophytes (amount of cover, biological type) for delimiting ecological zones, macrohabitats and also microhabitats, are recommended Haury and Baglinière, 1996. Methods for quantifying shelter and estimating their value for the fish also need to be further developed. In all cases, after an investigative phase, only easily measured parameters which guarantee reliable results should be retained.

(b) The modelling of relationships between habitat and trout population remains an important objective; the main types of model and their ecogeographical validity must be determined. The hierarchy and relationships between factors will certainly vary according to the type of water course, as has been shown for growth in trout. It appears necessary to develop the existing hydrological models with variables linked to the territorial behaviour of the trout, and the trophic aspect of the habitat. Macrophytes are, locally, good evidence of this trophic quality, but other factors should equally be taken into account including physical and chemical characteristics of water and invertebrates; ethological features should be taken into account to determine preferendum curves (Héland *et al.*, 1995).

(c) Studies following changes in habitat within seasonal cycles are undertaken to determine the most limiting factors for some periods and for each stage.

(d) The impact of human activities on fishery habitats linked to different uses and managements of water courses are still poorly understood. All activity on a watercourse which tends to render it more uniform is in conflict with the aims of management of aquatic environments (Mortensen, 1977). Detailed analysis of the impacts of arrangements would allow, via the development of 'practical ecological guides for use by managers', the limiting of unintentional effects of human intervention.

(e) In view of the identified limiting factors, experiments in improving the habitat (installing shelter, cleaning of rivers, fight against pollution) should aim, together with users and managers of watercourses, to achieve better management of the stocks via that of the habitat, the determining factor being the carrying capacity of a river.

However, knowledge of brown trout stocks across the whole extent of its habitat is still limited, in comparison to that of the Atlantic salmon. This fact can be attributed to the wide ecological range of the trout within the water course, possibly linked to strong genetic diversity and very varied demographic strategies.

BIBLIOGRAPHY

Alabaster J. S., Lloyd R., 1980. *Water Quality Criteria for Fresh Water Fish*, Butterworths, London, 297 pp.

Allen K. R., 1969. Limitations on production in Salmonid populations in streams, The University of British Columbia Institute of Fisheries. Symposium on Salmon and Trout in Streams, 1968, 3–18.

Amoros C., Petts G.E., 1993. *Hydrosystèmes fluviaux*. Masson, Paris, 300 pp.

Arrignon J., 1970. Aménagement piscicole des eaux intérieures. S.E.D.E.T.E.C. SA, Paris, 643 pp.

Arrignon J., 1972. Zonation piscicole de quelques cours d'eau normands (France). *Verh. Int. Verein. Limnol.*, **18**, 1135–1146.

Arrignon J. 1976. Aménagement écologique et piscicole des eaux douces, 3rd edn, Gauthier-Villars, Paris, 340 pp.

Bachman R. A., 1984. Foraging behavior of free-ranging wild and hatchery brown trout in a stream. *Trans. Am. Fish. Soc.*, **113**, 1–32.

Baglinière J. L., 1979. Les principales populations de poissons sur une rivière à Salmonidés de Bretagne-sud, le Scorff. *Cybium*, 3rd series, **7**, 53–74.

Baglinière J. L., 1093. *Impact des piscicultures sur les rivières à Salmonidés.* Rapp. Contr., Lab. Ecol. Hydrobiol. INRA, Rennes, 12 pp.

Baglinière J. L., 1991. L truite commune (*Salmo trutta* L.), son origine, son aire de répartition, ses intérêts économique et scientifique. In J. L. Baglinière and G. Maisse (Eds), *Lat Truite: Biologie et Écologie*, INRA Paris, 11–22.

Baglinière J. L., Champigneulle A., Nihouarn A., 1979. La fraie du saumon atlantique (*Salmo salar*) et de la truite commune (*Salmo trutta* L.) sur le bassin due Scorff. *Cybium*, 3rd series, **7**, 75–96.

Baglinière J. L., Arribe-Moutounet D., 1985. Microrépartition des populations de truite commune (*Salmo trutta* L.), de juvénile de saumon atlantique (*Salmo salar* L.) et des autres espèces présentes dans la partie haute du Scorff (Bretagne). *Hydrobiologia*, **120**, 229–239.

Baglinière J. L., Champigneulle A., 1982. Densité des populations de truite commune (*Salmo trutta* L.) et de juvéniles de saumon atlantique (*Salmo salar* L.) sur le cours principal du Scorff (Bretagne): préférendums physiques et variations annuelles (1976–1980). *Acta Oecol, Oecol. Appl.*, **3** (3), 241–256.

Baglinière J. L., Maisse G., Lebail P. Y., Prevost E., 1987. Dynamique de population de truite commune (*Salmo trutta* L.) d'un ruisseau breton (France): les géniteurs migrants. *Acta Oecol., Oecol. Appl.*, **8**, 201–215.

Baglinière J. L., Maisse G., 1989. Dynamique de la population de juvéniles de saumon atlantique (*Salmo salar* L.) sur un petit affluent du Scorff (Morbihan). *Acta Oecol., Oecol. Appl.*, **9**, 3–17.

Baglinière J. L., Maisse G., Lebail P. Y., Nihouarn A., 1989. Population dynamics of brown trout, *Salmo trutta* L., in a tributary in Brittany (France): spawning and juveniles. *J. Fish. Biol.*, **34**, 97–110.

Baglinière J. L., Maisse G., 1990. La croissance de la truite commune (*Salmo trutta* L.) sur le bassin du Scorff. *Bull. Fr. Pêche Piscic.*, **317**, Spec. Coll. (in press).

Baldes R. J., Vincent E. V., 1969. Physical parameters of microhabitats occupied by brown trout in an experimental flume. *Trans. Am. Fish. Soc.*, **98**(2), 230–238.

Baran P., Delacoste M., Lascaux J.M., Belaud A., 1993. Relations entre les caractéristiques de l'habitat et les populations de truites communes (*Salmo trutta* L.) de la vallée de la Neste d'Aure. *Bull. Fr. Pêche Piscic.*, **331**, 321–340.

Baran P., Delacoste M., Poizat G., Lascaux J.M., Lek, Belaud A., 1995a. Approche multi-échelle des relations entre les caractéristiques d'habitat et les populations de truite commune (*Salmo trutta* L.) dans les Pyrénées centrales. *Bull. Fr. Pêche Piscic.*, **337/338/339**, 399–406.

Baran P., Delacoste M., Lascaux J.M., Dauba F., Segura G., 1995b. La compétition interspécifique entre la truite commune (*Salmo trutta* L.) et la truite arc-en-ciel (*Oncorhynchus mykiss* Walbaum): influence sur les modèles d'habitat. *Bull Fr. Pêche Piscic.*, **337/338/339**, 283–290.

Bardonnet A., Heland M., 1994. The influence of potential predators on the habitat preferenda of emerging brown trout. *J. Fish Biol.*, **45** (Suppl. A.), 131–142.

Barila T., Williams R,. D., Stauffer J. R.Jr., 1981. The influence of stream order and selected stream bed parameters on fish diversity in Raystown Branch, Susquehanna River Drainage, Pennsylvania. *J. Appl. Ecol.*, **18**, 125–131.

Barnard S., Wyatt R.J., 1995. Analyse des modèles de prédition des populations de salmonidés en rivière. *Bull. Fr. Pêche Piscic.*, **337/338/339**, 365–373.

Barnard S., Wyatt R.J., Milner N.J., 1995. Le développement des modèles d'habitat pour les salmonidés d'eau courante et leur application à la gestion piscicole. *Bull. Fr. Pêche Piscic.*, **337/338/339**, 375–385.

Beaumont P., 1975. Hydrology. In *River Ecology*, B. A. Whitton (Ed), Blackwell Scientific Publications, Oxford, 1–38.

Binns N. A., Eiserman F. M., 1979. Quantification of fluvial trout habitat in Wyoming. *Trans. Am. Fish. Soc.*, **108**, 215–228.

Bjornn T. J., 1971. Trout and salmon movements in two Idaho streams as related to temperature, food, stream flow, cover and population density. *Trans. Am. Fish. Soc.*, **100** (3), 423–438.

Blondel J., 1979. *Biogéographie et écologie.* Collection d'écologie 15, Masson, Paris, 173 pp.

Bohlin T., 1977. Habitat selection and intercohort competition of juvenile sea-trout *Salmo trutta. Oïkos*, **29**, 112–117.

Bohlin T., 1978. Temporal changes in spatial distribution of juvenile sea-trout *Salmo trutta* in a small stream. *Oïkos*, **30**, 114–120.

Bournaud M., 1963. Le courant, facteur écologique et éthologique de la vie aquatique. *Hydrobiologia*, **21**, 125–165.

Boussu M. F., 1954. Relationship between trout populations and cover on a small stream. *J. Wildl. Mngt.*, **18**, **2**, 229–239.

Bovee K. D., Milhous R., 1978. *Hydraulic simulation in instream flow studies: theory and techniques.* Cooperative Instream Flow Service Group, Fish and Wildlife Service, U.S. Department of the Interior, FWS/OBS-78/33, 131 pp.

Bowlby J. N., Roff J. C., 1986. Trout biomass and habitat relationships in Southern Ontario streams. *Trans. Am. Fish. Soc.*, **115**, 503–514.

Bremond R., Vuichard R., 1973. Paramétres de la qualité des eaux, vol. 1, Ministère de la Protection de la Nature et de l'Environnement. La Documentation Française, Paris, 179 pp.

Brooker M. P., 1981. The impact of impoundment on the downstream fisheries and general ecology, in *Advances in Applied Biology*, 91–152, Academic Press, London.

Brown V. M., 1975. Fishes. In *River Ecology*, B. A. Winton (Ed.), p. 199–229, Blackwell Scientific Publications, Oxford, 725 pp.

Butcher R. W., 1933 Studies on the ecology of rivers. I. On the distribution of macrophytic vegetation in the rivers of Britain. *J. Ecol.*, **21**, 58–91.

Butler R. L., Hawthorne V. M., 1968. The reactions of dominant trout to changes in overhead artificial cover. *Trans. Am. Fish. Soc.*, **97**, 37–41.

Campbell J. S., 1977. Spawning characteristics of brown trout and sea trout *Salmo trutta* L. in Kirk Burn, River Tweed, Scotland. *J. Fish. Biol.*, **11**, 217–229.

Capra H., Valentin S., Breil P., 1995. Chroniques d'habitats et dynamique de populations de truite. *Bull. Fr. Pêche Piscic.*, **337/338/339**, 337–344.

Champigneulle A., 1978. Caractéristiques de l'habitat piscicole et de la population de juvéniles sauvages de saumon atlantique (*Salmo salar* L.) sur le cours principal de Scorff (Morbihan). Thèse 3e Cycle Biologie Animale, Univ. Rennes 1, 92 pp.

Chapman D. W., 1966. Food and space as regulators of salmonid populations in streams. *Am. Nat.*, **100**, (913), 345–357.

Chapman D. W., Bjornn T. C., 1969. Distribution of salmonids in streams, with special reference to food and feeding. In T. G. Northcote (Ed.), *Salmon and Trout in Streams*, 153–176. H. R. McMillan Lect. Fisheries Univ. Brit. Columbia (Vancouver).

Charlon N., 1969. Relation entre métabolisme respiratoire chez les poissons, teneur en oxygène et température. *Extrait Bull. Soc. Histoire Naturelle de Toulouse*, **105**, 1–2, 136–156.

Crisp D. T., 1963. A preliminary survey of brown trout (*Salmo trutta* l.) and bullheads (*Cottus gobio* L.) in high altitude becks. *Salmon Trout Mag.*, **167**, 45–59.

Crisp D. T., 1989. Some impacts of human activities on trout, *Salmo trutta*, populations. *Freshwater Biol.*, **21**, 21–33.

Crisp D. T., Matthews A. M., Westlake D. F., 1982. The temperature of nine flowing waters in Southern England. *Hydrobiologia*, **89**, 193–204.

Crisp D. T., Carling P. A., 1989. Observations on siting, dimensions and structure of salmonid reeds. *J. Fish. Biol.*, **34**, 119–134.

Cuinat R., 1971. Principaux caractères démographiques observés sur 50 rivières à truites françaises. Influence de la pente et du calcium. *Ann. Hydrobiol.*, **2** (2), 187–207.

Cuinat R., 1980. *Modification du lit des cours d'eau: conséquences écologiques et piscicoles*. 1 vol. ronéotypé, Coll. FAO, CECPI, Vichy avril 1980—Conseil Supérieur de la Pêche, 6e D.R., Clermont-Ferrand, 15 pp.

Cunjak R. A., Power G., 1986. Winter habitat utilization by stream resident brook trout (*Salvelinus fontinalis*) and brown trout (*Salmo trutta*). *Can. J. Fish. Aquat. Sci.*, **43**, 1970–1981.

Dawson F. H., 1978. Aquatic plant management in semi-natural streams: the role of marginal vegetation. *J. Environ. Mngt.*, **6**, 213–221.

Dawson F. H., Castellano E., Ladle M., 1978. Concept of species succession in relation to river vegetation and management. *Verh. Int. Verein Theor. Angew. Limnol.*, **20**. 1429–1434.

Decamps H., Capblancq J., 1980. *Recherches sur le bassin Lot-Dordogne et l'herbier d'Argentat*. 2 vol., Min. Environ. Cadre Vie, Com. Faune-Flore, Contr. 8046, Neuilly-sur-Seine, 94 pp.

Delacoste M., Baran P., Lek S., Lascaux J.M., 1995a. Classification et clé de détermination des faciès d'écoulements en rivières de montagne. *Bull. Fr. Pêche Piscic.*, **337/338/339**, 149–156.

Delacoste M., Baran P., Lascaux J.M., Segura G., Belaud A., 1995b. Capacité de la méthode des microhabitats à prédire l'habitat de reproduction de la truite commune. *Bull. Fr. Pêche Piscic.*, **337/338/339**, 345–353.

Dumas J., Haury J., 1995. Une rivière du Piémont pyrénéen: la Nivelle (Pays basque). *Acta Biologica Montana*, **11**, 113–146.

Dumont B., Rivier B., 1981. *Le Dévit de référence biologique*. 1 vol. ronéotype, Doc. CEMAGREF no,. 19, Aix en Provence, Section Qualité des Eaux, Pêche et Pisciculture, 19 pp.+ 11 pp. ann.

Dutartre A., Gross F., 1982. Evolution des végétaux aquatiques dans les cours d'eau recalibrés (Exemples pris dans le Sud-Ouest de la France). In J. J. Symoens, S. S.

Hooper and P. Compère (Eds), *Studies on Aquatic Vascular Plants*, 394–397, Bot. R. Soc. Belgium, Brussels.

Edington J. M., 1966. Some observations on stream temperature. *Oïkos*, **15**, (2), 265–273.

Egglishaw H. J., Shackley P. E., 1977. Growth, survival and production of juvenile salmon and trout in a Scottish stream, 1966–75. *J. Fish. Biol.*, **11**, 647–672.

Egglishaw H. J., Shackley P. E., 1982. Influence of water pH on dispersion of juvenile salmonids, *Salmo salar* L. and *S. trutta* L. in Scottish stream. *J. Fish Biol.*, **21**, 141–155.

Elliott J. M., 1984a. Growth, size, biomass and production of young migratory trout *Salmo trutta* in a Lake District stream, 1966–83. *J. Anim. Ecol.*, **53**, 979–994.

Elliott J. M., 1984b. Numerical changes and population regulation in young migratory trout *Salmo trutta* in a Lake District stream, 1966–83. *J. Anim. Ecol.*, **53**, 327–350.

Elliott J. M., 1985. Population regulation for different life-stages of migratory trout *Salmo trutta* in a Lake District stream, 1966–83. *J. Anim. Ecol.*, **54**, 617–638.

Elliott J. M., 1987. Population regulation in contrasting populations of trout *Salmo trutta* in two Lake District streams. *J. Anim. Ecol.*, **56**, 83–98.

Elliott J. M., 1989. Wild Brown trout *Salmo trutta*: an important national and international resource. *Freshwater Biol.*, **21**, 1–5.

Elser A. A., 1968. Fish populations of a trout stream in relation to major habitat zones and channel alterations. *Trans. Am. Fish. Soc.*, **97** (4), 389–397.

Euzenat G., Fournel F., 1976. Recherches sur la truite commune (*Salmo trutta* L.) dans une rivière de Bretagne, le Scorff. Thèse 3e Cycle Biologie Animale, Univ. Rennes 1, 213 pp.

Fausch K. D., 1984. Profitable stream positions for salmonids: relating specific growth rate to net energy gain. *Can. J. Zool.*, **62**, 441–451.

Fausch K. D., White R. J., 1981. Competition between brook trout (*Salvelinus fontinalis*) and brown trout (*Salmo trutta*) for positions in a Michigan stream. *Can. J. Fish. Aquat. Sci.*, **38**, 1220–1227.

Fausch K. D., White R. J., 1986. Competition among juveniles of coho salmon, brook trout, and brown trout in a laboratory stream, and implications for Great Lake tributaries. *Trans. Am. Fish. Soc.*, **115**, 363–381.

Fausch K.D., Hawkes C.L., Parsons M.G., 1988. Models that predict standing crop of stream fish from habitat variables: 1980–1985. Colorado State University, Dept. Fishery and Wildlife Biology, Fort Collins, USA, 52 pp.

Fragnoud E., 1987. Préférences d'habitat de la truite fario (*Salmo trutta fario* L., 1758) en rivière (Quelques cours d'eau du Sud-Est de la France). Thèse Doct. 3e Cycle Ecol. fond. appl. Eaux contin., Univ. Lyon I, C.E.M.A.G.R.E.F. Lyon, Lab. Hydroécol. quant., 435 pp.

Frost W. E., Brown M. E., 1967. *The Trout*, Collins, London, 286 pp.

Gaudin Ph., Heland M., 1984. Influence d'adultes de chabots (*Cottus gobio* L.) sur des alevins de truite commune (*Salmo trutta* L.): étude expérimentale en milieux semi-naturels. *Act Oecol., Oecol. Appl.*, **5** (1), 71–83.

Gaudin P., Heland M., 1995. Stratégies d'utilisation de l'habitat par les alevins post-émergents de truite commune (*Salmo trutta*) et de saumon Atlantique (*Salmo salar*). *Bull. Fr. Pêche Piscic.*, **337/338/339**, 199–205.

Gautier J. Y., Lefeuvre J. C., Richard G., Trehen P., 1978. *Ecoéthologie*, Masson, Paris, 166 pp.

Gillet C., Roubaud P., 1986. Survie embryonnaire précoce de 9 espèces de poissons d'eau douce après un choc de pH appliqué pendant la fécondation ou au cours des premiers stades de développement embryonnaire. *Reprod. Nutr. Dev.*, **26**, 1319–1333.

Ginot V., 1995. AVHA, un logiciel d'évaluation de l'habitat du poisson ous windows. *Bull. Fr. Pêche Piscic.*, **337/338/339**, 303–308.

Glova G. J., Duncan M. J., 1985. Potential effects of reduced flows on fish habitats in a large boarded river, New Zealand. *Trans. Fish. Am. Soc.*, **114**, (2), 165–181.

Gorman O. T., Karr J. R., 1978. Habitat structure and stream fish communities. *Ecology*, **59**, (3), 507–515.

Gourand V., Baglinière J.L. Sabaton C., Ombredane D., 1998. Application d'un modèle de dynamique de population (*Salmo trutta*) sur un bassin de Basse-Normandie: premières simulations. *Bull. Fr. Pêche Piscic.*, **350–351**, 675–691.

Grant J. W. A., Englert J., Bietz B. F., 1986. Application of a method for assessing the impact of watershed practices: effects of logging on salmonid standing crops. *North Am. J. Fish. Manage.*, **6**, 24–31.

Grant J. W. A., Noakes D. L. G., 1987. A simple model of territory size for drift feeding fish. *Can. J. Zool.*, **65**, 270–276.

Hansen E. A., Alexander G. R., Dunn W. H., 1983. Sand sediment in a Michigan trout stream. Part I. A technique for removing sand bedload from streams. *North Am. J. Fish. Manage.*, **3**, 355–364.

Haury J., 1985. Etude écologique des macrophytes sur Scorff (Bretagne-Sud). Thèse Doct. Ing. Ecol., Univ. Rennes 1, 196 pp.

Haury J., 1987. *Les macrophytes autour du Barrage de Rabodanges (Orne)—Impact du barrage. Etablissement d'un état de référence avant une augmentation du débit réservé.* E.N.S.A. Botanique et I.N.R.A. Ecol. Hydrobiol., Rennes, 39 pp.

Haury J., 1988a. Macrophytes due Scorff: distribution des espèces et bio-typologie. *Bull. Soc. Sci. Bretagne*, **59** (1–4), 53–66.

Haury J., 1988b. Macrophytes du Trieux (Bretagne-Nord): les ensembles floristiques. *Bull. Soc. Sci. Nat. Ouest Fr.*, Nouv. Sér., **10** (3), 135–150.

Haury J., 1990. Programme inter-agences, milieu et végétaux aquatiques fixés—Rapport d'expertise, Lab. INRA Ecol. Hydrobiol (Rennes). 58 pp.

Haury J., Baglinière J. L., 1990. Relations entre la population de truite commune (*Salmo trutta*), les macrophytes et les paramètres du milieu sur un ruisseau. *Bull. Fr. Pêche Piscic.*, **318**, 118–131.

Haury J., Baglinière J.L., 1996. Les macrophytes, facteurs structurant de l'habitat piscicole en rivière à salmonides. Etude de microrépartition sur un ecteur végétalisé du Scorff (Bretagne-Sud). *Cybium*, **20**(3) (suppl.), 107–122.

Haury J., Bernez I., Lahille V., 1996. Influence de la retenue de Rabodanges sur les peuplements macrophytiques de l'Orne. In Merot Ph. & Jigorel A. (Eds), Hydrologie dans les pays celtiques, Rennes 8–11 July 1996, *Colloques de l'I.N.R.A. 79*, 283–290.

Hawkes H. A., 1975. River zonation and classification. In *River Ecology*, B. A. Whitton (Ed), Blackwell Scientific Publications, Oxford, 312–374.

Hearn W. E., 1987. Interspecific competition and habitat segregation among stream-dwelling trout and salmon: a review. *Fisheries*, **12** (5), 24–31.

Heggberget T. G., Haukerbo T., Mork J., Staul G., 1988. Temporal and spatial segregation of spawning in sympatric populations of atlantic salmon, *Salmo salar* L., and brown trout, *Salmo trutta* L. *J. Fish. Biol.*, **33**, 347–356.

Heggenes J., 1988a. Substrate preferences of brown trout fry (*Salmo trutta*) in artificial stream channels. *Can. J. Fish. Aquat. Sci.*, **45**, 1801–1806.

Heggenes J., 1988b. Physical habitat selection by brown trout (*Salmo trutta*) in riverine systems. *Nordic. J. Freshwater Res.*, **64**, 74–90.

Heggenes J., 1988c. Effects of short-term flow fluctuations of displacement of, and habitat use by brown trout in a small stream. *Trans. Am. Fish. Soc.*, **117**, 336–344.

Heggenes I., Baglinière J.L., Cunjak R., 1995. Note de synthèse sur la sélection de niche spatiale et la compétition chez le jeune saumon atlantique (*Salmo salar*) et la truite commune (*Salmo trutta*) en milieu lotique. *Bull. Fr. Pêche Piscic.*, **337/338/339**, 231–239.

Heggenes I., Baglinière J.L. Cunjak R., 1998. Spatial niche variability for young Atlantic salmon (*Salmo salar*) an brown trout (*S. trutta*) in heterogeneous streams. *Ecol. Freshwat. Fish*, **8**, 1–21.

Heland M., 1971a. Observations sur les premières phases due comportement agonistique et territorial de la truite commune *Salmo trutta* L. en ruisseau artificiel. *Ann. Hydrobiol.*, **2** (1), 33–46.

Heland M., 1971b. Influence de la densité du peuplement initial sur l'acquisition des territoires chez la truite commune *Salmo trutta* L. en ruisseau artificiel. *Ann. Hydrobiol.*, **2** (1), 25–32.

Heland M., 1973. Observations préliminaires sur la compétition inter spécifique entre le vairon *Phoxinus phoxinus* (L.) et l'alevin de truite commune *Salmo trutta* L.. *Bull. Fr., Piscic.*, **250**, 5–16.

Heland M., 1980. La dévalaison des alevins de truite commune *Salmo trutta* L. I. Caractérisation en milieu artificiel. *Ann. Limnol.*, **16** (3), 233–245.

Heland M., Gaudin P., Bardonnet A., 1995. Mise en place des premiers comportements et utilisation de l'habitat après l'émergence chez les salmonidés d'eau courante. *Bull. Fr. Pêche Piscic.*, **337/338/339**, 191–197.

Huet M., 1949. Aperçu des relations entre la pente et les populations piscicoles des eaux courantes. *Rev. Suisse d'Hydrol.*, **11**, (3/4), 332–351.

Huet M., 1954. Biologie, profils en long et en travers des eaux courantes. *Bull. Fr. Piscic.*, **175**, 41–53.

Huet M., 1962. Influence du courant sur la distribution des poissons dans les eaux courantes. *Rev. Suisse d'Hydrol.*, **24**, (2), 412–432.

Huet M., Timmermans J. A., 1976. Influence sur les populations de poissons des amémangements hydrauliques des petits cours d'eau assez rapides. *Trav. Stn. Rech. Eaux For.*, Ser. D, **46**, 27 pp.

Hunt R. L., 1969. Effects of habitat alteration on production, standing crops and yield of brook trout in Lawrence Creek, Wisconsin. In T. G. Northcote, *Salmon and Trout in Streams*, 281–312. H. R. McMillan Lect. Fisheries Univ. Brit. Columbia (Vancouver).

Hunt R. L., 1976. A long term evaluation of trout habitat development and its relation to improving management-related research. *Trans. Am. Fish. Soc.*, **105** (3), 361–364.

Jenkins T. M. Jr., 1969. Social structure, position choice and micro-distribution of two trout species (*Salmo trutta* and *Salmo gairdneri*) resident in mountain streams. *Anim. Behav. Monog.*, **2** (2), 57–123.

Johnson I.W., Elliot C.R.N., Gustard A., 1995. Mise en oeuvre de la méthode IFIM (Instream Flow Incremental Methodology) pour modéliser l'habitat des salmonidés dans la rivière Allen, Angleterre. *Bull. Fr. Pêche Piscic.*, **337/338/339**, 329–336.

Jones J. W., Ball J. N., 1954. The spawning behaviour of brown trout and salmon. *Br. J. Anim. Behav.*, **2**, 103–104.

Jones A. N., 1975. A preliminary study of fish segregation in salmon spawning streams. *J. Fish Biol.*, **7**, 95–104.

Kaeding L. R., 1980. Observations on communities of brook and brown trout separated by an upstream-movement barrier on the Firehole river. *Prog. Fish Cult.*, **42** (3), 174–176.

Kalleberg H., 1958. Observations in a stream tank of territoriality and competition in juvenile salmon and trout (*Salmo salar* L. and *Salmo trutta* L.). *Rep. Int. Freshwater Res. Drottningholm*, **39**, 55–98.

Karlstrom O., 1977. Habitat selection and population densities of salmon (*Salmo salar* L.) and trout (*Salmo trutta* L.) parr in Swedish rivers with some reference to human activities. *Acta Univ. Upsaliensis*, **404**, 3–12.

Kennedy G. J. A., Strange C. D., 1982. The distribution of salmonids in upland streams in relation to pH and gradient. *J. Fish Biol.*, **20**, 579–591.

Kennedy G. J. A., Strange C. D., 1986. The effects of intra and interspecification competition on the survival growth of stocked juvenile Atlantic salmon, *Salmo salar* L., in a upland stream. *J. Fish Biol.*, **28**, 479–489.

Khalanski M., Bonnet M., Gregoire A., 1987. *Evaluation quantitative de la biomasse végétale à l'aval de Serre-Ponçon.* E.D.F. Dir. Etud. Rech., HE/32-87.05, 88 pp. + 8 ann. + 15 figs + 17 tables.

Lamouroux N., 1995. Les modèles statistiques de description de l'habitat hydraulique: des outils pour l'écologie. *Bull. Fr. Pêche Piscic.*, **337/338/339**, 157–163.

Lanka R. P., Hubert W. A., Wesche T. A., 1987. Relations of geomorphology to stream habitat and trout standing stock in small rocky mountain streams. *Trans. Am. Fish. Soc.*, **116**, 21–28.

Le Coarer Y., Dumont B., 1995. Modélisation de la morphodynamic fluviale pour la recherche des relations habitat/faune aquatique. *Bull. Fr. Pêche Piscic.*, **337/338/339**, 309–316.

Leger L., 1910. Principes de la méthode rationnelle du peuplement des cours d'eau à Salmonidés. *Bull. Soc. cent. Aquiculture Pêche*, **22**, 241–269.

Lewis S. L., 1969. Physical factors influencing fish populations in pools of a trout stream. *Trans. Am. Fish. Soc.*, **98** (1), 14–19.

Lewis W. M., Morris D. P., 1986. Toxicity of nitrite to fish: a review. *Trans. Am. Fish. Soc.*, **115**, 183–195.

Liebig H., Lim P., Belaud A., 1998. Influence du débit de base et de la durée des éclusées sur la dérive d'alevins de truite commune: expérimentations en canal semi-naturel. *Bull. Fr. Pêche Piscic.*, **337/338/339**, 337–347.

Lindroth A., 1955. Distribution territorial behavior and movements of sea trout fry in the river Indalsälven. *Rep. Int. Freshwater Res. Drottningholm*, **36**, 104–119.

Maisse G., Baglinière J. L., 1990. Biologie de la truite commune (*Salmo trutta*) dans les rivières françaises. In J. L. Baglinière and G. Maisse (Ed.), *La Truite: Biologie et Écologie*, INRA, Paris, 25–46.

Malavoi J. R., 1988. *Protocole de description des composantes morpho-dynamiques d'un cours d'eau à fond caillouteux.* C.E.M.A.G.R.E.F. Lyon, Lab. Hydroécol. quant., Lyon, 21 pp.

Malavoi J. R., 1989. Typologie des facies d'écoulement ou unités morpho-dynamiques des cours d'eau à haute énergie. *Bull. Fr. Pêche Piscic.*, **315**, 189–210.

Maridet L., Souchon Y., 1995. Habitat potentiel de la truite fario (*Salmo trutta fario*, L. 1758) dans trois cours d'eau du Masssif central. *Bull. Fr. Pêche Piscic.*, **336**, 1–18.

Martin G., 1979. Le problème de l'azote dans les eaux. Technique et Documentation, Paris, 279 pp.

Massa F., Grimaldi C., Baglinière J.L. Prunet P., 1998. Evolution des caractéristiques physico-chimiques de deux zones de frayères à sédimentation contrastée et premiers résultats de survie embryo-larvaire de truite commune (*Salmo trutta* L.). *Bull. Fr. Pêche Piscic.*, **350–351**, 359–376.

Melhaoui M., 1985. Eléments d'écologie de la truite de lac (*Salmo trutta* L.) du Léman dans le système lac-affluent. Thèse 3e cycle de Biol. Animale, Université Pierre et Marie Curie, Paris 6, 127 pp.

Meriaux J. L., Verdevoye P., 1983. Données sur le *Callitrichetum obtusangulae* Serbert 1962. Synfloristique, syntaxonomie, synécologie et faune associée. In J. M. Gehu, *Les Végétations Aquatiques et Amphibies*, 45–68, J. Cramer, Vaduz.

Mesick C. F., 1988. Effects of food and cover on numbers of apache and brown trout establishing residency in artificial stream channels. *Trans. Am. Fish Soc.*, **117**, 421–431.

Mills D. H., 1969. A study of trout and young salmon populations in forest streams with a view to management. *Forestry*, **40** (1) Supp., 85–90.

Mills D. H., 1971. *Salmon and Trout Resource, its Ecology, Conservation and Management.* Oliver & Boyd, Edinburgh, 351 pp.

Milner N. J., Gee A. S., Hemsworth R. J., 1978. The production of brown trout, *Salmo trutta* in tributaries of the Upper Wye, Wales. *J. Fish. Biol.*, **13**, 599–612.

Milner N. J., Hemsworth R. J., Jones B. E., 1985. Habitat evaluations as a fisheries management tool. *J. Fish. Biol.*, **27** (supplement A), 85–108.

Mortensen E., 1977. Density-dependence mortality of trout fry (*Salmo trutta* L.) and its relationship to management of small streams. *J. Fish. Biol.*, **11**, 613–617.

Neveu A., 1981. Densité et microrépartition des différentes espèces de poissons dans la Basse Nivelle, petit fleuve côtier des Pyrénées atlantiques. *Bull. Fr. Piscic.*, **280**, 86–103.

Nihouarn A., 1983. Etude de la truite commune (*Salmo trutta* L.) dans le bassin du Scorff

(Morbihan): démographie, reproduction, migrations. Thèse 3e cycle Ecologie, Univ. Rennes 1, 64 pp.

Ombredane D., Haury J., Thibault M., 1988. *Etude des peuplements piscicoles de l'Elorn en relation avec les habitats aquatiques en octobre 1987.* Stn Physiol. Ecol. Poissons INRA, Départ. Halieutique ENSA, Rennes, 24 pp.

Ombredane D., 1989. Les peuplements et habitats piscicoles de l'Elorn en 1988, INRA-ENSAR (Rennes), 14 pp.

Ombredane D., Haury J., Chapon P.M., 1995. Heterogeneity and typology of fish habitat in the main stream of a Breton coastal river (Elorn-Finistère, France). *Hydrobiologia*, **300–301**, 259–268.

Ombredane D., Baglinière J.L., Marchand F., 1998. The effects of Passive Integrated Transponder tags on survival and growth of juvenile brown trout (*Salmo trutta* L.). *Hydrobiologia*, **371/372**, 99–106.

Ottaway E. M., Carling P. A., Clarke A., Reader N. A., 1981. Observations on the structure of brown trout, *Salmo trutta* Linnaeus, redds. *J. Fish. Biol.*, **19**, 593–607.

Ottaway E. M., Clarke A., 1981. A preliminary investigation into the vulnerability of young trout (*Salmo trutta* L.) and Atlantic salmon (*S. salar* L.) to downstream displacement by high water velocities. *J. Fish. Biol.*, **19**, 135–145.

Paris P., 1989. *Etude des végétaux aquatiques fixés en relation avec la qualité du milieu—Rapport d'avancement des travaux.* 2 vol., Loisirs et Détente Joinville-le-Pont (94), Agence de l'Eau Rhin-Meuse Rozerieulles (57), 12 pp. + 12 pp. ann.

Peltre M.C., Muller S., Dutartre A., Barbe J. *et al.*, GIS Macrophytes des eaux continentales 1998. Biologie et écologie des espèces végétales proliférantes en France. Synthèse bibliographique. Les études de l'Agence de l'eau 68, 199 pp.

Peters J. C., 1967. Effects on a trout stream of sediment from agricultural practices. *J. Wildl. Mngt.*, **31** (4), 805–812.

Pouilly M., Souchon Y., 1995. Méthode des microhabitats: validation et perspectives. *Bull. Fr. Pêche Piscic.*, **337/338/339**, 329–336.

Piegay H., Maridet L., 1994. Formations végétales arborées riveraines des cours d'eau potentialités piscicoles (revue bibliographique). *Bull. Fr. Pêche Piscic.*, **333**, 125–147.

Ramade F., 1982. Elément d'écologie appliquée. 1 vol., McGraw-Hill, Paris, 452 pp.

Rousel J.M., 1998. Utilisation de l'espace par la Truite commune (*Salmo trutta* L.) au cours du nycthémère. Un exemple du rôle fonctionnel de l'habitat en ruisseau. Thèse Doct. Ec. Nat. Sup. Agron. Rennes, Sciences de l'Environnement, 186 pp.

Roussel J.M., Bardonnet A., 1995. Activité nycthémérale et utilisation de la séquence radier-profond par les truitelles d'un an (*Salmo trutta* L.). *Bull. Fr. Pêche Piscic.*, **337/338/339**, 221–230.

Roussel J.M., Bardonnet A., 1996. Changements d'habitat de la truite (*Salmo trutta*) et du chabot (*Cottus gobio*) au cours du nycthémère. Approches multivariées à différentes échelle spatiales. *Cybium*, **20** (3) (suppl.), 43–53.

Roussel J.M., Bardonnet A., 1997. Diet and seasonal patterns of habitat use by fish in a natural salmonid brook: an approach to the functional role of the riffle-pool sequence. *Bull. Fr. Pêche Piscic.*, **346**, 573–588.

Roussel J.M., Bardonnet A., 1999. Ontogeny of diet pattern of stream-margin habitat use by emerging brown trout, *Salmo trutta*, in experimental channels: influence of food and predator presence. *Environmental biology of fishes* (in press).

Roussel J.M., Bardonnet A., Haury J., Baglinière J.L., Prévost E., 1998. Végétation aquatique et peuplement pisciaire: approche expérimentale de l'enlévement des macrophytes dans les radiers d'un cours d'eau breton. *Bull. Fr. Pêche Piscic.*, **350–351**, 693–709.

Saunders J. W., Smith M. W., 1962. Physical alterations of stream habitat to improve brook trout production. *Trans. Am. Fish. Soc.*, **92** (2), 185–189.

Shirvell C. S., Dungey R. G., 1983. Microhabitats chosen by brown trout for feeding and spawning in rivers. *Trans. Am. Fish. Soc.*, **112**, 355–367.

Souchon Y, Trocherie F., Fragnoud E., Lacombre C., 1989. Les modèles numériques des microhabitats des poissons: application et nouveaux développements. *Rev. Sci. Eau*, **2**, 807–830.

Taube C. M., 1974. Stability of residence among brown trout and rainbow trout in experimental sections of Platte River. Fisheries Research Report no. 1817, Michigan Depart. of Natural Resources—Fisheries Division, 20 pp.

Thevenet A., 1998. Intérêt des débris ligneux grosiers pour les poissons des grandes riviéres. Pour une prise en compte de leur dimension écologique dans la gestion des cours d'eau. Thèse Univ. Claude Bernard-Lyon I, 111 pp. + annex.

Tiberghien G. (Ed.) 1985. *Le Scorff: système de référence floristique et faunistique de la qualité des eaux courantes*. Rapp. Contr., Lab. Ecol. Hydrobiol. INRA, Rennes, 176 pp. + 44 pp. ann.

Mimmermans J. A., 1960. Observations concernant les populations de truite commune (*Salmo trutta fario* L.) dans les eaux courantes. *Trav. Stn. Rech. Eaux For. Groendaal-Hoeilaart*, Sér. D (28), 25 pp.

Vander Borght P., Ska B., Schmitz A., Wollast R., 1982. Eutrophisation de la rivière Semois: le développement de *Ranunculus* et ses conséquences sur l'écosystème aquatique. In J. J. Symoens, S. S. Hooper and P. Compere (Eds), *Studies on Aquatic Vascular Plants*, 340–345, Bot. R. Soc. Belgium, Brussels.

Vannote R. L., Minshall G. W., Cummins K. W., Sedell J. R., Cushing C. E., 1980. The river continuum concept. *Can. J. Fish. Aquat. Sci.*, **37**, 130–137.

Verneaux J., 1973. Les principales méthodes biologiques de détermination du degré de pollution des eaux courantes. *Econ. Méd. Anim.*, **14** (1), 11–19.

Verneaux J., 1977a. Biotypologie de l'écosystème 'eau courante'. Déterminisme approché de la structure biotypologique. *C.R. Acad. Sci. Paris*, **284**, Series D, 77–79.

Verneaux J., 1977b. Biotypologie de l'écosystème à 'eaux courante'. Détermination approchée de l'appartenance typologique d'un peuplement icthyologique. *C.R. Acad. Sci. Paris*, **284**, Series D, 675–678.

Verneaux J., 1980. Fondements biologiques et écologiques de l'étude de la qualité des eaux continentales. Principales méthodes biologiques. In P. Pesson (Ed.), *La Polution des Eaux Continentales*, 289–345. Gauthier-Villars, Paris, 345 pp.

Wasson G., Dumont B., Trocherie F., 1981. *Protocole de description des habitats aquatiques et de prélèvement des invertébrés benthiques dans les cours d'eau.* Etude no. 1, Centre d'Etude du Machinisme Agricole, du Génie Rural, des Eaux et des Forêts, Div. Qualité des Eaux, Pêche et Pisciculture, Lyon, 17 pp.

Welcomme R. L., 1985. River fisheries, *F.A.O. Fish. Tech. Pap.*, 262, 330 pp.

Westlake D. F., 1975. Aquatic macrophytes. In *River Ecology*, B. A. Whitton (Ed), Blackwell Scientific Publications, Oxford, 106–128.

Westlake D. F., Casey H., Dawson F. H., Ladle M., Mann R. M. K., Marker A. F. H., 1972. The chalk stream ecosystem. In *Productivity Problems of Freshwater*, Z. Zajak and A. Hillbricht (Eds), Ibkowska, 615–635.

White R. J., 1975. Trout population responses to streamflow fluctuation and habitat management in Big Roche-a-Cri Creek, Wisconsin. *Verh. Int. Verein. Limnol.*, **19**, 2469–2477.

Witzel L. D., MacCrimmon H. R., 1983. Redd site selection by brook trout and brown trout in south eastern Ontario streams. *Trans. Am. Fish. Soc.*, **112**, 760–771.

3

Feeding strategy of the brown trout (*Salmo trutta* L.) in running water

A. Neveu

I. INTRODUCTION

The approach to the study of an aquatic ecosystem involves, above all, the analysis of the partitioning of energy between the different food chains. The objective is to bring together the elements for a better understanding of the interactions in order to maximize exploitation of the resources either for fishing or for aquaculture.

Studies of energy flow of systems tend more and more to consider an entire hydrographic catchment as a complete ecological unit. The interdependence of the different elements is underlined by the concept of the 'river continuum' of Vannote *et al.* (1980) associated with that of the spiral cycle of nutrients of Newbold *et al.* (1981). Numerous works show also that organic detritus, some autochthonously derived, but mostly allochthonous, represents the principal source of energy and that it is possible to apply the idea of functional trophic groups at the invertebrate level (Cummins, 1973; Minshall *et al.*, 1982).

However, much remains to be learnt about the fishes, which are generally at the top of the food chains. Energy flow towards them is poorly understood; estimations of consumption rates are very limited at the wild population level, even for the trout, the most studied species.

Fish require highly protein-rich food (of the order of 40%), which is used not only for tissue production, but also as an energy source. This aptitude is linked to their capacity to easily excrete ammonia products and to their relatively limited energetic requirements (poikilothermy, support by water).

In this context, the trout is carnivorous, like the majority of fishes. This diet can be considered as an advantage where grazing (of algae, plants etc.) is a laborious, very inefficient way of acquiring proteins.

II. FOOD RESOURCES

In salmonid water courses, food level is generally slightly raised, in relation to a location at the head of the basin; primary production is reduced, to the periphyton, to bryophytes and some hydrophytes. This autochthonous primary production is limited; the principal source of energy necessary for the 'economy' of the water course comes from the imported allochthonous organic detritus which serves as food for the invertebrates (Kaushik and Hynes, 1971; Minshall, 1978; Anderson and Sedell, 1979).

1. Food sources

Taking into account the conditions specific to this type of water course, zooplankton is negligible and is supplied mainly from outside sources. For example, when a lake is present in the basin, its zooplankton production can benefit the local water course through drift (Neveu and Echaubard, 1975; Crisp *et al.*, 1978).

The main source of food for the trout is the assembly of benthic invertebrates, the density of which varies according to local conditions. These invertebrates can be present in water through their entire life-cycle (crustaceans, molluscs) or present only at certain times, usually during larval development (insects). Most of the time, these different constituents of the benthos are accessible to a varying degree in the substrate, but they occasionally become very vulnerable (metamorphosis or part of the drift).

Another source of food is represented by the import of exogenous fauna, terrestrial invertebrates which fall into the water and constitute the surface drift which is often very important for the trout (Chaston, 1969; Tusa, 1969; Metz, 1974). This import is related to the flora of the banks and current climatic conditions (Hunt, 1975).

A final option for the trout is the capture of other, smaller species of fish (minnow, loach) or juveniles of its own species. In most flowing waters, such fish-eating does not seem to be greatly developed; however, it presents a nutritional advantage (proteins), especially for the largest individuals.

2. Estimation of stocks and production

The quantification of available food is made using various methods depending on whether benthos and drift are both being assessed.

The most common method consists of regular and controlled fauna sampling using fine-meshed (0.25–0.30 mm) nets (SURBER type). By exhaustive sorting and the determination of the principal species and their numbers, it is possible to estimate stocks (mean biomass). Following developmental cycles, productivity can be very different for the same biomass; this knowledge is required for different methods of calculation reviewed by Waters (1977), improved recently for species with continuous reproduction, by Kimmerer (1987) and for their confidence limits, by Morin *et al.* (1987).

The complexity of this work means data for productivity of the macrobenthos for an entire water course remain scarce; they are too often limited to a certain few species. Some figures can, however, be presented: 800 kg/ha/yr for a river in the foothills of the Pyrenees (Neveu *et al.*, 1979), 1188 kg/ha/yr for a little river in the plains (Maslin and Pattee, 1981), between 325 and 1324 kg/ha/yr for various water courses in Minnesota

(Krueger and Waters, 1983). In other words, these values are of the same order, close to 100 kg/ha/yr for temperate zones.

Note that the establishment of the ratio P/B for a particular region, can later be used for quick estimates of productivity (P) from mean biomasses (B).

Estimates of energetic relationships can be made either from calorimetric analysis of the stock or from direct measurements on samples (see methodological review: Richman, 1971), or by the conversion of biomass to energy using tables (Cummins and Wuychek, 1971; Wissing and Hasler, 1968, 1971; Caspers, 1975a, b; Alimov and Shadrin, 1977; Penczak *et al.*, 1984). It is also possible to make estimates of real values using the knowledge that, on average, 1 g of dry matter = 6 g live matter = 5 kilocalories = 21 kilojoules = 0.90 g ash-free dry matter = 0.50 g carbon (Waters, 1977).

Chemical analysis of the components also allows nutritional value to be assessed, e.g. protein content. Thus Yurkowski and Tabachek (1978) showed the wide variation in amino acid content of various invertebrates, while Gardner *et al.* (1985) found seasonal variation in lipid content. Nutritional value data can be complemented with digestibility studies of different foods, using the methods described in Utne (1978).

3. Variability and accessibility of stocks

(a) Spatially

The quantity of available prey varies in relation to the size of the water course, habitat type (influence of flow, substrate particle size etc.) and the form of the banks (wooded or not, etc.). Different measurements in various water courses have shown that there is less surface drift in forested areas (10–24% of total drift) and in arable land (11–14%) than in natural mountain prairie (28–70%). In small streams, stocks can vary between 1 and 3 g/m^2 in pool areas up to 9–20 g/m^2 in flowing areas; the annual mean for a small Pyrenean river is 16 g/m^2 (Neveu *et al.*, 1979). In general, sunny areas are more productive (Behmer and Hawkins, 1986) and the addition of sediments reduces the stocks (Neveu, 1980).

Occasional organic inputs (organic pollution) can alter the original arrangement by increasing the stocks; thus in the French Massif Central, a flood of outflow from dairies, while changing the biological balance, increased the stock from 9–27 g/m^2 in natural zones (prairies) to 60–371 g/m^2 in the affected area (Neveu and Echaubard, 1983).

Not all the available stocks in a stream can be exploited by the fish; some of it remains inaccessible. While terrestrial animals falling into the water are very vulnerable, the accessibility of benthic species depends largely on the substrate and vegetation and also their means of protection (camouflage, burrowing, etc.). Much also depends on the coincidental meeting of prey and fish, which may occur in inaccessible areas (clumps of dead leaves, shallows, under banks etc.).

(b) Temporally

In the course of a year, stocks vary according to the developmental cycles of invertebrates; eggs, especially if they develop slowly, cannot be exploited by fish; the same is true for young larval stages.

From one year to the next, the influence of key factors on invertebrate development (temperature limits, spates, low waters etc.) can change the availability of food and reduce trout growth.

Over the course of a year, access to prey can change, particularly in relation to the drift, which is generally at night for larvae and mostly diurnal for adults and exogenous fauna. The latter depends a lot on the weather (wind and rain). Variations in water levels (spates) and in temperature can also change the rhythm of the drift, and its intensity, at a time when the trout is foraging for its prey.

This accessibility of prey can also depend on developmental stage; for example, vulnerability of imagos and subimagos at hatching, displacement of larvae during their activities such as moulting and feeding. Changes in water levels can cause migrations within the substrate–covering over of prey items.

III. EXPLOITATION OF FOOD RESOURCES

1. Study methods

The different methods of estimating food consumption by fishes have been the subject of several methodological reviews (Mann, 1978; Windell, 1978a; Neveu, 1979; Hyslop, 1980). The methods are based on two principles: either the analysis of stomach contents, or the extrapolation to the wild of results obtained experimentally in the laboratory.

Analysis of stomach contents is the method most frequently employed to provide data on feeding behaviour and consumption rate. The main obstacle is the need to sacrifice the fish. Certain studies have been carried out on living animals by using stomach washes (Seaburg and Moyle, 1962; Baker and Fraser, 1975; Mechan and Miller, 1978; Giles, 1980; Light *et al.*, 1983; Georges and Gaudin, 1984). However, the efficiency of this technique varies depending on the size of the stomach and prey type and is less effective for the posterior part of the stomach (Neveu and Thibault, 1977).

2. Feeding behaviour

Capture results from interactions of parameters which depend on both predator and prey.

(a) The chase

For the trout, the strategy of taking food is characterized by a series of stages, often lasting only several tenths of a second, as summarized by Nilsson (1978).

The initial motivation resulting in searching for food is above all physiological with the appearance of hunger brought on by the emptying of the stomach (Ivlev, 1961; Ware, 1972; Colgan, 1973) but also with other physiological parameters (Colgan, 1973).

Encountering the prey determines the second sequence where certain choices may arise, depending on both predator and prey. For example, the reaction distance increases with trout size (Ware, 1972, 1973) as in salmon (Wankowski, 1979) but also with the size of the prey (Ware, 1972, 1973). This aspect, basically visual, can be affected by decreased light levels, which in turn are influenced by turbidity (Berg and Northcote, 1985) or rain (Bachman, 1984). Elsewhere, the attack duration is very variable between individuals (Ringler, 1979; Ringler and Brodowski, 1983).

The third sequence results in the capture which may or may not be followed by ingestion. The stimuli triggering the attack can be visual, olfactory or gustatory (taste), according to species, but ingestion relates directly to taste. This influence of taste is clearly detectable experimentally (Hara, 1975; Adron and Mackie, 1978; Appelbaum, 1980). It is difficult to determine its exact role in wild conditions. Amino acid molecules appear to play the most important role.

(b) Prey selection

A large number of factors come into play in prey selection by a predator. The study of choice in fishes has often been carried out in the laboratory, with all the inherent difficulties of extrapolation to the natural environment (Ivlev, 1961; Ware, 1971).

In the wild, studies can be carried out by comparing stomach contents and the fauna *in situ*, using different indices (Ivlev, 1961; Jacobs, 1974; Herrera, 1976; Chesson, 1978; Strauss, 1979) or by comparing ranks in the trophic spectra (Neveu, 1980, 1981a). In all cases, great care must be taken in sampling the food stocks, and a correct size of net used (O'Brien and Vinyard, 1974).

Prey movements play a big role, whether they are active or caused by the current (Waters, 1969; Serebrov, 1983; Ware, 1973; Rimmer and Power, 1978; Irwine and Northcote, 1983). Selection of a prey depends on its relative density (Ware, 1973; Ringer and Brodowski, 1983). An increase, even if only instantaneous, of a species, can increase its consumption (Murdoch *et al.*, 1975) even though this capacity varies according to the individuals (Ringler, 1985).

The size of prey captured is proportional to the size of the trout (Dahl, 1962; Nilsson, 1965; Egglishaw, 1967; Ware, 1973; Vinyard and O'Brien, 1975), relating to variations in mouth size (Bannon and Ringler, 1986). This is particularly clear in the case of the trout in a French Pyrenean stream (Neveu and Thibault, 1977). The impression that the trout were actively selecting prey of certain sizes (Thomas, 1962), could be linked to selective capture of certain prey (Ringler, 1979; Fahy, 1980; Newman and Waters, 1984) as in the case of the brook charr (Allen, 1981). This selection is best seen in the case of the consumption of the mollusc *Potamopyrgus jenkinsi* (Table 1).

Prey vulnerability also depends on their means of protection (colour, shell, case etc.) (Ware, 1972, 1973; Ginetz and Larkin, 1973).

The role of experience in the predator is important, but this experience is variable (Ware, 1971; Ringler, 1979) in as much as learning capacity varies according to individuals (Ringler, 1983) and can also lead to errors such as the capture of pieces of wood (Rocha and Mills, 1984). This last phenomenon is often observed in animals used for restocking from fish farms (Neveu, unpublished report).

Despite all this, certain individuals persist in specialist feeding (Bryan and Larkin, 1972).

3. Capture rate

(a) Intensity of consumption

This is the quantity of food ingested per unit time. It depends on concentration of food, its spatial distribution and also the hunger of the predator.

Table 1. Size frequency distribution for the mollusc *Potamopyrgus jenkinsi* (*L*: total shell length in mm) found in brown trout stomachs and in the benthos of a Pyrenean stream (unpublished data)

	Trout length (cm)				
L (mm)	5–9.9	10–14.9	15–19.9	⩾ 20.0	Benthos
0.40–0.79	10	4			2
0.80–1.19	44	4			12
1.20–1.59	23	18	2		14
1.60–1.99	18	9	12	19	14
2.00–2.39	5	4	5	11	8
2.40–2.79		10	5	4	4
2.80–3.19		8	10	26	8
3.20–3.59		11	27	7	12
3.60–3.99		20	22	19	8
4.00–4.39		11	12	15	14
4.40–4.79		2	5		3
4.80–5.19					2

Since the experimental work of Ivlev (1961) on different species, it has been confirmed that consumption intensity depends directly on the number of prey items available in the environment; a reduction of these leads to a lowering in consumption by trout (Neveu, 1980).

In general terms, hunger and consumption intensity increase with the emptying of the stomach. Fortmann *et al.* (1961) noticed that anglers catch more trout with empty stomachs. Elliott (1975a) was able to estimate the time required to reach satiation in the trout, as a function of temperature and shows that stomach filling slows down consumption progressively. Increase in appetite is thus at a maximum when the stomach contents fall below 10% of the maximum capacity at 15°C (that is 10 h after a meal); if the deprivation of food lasts longer, appetite (expressed as rate of consumption) remains maximal (Elliott, 1975b).

In the wild, it is difficult to measure consumption rate; it is probably lower than in the laboratory, as the animal has to search for food, which cannot be abundant enough to fill its stomach and eliminate hunger. In fact, consumption intensity is very variable with time; this results in marked feeding rhythms.

(b) Feeding rhythms

In the aquatic environment, many parameters show periodic variations whether at the primary level (light, temperature, minerals) or at the secondary level (oxygen, pH, etc.). Even biotic factors can have rhythms: these include migration, benthic drift and the hatching cycle of imagos. Fish are at the top of the food chain and are reliant on these fluctuations (poikilothermy, availability of food). Rhythms vary greatly according to species, arrhythmia being very rare in nature (Neveu, 1981b).

It is possible to distinguish seasonal rhythms, which are increasingly accentuated towards the north; they depend above all on water temperature. But the most fundamental are the daily rhythms, which depend on the succession of light and dark.

The most important factor is light which acts on the periodicity of these rhythms; the sensitivity threshold can be as low as 10^{-3} to 3×10^{-2} lux (starlight 10^{-3}, moonlight from 10^{-2} to 10^{-1} lux) for salmonids (Kalleberg, 1958; Tanaka, 1970; Brett, 1971). But temperature appears to be the factor orchestrating the number of meals for the trout, at least at the experimental level (Elliott, 1975b). The temperature threshold to which they react lies within a few degrees, for the salmonids. A succession of several hot days (more than 20°C) can inhibit feeding (Hoar, 1942; Baldwin, 1957; Elliott, 1975b). In fact, it is difficult to separate the effects of light and temperature; it is the combined effect of the two which allows the development of the annual pattern, in particular close to the Arctic circle, where salmonids are diurnal in winter and nocturnal in summer, with a transitory biomodal phase (Kalleberg, 1958; Erickson, 1973).

Other factors can influence feeding rhythms in nature. It has often been considered that invertebrate drift sets the rhythm for the trout (Elliott, 1967, 1970, 1973; Chaston, 1969; Jenkins *et al.*, 1970; Tanaka, 1970). But analyses of the composition of stomach contents and the drift cannot prove this definitively; some still lean rather in favour of benthic feeding (Neveu, 1980), above all for young trout (Table 2).

Table 2. Comparison using Spearman's rank coefficient (significant to the 5% level) of the composition of the stomach contents of brown trout of different sizes (*n*: number of individuals) and that of the benthos and drift in a French Pyrenean stream (unpublished data)

Size class (cm)	June			October		
	n	Benthos	Drift	*n*	Benthos	Drift
5–9.9	26	0.65*	0.33	83	0.73*	0.52
10–14.9	58	0.50	– 0.21	92	0.69*	0.42
14–19.9	44	0.24	– 0.50	56	0.11	0.07
⩾ 20.0	25	0.38	– 0.43	0		
Benthos		1.00	0.14		1.00	0.36

Digestion rate and metabolic rate are both influenced by temperature, which acts directly on food consumption. Poor-quality food can be compensated for by increased rate of capture (Elliott, 1976c; Grove *et al.*, 1978). Similarly, while learning may have a role experimentally (Ware, 1971), this does not appear to be the case in nature (Bryan, 1973), except under certain conditions such as the regular input of organic pollution (Ellis and Gowing, 1957).

Many toxins can affect feeding rates by disrupting general behaviour, for example copper and mercury in the trout (Drummond *et al.*, 1973; Hara *et al.*, 1976).

It should also be noted that rhythms change with age. Small trout in flowing waters feed more on the endogenous nocturnal drift than adult trout from the deep waters which attack the exogenous diurnal drift (Table 3).

The study of feeding rates remains indispensable to the knowledge of consumption rates, as it allows the number of food items taken per day to be determined.

(c) Consumption rate

Different methods have been employed to estimate daily consumption; these have been reviewed by Windell (1978a):

- an energetic approach (Winberg, 1956; Mann, 1965; Morgan, 1974; Elliott, 1976c);
- an approach based on nitrogenous requirements (Smith and Thorpe, 1976; Braaten, 1979);
- an approach based on growth rates (Allen, 1951; Horton, 1961; Warren *et al.*, 1964; Elliott, 1975c, d).

All these methods demand significant extrapolations from the laboratory to the wild, assume a certain number of simplifying hypotheses, and are often only applicable to mean values over long time periods.

It is possible to apply a more direct methodology over set time periods, by the analysis of variations in stomach contents based on a knowledge of gastric transit times.

The speed of passage through the stomach must be evaluated in the laboratory from a 'standard' meal. Most evacuation rates are exponential (Windell, 1978b), in particular for the brown trout (Elliott, 1972). In the latter, the evacuation rate is not affected by fish size, meal size and only a little by prey size and meal frequency; it depends above all on temperature (Elliott, 1972). However, Jobling (1981) cast doubt on the influence of fish size and meal size, but especially on food composition, with regard to lipid content (which slows transit). A long period of fasting can also slow down transit, as can any form of stress (Pickering *et al.*, 1982).

In fact, the exponential model mainly fits the evacuation of small food particles (most benthic invertebrates). The different relationships between surface/volume for the small and large particles and their degree of friability, affects the type of evacuation, which comes closer to a linear model for the piscivorous older trout (Jobling, 1986, 1987). The model must therefore be adapted according to the type of food the trout is eating—invertebrates or fish.

Bajkov (1935) carried out studies on consumption from knowledge of stomach contents and transit times. Her method has often been employed, despite its imperfections; she assumes, amongst other things, continuous feeding and constant evacuation rate.

For the trout, the best method of calculation at present seems to be that of Elliott and Persson (1978) with:

$$D = \sum C_t \qquad C_t = \frac{\left(S_t - S_0 e^{-R_t}\right) R_t}{1 - e^{-R_t}}$$

Table 3. Frequency (F) of aquatic and terrestrial fauna in brown trout stomachs in a Pyrenean stream, in relation to time of capture in June (expressed as a percentage of the total number of prey) and diversity (D) of the food source expressed by the number of invertebrate families captured, for each size class of trout (*n*: total number, mean and maximum, of families in the stomach; *b*: number of families in the benthos; *d*: number in the drift) (unpublished data)

	F						D				
	Time of capture						*n*				
Trout length (cm)	11 h 30 12 h 00	15 h 30 16 h 00	19 h 30 20 h 00	23 h 30 24 h 00	3 h 30 4 h 00	7 h 30 8 h 00	Total	Mean	Max.	*d*	*b*
Aquatic fauna										19	34
5–9.9	99	96	97	100	91	96	10	6.1	8		
10–14.9	52	3	92	59	56	75	18	10.6	15		
15–19.9	28	3	52	80	59	62	17	9.5	10		
⩾ 20.0	85	27	81	98	76	20	12	5.8	8		
Terrestrial fauna										7	0
5– 9.9	1	4	3	0	5	4	4	0.8	2		
10–14.9	48	97	8	27	36	8	13	6.1	8		
15–19.9	69	97	48	13	26	36	11	7.0	9		
⩾ 20.0	7	72	13	2	19	56	9	5.3	7		

where D is daily consumption (dry matter), C_t is instantaneous consumption (measured over t hours), S_0 is the amount of food in stomach at time t_0, S_t is the amount of food in stomach at time t, and R_t is the gastric evacuation rate at study temperature.

Measurements of C_t must be made over periods that are as short as possible, usually between 1 and 3 hours.

If $S_0 = S_{24}$, that is if the amount of food is the same after 24 hours, the formula becomes $D = 24SR$, S being the mean amount of food in the stomach during the course of the day.

The application of these calculations to trout in a French Pyrenean stream was carried out at two different times of year; June and October (Table 4). The results allow the change in consumption during the day to be followed, and the estimation of consumption by different size classes of fish. In June, consumption is higher than in October, especially for the larger individuals with associated low availability of autumnal food (invertebrates at early stages of development). This consumption represents between 0.4 and 6.4% of trout body weight in June compared to 0.1 to 1.2% in October. Application of the formula $D = 24SR$ gives comparable results (respectively 14.8, 14.1, 13.1, 9.1 mg/g/day in June, 10.1, 4.2, 3.7 mg/g/day in October).

Very little information is available at present for wild trout in rivers (Elliott, 1973; Neveu, 1980; Mortensen, 1985).

Existing data remain patchy in time and require complementary measurements to be made for annual calculations, as for those of Lien (1978) on a lake.

IV. ENERGY BUDGET FOR THE TROUT

To understand the growth, and hence the production of a species, it is necessary to understand the way in which the energetic content of food is utilized. In particular, a balance should be set up between input and expenditure, especially at the metabolic level. In principle, many parameters of an energy budget can be measured experimentally, but the animal in the wild cannot be placed inside a calorimeter; it is therefore necessary to devise indirect approaches with complete complementarity between the laboratory and the wild.

The energy budget can be written as follows:

$$C = \delta B + R + U + F$$

where C is the energetic input of food, δB is the accumulation of energy in the tissues, R is the energy loss through general metabolism, U is the loss through nitrogenous excretion, and F is the loss through faeces.

1. Accumulation of energy δB

When energetic inputs from food are sufficient, part of it is stored in the form of somatic and germinal growth.

Measurement of this growth is carried out by numerous direct (weight, length, etc.) or indirect (osteochronological back calculation, RNA-DNA ratios) methods. In all these cases, interest is directed at the energetic content of the tissues which can vary greatly in the same species, as a function of size, maturity, type of food, season and fish activity.

Table 4. Feeding pattern and daily consumption of different sizes of brown trout in a Pyrenean stream in June and October (T°C: water temperature, R: gastric evacuation rate; Ct: consumption calculated over 4 hours; ΣCt: daily consumption in mg of dry matter per gram of trout) (unpublished data)

	Time of capture												
June	11 h 30 12 h 00	Ct	15 h 30 16 h 00	Ct	19 h 30 20 h 00	Ct	23 h 30 24 h 00	Ct	3 h 30 4 h 00	Ct	7 h 30 8 h 00	Ct	ΣC_t
	Night												
T°C	16.0		17.0		15.5		13.2		12.2		14.0		
R		0.318		0.356		0.301		0.232		0.208		0.254	
Trout length (cm)													
5– 9.9		4.31		0.00		4.43		2.98		1.59		2.07	15.38
10–14.9		4.59		0.88		4.10		2.96		0.71		1.50	14.74
15–19.9		3.65		4.27		2.22		1.66		0.81		1.63	14.24
⩾ 20.0		1.84		2.19		0.42		0.73		0.73		4.01	9.92
October	14 h 00 14 h 30		18 h 00 18 h 30		22 h 00 22 h 30		2 h 00 2 h 30		6 h 00 6 h 30		10 h 00 10 h 30		
	Night												
T°C	15.9		15.9		15.1		14.8		14.8		15.6	0.304	
R		0.316		0.316		0.288		0.278		0.278			
Trout length (cm)													
5–9.9		2.37		2.24		1.70		1.29		0.68		1.92	10.21
18–14.9		0.63		0.43		0.72		1.06		0.81		0.54	4.19
⩾ 15.0		0.00		0.77		0.77		0.13		0.09		1.13	2.89

This is the case for the trout where these different factors have often been studied (Elliott, 1976a; Staples and Nomura, 1976; Weatherley and Gill, 1983). This variability is linked to changes in lipid level, which results in variations in stoutness, which can be estimated from the condition factor, obtained from weight–length measurements.

Growth of the gonads can be followed particularly at certain periods of the year when energetic reserves are focused on this phenomenon, to the detriment of somatic growth.

Relationships between the input of energy and its use for growth have been reviewed by Jobling (1985) and Wootton (1985).

2. Losses through excretion, $U + F$

Losses cannot be measured directly because of the rapid dispersal of particles and dissolved elements in water; only laboratory analyses are possible at present.

Winberg (1956) estimated faecal loss at 15% of energetic output and losses through all the other diverse excretory products, at 3%. This is why, to avoid complex analysis, many authors use an approximation of 20% for $U + F$.

Elliott (1976b) reviewed these figures for the trout and showed that 25–30% is a more realistic figure. In effect, faecal losses (F) are a function of temperature and food ration. They vary from 31% of the maximum ration at 3°C, to only 11% at a reduced ration. They are no greater than 10% at 22°C for the maximum ration. Losses through excretory products (U) vary from 3.6% at 3°C at maximum ration, to 15.1% for a reduced ration at 22°C. In addition, when the lipid/protein ratio increases, F increases while U decreases.

3. Metabolic losses, R

Metabolic losses, both basal metabolism and that of various activities (feeding, swimming, digestion) represent a large part of the energy flux. Measuring these outputs cannot be done *in situ*. They must be estimated in the laboratory, under conditions as close as possible to nature and with certain simplifying hypotheses.

Classically, the different metabolic outputs are separated:

$$R = R_s + R_d + R_a$$

where R_s is basal metabolism, R_d is loss through feeding, and R_a is the loss through locomotion.

Basal metabolism, R_s is estimated from individuals without food, showing minimal activity, by analysis of their oxygen consumption, in relation to their weight and the temperature. It can also be estimated from tissue loss and excretion from these same fish (Elliott, 1976c).

Capture of prey results in an increase in heat and oxygen consumption. These losses, R_d, are linked to activity expenditure for hunting, ingestion, digestion and absorption, especially of proteins (specific dynamic action (SDA)). These expenditures depend on the level of consumption, on temperature and on fish weight (Soofiani and Hawkins, 1985).

Expenditure related to fish movements (R_a) (static swimming, flight, migration, searching for food) are difficult to estimate as they are very variable under natural conditions. It is, however, possible to study the movements of trout on video, by counting tail beats (Feldmeth and Jenkins, 1973). But the method requires careful interpretation. It is also possible to use telemetry to monitor heart beats (Priede and Tytler, 1977; Priede and Young, 1977) or gill movements (Oswald, 1978).

Faced with these difficulties, a global estimation can be made from $R_a + R_d = R - R_s$, R being estimated as a function of trout weight and temperature (Elliott, 1976c).

4. Calculation of the energy budget

There are very few data on the evaluation of the energy budget of trout in the wild, taking into account the complexity of this problem and its temporal variability.

Some approaches have, however, been tried, either in rivers (Mortensen, 1986; Cunjak and Power, 1987) or in lakes (Lien, 1978) using the experimental results obtained by Elliott (1976a,b,c).

This last methodology can be used on data from trout from a French Pyrenean stream (Table 5).

Elliott's formulae are applied to individuals between 11 and 250 g in weight. They are also valid for fry of 2–3 g. The calculations are made on the basis of 5.2 kcal/g of dry weight of invertebrates in June and 5 kcal/g in October; this relates to the faunal consumption. The calorific value of the trout is rounded up from Elliott (1976a), in the absence of direct analysis. Similarly, metabolism, for a ration between maintenance and maximum, is estimated from Elliott's (1976c) experiments (mean temperature 15°C, trout 12–91 g).

It is possible to calculate:

$$\frac{R}{R_{\text{main}}} = 1.37\frac{C}{C_{\text{max}}} + 0.67 \qquad (r = 0.97,\ \text{for } n = 9)$$

The calculation of R_{main} and C/C_{max} allows R to be calculated.

In June, apart from the fry, consumption is close to the maximum ration, which gives a greatly increased efficiency ($\delta B/C_{\text{max}} = 0.29$) and high growth. The losses $F + U$ represent 27–31% of inputs and metabolism 39–59% (expenditure through activity and digestion $R_a + R_d$ can be estimated at 45–63% of metabolism).

In October, consumption is lower. The relatively high temperature increases expenditure which is greater than consumption, and so utilization of reserves occurs. This results in a lowering of condition factor between June and October from 1.32 to 1.14 for the males and 1.24 to 1.05 for the females.

Note that the calculations are carried out on 'average' animals for each age class, while in October certain individuals have empty digestive tracts. For these $C = 0$, $-\delta B = F + U + R$, with R close to R_s, the calculation of $F + U$ and R_s (Elliott, 1976b) allows δB to be estimated at –91.16, –333.87 and –871.03 cal/day/trout respectively, or a ratio $\delta B/W$ of –0.023, –0.013 and –0.008 respectively, which correspond to a faster rate of weight loss and consequent decrease in condition factor in the trophically inactive individuals.

These results are linked to the food available in this stream which is very reduced at this pre-autumnal period, most invertebrates being at the egg or very young stages and the water level being very low (Neveu *et al.*, 1979).

Table 5. Energy budget of different sizes of trout in a Pyrenean stream at two times of year (calculations made according to the formulae of Elliott, in cal/trout/day). K: calorie content of trout; C: actual consumption; C_{max}: maximum consumption; F: output through faeces; U: output through ammonium excretion; R_{main}: output through maintenance; R: total metabolic output; R_s: basal metabolism; Ra; output through locomotion; Rd: output linked to digestion; δB: energy available for growth (unpublished data)

Length in cm	Weight Wg	K Cal/g	C	Cmax	C/Cmax	F	U	Rmain	R/Rmain	R	Rs	Ra+Rd	δB	δB/C	δB/W
								June T°C = 14.6							
5– 9.9	2.46	1200	195	391	0.50	31	21	87	1.34	116	63	53	27	0.14	0.008
10.14–9	25.36	1300	1939	2342	0.82	382	186	466	1.79	835	348	487	536	0.28	0.016
15–19.9	59.14	1400	4581	4484	1.02	1008	417	859	2.07	1778	645	1133	1378	0.30	0.017
⩾ 20.0	126.36	1500	6269	8027	0.78	1197	608	1485	1.74	2584	1124	1460	1879	0.29	0.010
								October T°C = 15.3							
5– 9.9	3.48	1200	178	478	0.37	26	20	119	1.18	140	87	53	–008	–0.04	–0.002
10–14.9	20.44	1300	428	1857	0.23	57	50	427	1.00	427	317	109	–106	–0.24	–0.004
⩾ 15.0	75.22	1400	1087	5045	0.21	143	129	1092	1.00	1092	824	268	–277	–0.25	–0.003

V. GENERAL CONCLUSIONS AND PROSPECTS

Many aspects of energy budgets remain to be quantified, especially concerning biotic factors such as the cost of behaviour, especially social behaviour, with the advantages of different social strategies and their ecological effects. In general, benthic feeders are more territorial and more aggressive than plankton feeders. In the trout, dominant individuals eat more and grow faster. They therefore mature earlier and have a higher fecundity.

Similarly, the migratory strategy (migration of salmonids at sea) can be considered as a means of increasing growth, survival and density, that is to say the productivity of the population. However, this migration is often very costly in energetic terms (Wootton, 1985). It must therefore present major advantages, such as the return of much larger adults with high fecundity, which increase the probability of survival.

The scarcity of studies concerning trophic analyses in the wild is linked to two major difficulties: on the one hand the reliable estimation of consumption and on the other hand, the estimation of available prey.

Allen (1951) estimated the consumption of a trout population to be 40–50 times the available benthic food stock and Horton (1961), 8–26 times. This appears to be a paradox, arising from a lack of knowledge about gastric transit times, feeding patterns and their variations and on the efficiency of transformation into growth (Mann, 1978). There is also misunderstanding of the renewal rate of benthic stocks which, for example, is close to 5 in the case of a French Pyrenean river (Neveu *et al.*, 1979). Similarly, the use of a net with too large a mesh size can result in an underestimation of the stock (Macan, 1958). This reflects the general problem in obtaining reliable estimates of invertebrate numbers, some of which can be inaccessible in the substrate (Hynes *et al.*, 1976).

Such predation, if it is so important, will have an effect on the benthic population. Griffith (1981) observed a reduction in invertebrates after the introduction of trout in a stream where they were not previously present. Mann (1978) also noted high predation by fish on the benthos; similarly, trout can significantly reduce zooplankton numbers in still waters (Macan, 1966). In contrast, many observations show no significant difference between the aquatic fauna of a water course where fish are present or absent (Zelinka, 1974; Allan, 1982). This agrees with Mundie (1974) who considered the drift to be sufficient to feed more salmonids than are present naturally, although this drift is only a small part of the benthos.

The effect of the predation of trout on the food supplies is therefore difficult to demonstrate, all the more so as the role of invertebrate predators is significant and often ignored. In addition, invertebrates show a range of adaptations which allow them to avoid predation. These adaptations are either behavioural, with the drift more nocturnal for later larval stages (Stein, 1972; Allan, 1978; Ringler, 1979), or morphological, as in the protection of Trichoptera (caddis fly) by cases (Otto and Svensson, 1980). The activity of the predator can also sometimes decrease prey activity (Stein and Magnusson, 1976).

Even though some experiments show a direct relationship between the stock of available food and growth in trout (Warren *et al.*, 1964), it remains very difficult to establish whether such relationships exist in the wild. Despite evidence of intense consumption of benthic stocks, these do not appear to be affected. All these apparently contradictory facts demonstrate the inability of the present methods for obtaining accurate estimates of both the consumption by wild trout and the quantity of food available.

BIBLIOGRAPHY

Adron J. W., Mackie, A. M., 1978. Studies on the chemical nature of feeding stimulants for rainbow trout, *Salmo gairdneri* Richardson. *J. Fish Biol.*, **12**, 303–310.

Alimov A. F., Shadrin, N. W., 1977. The calorific value of some representatives of freshwater benthos. *Hydrobiol. J.*, **13**, 68–73.

Allan J. D., 1978. Trout predation and the size composition of stream drift. *Limnol. Oceanogr.*, **23**, 1231–1237.

Allan J. D., 1981. Determinant of diet of brook trout (*Salvelinus fontinalis*) in a mountain stream. *Can. J. Fish. Aquat. Sci.*, **38**, 184–192.

Allan J. D., 1982. The effects of reduction in trout density on the invertebrate community of a mountain stream. *Ecology*, **63**, 1444–1455.

Allen K. R., 1951. The Horokiwi stream. *Bull. Mar. Dep. N.Z. Fish.*, **10**, 1–231.

Anderson N. H., Sedell, J. R., 1979. Detritus processing by macroinvertebrates in stream ecosystems. *Ann. Rev. Entomol.*, **24**, 351–377.

Appelbaum S., 1980. Zur der Regenbogen forellen (*Salmo gairdneri* R.) bei der Aufnahme von aromatisiertem bzw. nicht aromatisiertem Futter. *Arch. Fischereiwiss*, **31**, 21–27.

Bachman R. A., 1984. Foraging behavior of free ranging wild and hatchery brown trout in a stream. *Trans. Am. Fish. Soc.*, **113**, 1–32.

Bajkov A. D., 1935. How to estimate the daily feed consumption of fish under natural conditions. *Trans. Am. Fish. Soc.*, **65**, 288–289.

Baker A. M., Frazer, D. F., 1975. A method for securing the gut contents of small, live fish. *Trans. Am. Fish. Soc.*, **105**, 520–522.

Baldwin N. S., 1957. Food consumption and growth of brook trout at different temperatures. *Trans. Am. Fish. Soc.*, **86**, 323–328.

Bannon E., Ringler N. H., 1986. Optimal prey size for stream resident brown trout (*S. trutta*); tests of predictive models. *Can. J. Zool.*, **64**, 704–713.

Behmer D. J., Hawkins C. P., 1986. Effects of overhead canopy on macroinvertebrate production in a Utah stream. *Freshwater Biol.*, **16**, 287–300.

Berg L., Northcote T. G., 1985. Changes in territorial, gill-flaring, and feeding behavior in juvenile Coho salmon (*Oncorhynchus kisutch*) following short-term pulses of suspended sediment. *Can. J. Fish. Aquat. Sci.*, **42**, 1410–1417.

Braaten B. R., 1979. Bioenergetics—A review on methodology. In Halver J. E. and Tiews K. (Eds). *Proc. World Symp. on Finfish Nutrition and Fishfeed Technology*, Vol. II, Berlin, Heinemann, 461–504.

Brett J. R., 1971. Satiation time, appetite and maximum food intake of sockeye salmon (*Oncorhynchus nerka*). *J. Fish. Res. Board Can.*, **28**, 409–415.

Bryan J. E., 1973. Feeding history, parental stock and food selection in rainbow trout. *Behaviour*, **45**, 123–153.

Bryan J. E., Larkin P. A., 1972. Food specialization by individual trout. *J. Fish. Res. Board Can.*, **29**, 1615–1624.

Caspers N., 1975a. Kalorische Werte der dominierenden Invertebraten zweier Waldbäche des Naturparkes Kottenforst-Ville. *Arch. Hydrobiol.*, **75**, 484–489.

Caspers N., 1975b. Untersuchungen über Individuendichte, Biomasse und kalorische Aquivalente des Makrobenthos eines Waldbaches. *Int. Revuew. Ges. Hydrobiol.*, **60**, 557–566.

Chaston L., 1969. Seasonal activity and feeding pattern of brown trout (*Salmo trutta*) in a Dartmoor stream in relation to availability of food. *J. Fish. Res. Board Can.*, **26**, 2165–2171.

Chesson J., 1978. Measuring preference in selective predation. *Ecology*, **59**, 211–215.

Colgan P., 1973. Motivation analysis of fish feeding. *Behaviour*, **45**, 38–66.

Crisp D. T., Mann R. H. K., McCormack J. C., 1978. The effects of impoundment and regulation upon the stomach contents of fish at Cow Green, Upper Teesdale. *J. Fish Biol.*, **12**, 287–301.

Cummins K. W., 1973. Trophic relations of aquatic insects. *Ann. Rev. Entomol.*, **18**, 183–206.

Cummins K. W., Wuycheck J. C., 1971. Caloric equivalents for investigations in ecological energetics. *Mitt. int. Ver. Limnol.*, **18**, 1–158.

Cunjak R. A., Powers G., 1987. The feeding and energetics of stream resident trout in winter. *J. Fish Biol.*, **31**, 493–511.

Dahl J., 1962. Studies on the biology of Danish stream fishes. I. The food of grayling (*Thymallus thymallus*) in some inland streams. *Medd. Dan. Fisk. Havunders.*, **3**, 199–264.

Drummond R. A., Spoor W. A., Olson G. F., 1973. Some short-term indicators of sublethal effects of copper on brook trout. *J. Fish. Res. Board Can.*, **30**, 698–701.

Egglishaw H. J., 1967. The food, growth and population structure of salmon and trout in two streams in the Scottish Highlands. *Freshwater Salm. Fish. Res.*, **38**, 1–32.

Elliott J. M., 1967. The food of trout (*Salmo trutta*) in a Dartmoor stream. *J. Appl. Ecol.*, **4**, 59–61.

Elliott J. M., 1970. Diet changes in invertebrate drift and the food of trout *Salmo trutta* L. *J. Fish Biol.*, **2**, 161–165.

Elliott J. M., 1972. Rates of gastric evacuation in brown trout *Salmo trutta* L. *Freshwater Biol.*, **2**, 1–18.

Elliott J. M., 1973. The food of brown and rainbow trout (*Salmo trutta* and *S. gairdneri*) in relation to the abundance of drifting invertebrates in a mountain stream. *Oecologia*, **12**, 329–347.

Elliott J. M., 1975a. Weight of food and time required to satiate brown trout *Salmo trutta* L. *Freshwater Biol.*, **5**, 51–64.

Elliott J. M., 1975b. Number of meals in a day, maximum weight of food consumed in a day and maximum rate of feeding for brown trout *Salmo trutta* L. *Freshwater Biol.*, **5**, 287–303.

Elliott J. M., 1975c. The growth rate of brown trout (*Salmo trutta* L.) fed on maximum rations. *J. Anim. Ecol.*, **44**, 561–580.

Elliott J. M., 1975d. The growth rate of brown trout (*Salmo trutta* L.) fed on reduced rations. *J. Anim. Ecol.*, **44**, 823–842.

Elliott J. M., 1976a. Body composition of brown trout (*Salmo trutta* L.) in relation to temperature and ration size. *J. Anim. Ecol.*, **45**, 273–283.

Elliott J. M., 1976b. Energy losses in the waste products of brown trout (*Salmo trutta* L.). *J. Anim. Ecol.*, **45**, 561–580.

Elliott J. M., 1976c. The energetics of feeding, metabolism and growth of brown trout (*Salmo trutta* L.) in relation to body weight, water temperature and ration size. *J. Anim. Ecol.*, **45**, 923–948.

Elliott J. M., Persson L., 1978. The estimate of daily rates of food consumption for fish. *J. Anim. Ecol.*, **47**, 977–991.

Ellis R. J., Gowing H., 1957. Relationship between food supply and condition of wild brown trout, *Salmo trutta* L., in a Michigan stream. *Limnol. Oceanogr.*, **2**, 299–308.

Ericksson L. D., 1973. Spring inversion of the diet rhythm of locomotor activity in young sea-going brown trout, *Salmo trutta trutta* L. and atlantic salmon, *Salmo salar* L. *Aquilo Ser. Zool.*, **14**, 68–79.

Fahy E., 1980. Prey selection by young trout fry (*Salmo trutta*). *J. Zool.*, **190**, 27–37.

Feldmeth C. R., Jenkins T. M., 1973. An estimate of energy expenditure by rainbow trout (*Salmo gairdneri*) in a small mountain stream. *J. Fish. Res. Board Can.*, **30**, 1755–1765.

Fortmann H. R., Hazzard A. S., Bradford A. D., 1961. The relation of feeding before stocking to catchability of trout. *J. Wildl. Mngt.*, **25**, 391–397.

Gardner W. S., Nalepa T. F., Frez W. A., Cichocki E. A., Landum P. F., 1985. Seasonal patterns in lipid content of lake Michigan macroinvertebrates. *Can. J. Fish. Aquat. Sci.*, **42**, 1827–1882.

Georges J. P., Gaudin P., 1984. Le tubage gastrique chez les poissons: expérimentation chez la truitelle (*Salmo trutta* L.). *Arch. Hydrobiol.*, **101**, 483–460.

Giles N., 1980. A stomach sampler for use on live fish. *J. Fish Biol.*, **16**, 441–444.

Ginetz R. M., Larkin P. A., 1973. Choice of colors of food items by rainbow trout (*Salmo gairdneri*). *J. Fish. Res. Board Can.*, **30**, 229–234.

Griffith R. W., 1981. The effect of trout predation on the abundance and production of stream insects. M.S. Thesis, University of British Columbia, Vancouver, 250 pp.

Grove D. J., Loizides L. G., Nott J., 1978. Satiation amount, frequency of feeding and gastric emptying rate in *Salmo gairdneri*. *J. Fish Biol.*, **12**, 507–516.

Hara T. J., 1975. Olfaction in fish. *Prog. Neurobiol.*, **5**, 271–335.

Hara T. J., Law Y. M. C., McDonald S., 1976. Effects of mercury and copper on the olfactory response in rainbow trout, *Salmo gairdneri*. *J. Fish Res. Board Can.*, **33**, 1568–1573.

Herrera C. M., 1976. A trophic diversity index for presence–absence food data. *Oecologia*, **25**, 187–191.

Hoar W. S., 1942. Diurnal variations in feeding activity of young salmon and trout. *J. Fish. Res. Board Can.*, **6**, 90–101.

Horton P. A., 1961. The bionomics of brown trout in a Dartmoor stream. *J. Anim. Ecol.*, **30**, 311–338.

Hunt R. L., 1975. Use of terrestrial invertebrates as food by salmonids. *Ecol. Stud.*, **10**, 137–151.

Hynes H. B. N., Williams D. D., Williams N. E., 1976. Distributions of the benthos within the substratum of a Welsh mountain stream. *Oïkos*, **27**, 307–310.

Hyslop E. J., 1980. Stomach content analysis. A review of methods and their application. *J. Fish Biol.*, **17**, 411–429.

Irvine J. R., Northcote T. G., 1983. Selection by young rainbow trout (*Salmo gairdneri*) in simulated stream environments for live and dead prey of different sizes. *Can. J. Fish. Aquat. Sci.*, **40**, 1745–1749.

Ivlev V. S., 1961. *Experimental Ecology of the Feeding of Fishes*. Transl. D. Scott. Yale Univ. Press, New Haven, Conn., 302 pp.

Jacobs J., 1974. Quantitative measurement of food selection. A modification of the forage ratio and IVLEV'S selectivity index. *Oecologia*, **14**, 413–417.

Jenkins T. M., Feldmeth C. R., Elliott G. V., 1970. Feeding of rainbow trout (*Salmo gairdneri*) in relation to abundance of drifting invertebrates in a mountain stream. *J. Fish. Res. Board Can.*, **27**, 2356–2361.

Jobling M., 1981. Mathematical modes of gastric emptying and the estimation of daily rates of food consumption for fish. *J. Fish Biol.*, **19**, 245–257.

Jobling M., 1985. Growth. In *Fish Energetics. New Perspectives*, Tytler P. and Carlow P., Croom Helm, London, 213–230.

Jobling M., 1986. Mythical models of gastric emptying and implications for food consumption studies. *Environ. Biol. Fish.*, **16**, 35–50.

Joblin M., 1987. Influences of food particle size and dietary energy content on patterns of gastric evacuation in fish: test of a physiological model of gastric emptying. *J. Fish. Biol.*, **30**, 289–314.

Kalleberg H., 1958. Observations in a stream tank of territoriality in juvenile salmon and trout (*Salmo salar* L. and *S. trutta* L.). *Rep. Inst. Freshwater Res. Drottningholm*, **39**, 55–98.

Kaushik N. K., Hynes H. B. N., 1971. The fate of dead leaves that fall into streams. *Arch. Hydrobiol.*, **68**, 465–515.

Kimmerer W. J., 1987. The theory of secondary production calculation for continuously reproducing populations. *Limnol. Oceanogr.*, **32**, 1–13.

Krueger C. C., Waters T. F., 1983. Annual production of macroinvertebrates in three streams of different water quality. *Ecology*, **64**, 840–850.

Lien C., 1978. The energy budget of the brown trout population of Ovre Heindalsvatn. *Holarctic Ecol.*, **1**, 279–300.

Light R. W., Adler P. H., Arnold D. E., 1983. Evaluation of gastric lavage for stomach analysis. *North Amer. J. Fish. Mngt.*, **3**, 81–85.

Macan T. T., 1958. Causes and effects of short emergence periods in insects. *Verh. Int. Verein. Limnol.*, **13**, 845–849.

Macan T. T., 1966. The influence of predation on the fauna of moorland fishpond. *Arch. Hydrobiol.*, **61**, 432–452.

Mann K. H., 1965. Energy transformation by a population of fish in the River Thames. *J. Anim. Ecol.*, **34**, 253–275.

Mann K. H., 1978. Estimating the food consumption of fish in nature. In Gerking S. (Ed.), *Ecology of Freshwater Fish Production*, Blackwell Scientific, London, 250–273.

Maslin J. L., Pattee E., 1981. La production du peuplement benthique d'une petite rivière: son evaluation par la méthode de Hynes Coleman et Hamilton. *Arch. Hydrobiol.*, **92**, 321–345.

Mechan W. R., Miller R. A., 1978. Stomach flushing: effectiveness and influence on survival and condition of juvenile salmonids. *J. Fish. Res. Board Can.*, **35**, 1359–1363.

Metz J. P., 1974. Die Invertebratendrift an der Oberfläche eines Voralpenflusses und ihre selektive Ausnutzung durch die Regenbogenforellen (*Salmo gairdneri*). *Oecologia*, **14**, 247–267.

Minshall G. W., 1978. Autotrophy in stream ecosystems. *Bioscience*, **28**, 767–771.

Minshall G. W., Brock J. T., Lapoint T. W., 1982. Characterization and dynamics of benthic organic matter and invertebrate functional feeding group relationships in the Upper Salmon River (USA). *Int. Rev. Ges. Hydrobiol.*, **67**, 793–820.

Morgan R. I. G., 1974. The energy requirement of trout and perch population in Loch Leven, Kinross. *Proc. R. Soc. Edinburgh*, **74**, 333–345.

Morin A., Mousseau T. A., Roff D. A., 1987. Accuracy and precision of secondary production estimates. *Limnol. Oceanogr.*, **32**, 1342–1352.

Mortensen E., 1985. Population and energy dynamics of trout *Salmo trutta* in a small Danish stream. *J. Anim. Ecol.*, **54**, 869–882.

Mundie J. H., 1974. Optimization of the salmonid nursery stream. *J. Fish. Res. Board Can.*, **31**, 1827–1837.

Murdoch W. W., Avery S., Smyth M. E. B., 1975. Switching in predatory fish. *Ecology*, **56**, 1094–1105.

Neveu A., 1979. Les problèmes posés par l'étude de l'alimentation naturelle des populations sauvages de poissons. *Bull. Cent. Etud. Rech. Sci. Biarritz*, **12**, 501–502.

Neveu A., 1980. Influence d'une fine sédimentation dans un canal expérimental sur la densité du macrobenthos, sa composition et sa consommation par les Salmonidés. *Bull. Fr. Pisc.*, **276**, 104–122.

Neveu A., 1981a. Relations entre le benthos, la dérive, le rythme alimentaire et le taux de consommation de truites communes (*S. trutta* L.) en canal expérimental. *Hydrobiologia*, **76**, 217–228.

Neveu A., 1981b. Les rythmes alimentaires en milieu naturel. In *La Nutrition des Poissons*. CNERNA, CNRS, Paris, 339–354.

Neveu A., Echaubard M., 1975. La dérive estivale des invertébrés aquatiques et terrestres dans un ruisseau du Massif central, la Couze Pavin. *Ann. Hydrobiol.*, **6**, 1–26.

Neveu A., Echaubard M., 1983. L'action de l'homme sur la faune des torrents et ruisseaux. Un exemple de torrent pollué: la Couze Pavin. *Ann. Stn Biol. Besse*, **17**, 92–111.

Neveu A., Lapchin L., Vignes J. C., 1979. La macrobenthos de la basse Nivelle, petit fleuve côtier des Pyrénées atlantiques. *Ann. Zool. Ecol. Anim.*, **11**, 85–111.

Neveu A., Thibualt M., 1977. Comportement alimentaire d'une population sauvage de truites fario (*S. trutta* L.). *Ann. Hydrobiol.*, **8**, 111–128.

Newbold J. D., Elwood J. W., O'Neil R. V., Van Winkle W., 1981. Measuring nutrient spiralling in streams. *Can. J. Fish. Aquat. Sci.*, **38**, 860–863.

Newmann R. M., Waters T. F., 1984. Size selective predation on *Gammarus pseudolimnaeus* by trout and sculpins. *Ecology*, **65**, 1535–1545.

Nilsson N. A., 1965. Food segregation between salmonid species in North Sweden. *Rep. Inst. Freshwater Res. Drottningholm*, **46**, 58–78.

Nilsson N. A., 1978. The role of size-biased predation in competition and interactive segregation in fish. In Gerking S. D. (Ed.), *Ecology of Freshwater Fish Production*, Blackwell Scientific, Oxford, 303–325.

O'Brien W. J., Vinyard G. L., 1974. Comment on the use of Ivlev's selectivity index with planktivorous fish. *J. Fish. Res. Board Can.*, **31**, 1427–1429.

Oswald R. L., 1978. The use of telemetry to study light synchronization with feeding and gill ventilation rates in *Salmo trutta. J. Fish Biol.*, **13**, 729–732.

Otto C., Svensson B. S., 1980. The significance of case material selection for the survival of caddis larvae. *J. Anim. Ecol.*, **49**, 855–866.

Penczak T., Kusto E., Krzyzanowska D., Molinski M., Suszycka E., 1984. Food consumption and energy transformation by fish populations in two small lowland rivers in Poland. *Hydrobiologia*, **108**, 135–144.

Pickering A. D., Pottinger T. G., Christie P., 1982. Recovery of the brown trout (*S. trutta* L.) from acute handling stress: a time-cause study. *J. Fish Biol.*, **20**, 229–235.

Priede I. G., Tytler P., 1977. Heart rate as a measure of metabolic rate in teleost fishes: *Salmo gairdneri*, *Salmo trutta* and *Gadus morhua. J. Fish Biol.*, **10**, 231–242.

Priede I. G., Young A. H., 1977. The ultrasonic telematry of cardiac rhythms of wild brown trout (*Salmo trutta* L.) as an indicator of bioenergetics and behaviour. *J. Fish Biol.*, **10**, 299–318.

Richman S., 1971. Calorimetry. In 'a manual on methods for the assessment of secondary productivity in fresh waters'. IBP handbook no. 17, Blackwell Sci. Publ., Oxford and Edinburgh, 146–149.

Rimmer D. M., Power G., 1978. Feeding response of Atlantic salmon (*S. salar* L.) alevins in flowing and still water. *J. Fish. Res. Board Can.*, **35**, 329–332.

Ringler N. H., 1979. Selective predation by drift feeding brown trout (*S. trutta*). *J. Fish. Res. Board Can.*, **36**, 392–403.

Ringler N. H., 1983. Variation in foraging tactics of fishes. In *Predators and Preys in Fishes*. Noakess D. L. G. *et al.* (Eds), Dr W. Junk, The Hague, 159–171.

Ringler N. H., 1985. Individual and temporal variation in prey switching by brown trout, *Salmo trutta. Copeia*, **4**, 918–926.

Ringler N. H., Brodowski D. F., 1983. Functional responses of brown trout (*Salmo trutta*) to invertebrate drift. *J. Freshwater Ecol.*, **2**, 45–57.

Rocha A. J., Mills D. H., 1984. Short communication 'Utilization of fish pellets and wood fragments by brown trout (*S. trutta*) in Cobbinshaw Reservoir, West Lothian, Scotland. *Fish Mngt.*, **15**, 141–142.

Seaburg K. G., Moyle J. B., 1964. Feeding habits, digestion rates and growth of some Minnesota warmwater fishes. *Trans. Am. Fish. Soc.*, **93**, 269–285.

Serebrov L. I., 1983. Effect of a current on the intensity of feeding in certain fishes. *Hydrobiol. J.*, **9**, 68–70.

Smith M. A. K., Thorpe A., 1976. Nitrogen metabolism and trophic input in relation to growth in freshwater and saltwater *Salmo gairdneri. Biol. Bull.*, **150**, 139–151.

Soofiani N. M., Hawkins A. D., 1985. Field studies of energy budgets. In P. Tytler and P. Calow (Eds), *Fish Energetics. New Perspectives*, Croom Helm, London, 283–308.

Staples J., Nomura N., 1976. Influence of body size and food ration on the energy budget of rainbow trout *Salmo gairdneri* Richardson. *J. Fish Biol.*, **9**, 28–43.

Stein R. A., Magnuson J. J., 1976. Behavioral response of crayfish to a fish predator. *Ecology*, **57**, 751–761.

Steine I., 1972. The number and size of drifting nymphs of Ephemeroptera, Chironomidae and Simuliidae by day and night in the river Stranda, Western Norway. *Norw. J. Ent.*, **19**, 127–131.

Strauss R. E., 1979. Reliability estimates for Ivlev's electicity index, the forage ratio, and a proposed linear index of food selection. *Trans. Am. Fish. Soc.*, **108**, 344–352.

Tanaka H., 1970. On the nocturnal feeding activity of rainbow trout (*Salmo gairdneri*) in streams. *Bull. Freshwater Fish. Res. Lab.*, Tokyo, **20**, 73–82.

Thomas J. D., 1962. The food and growth of brown trout (*Salmo trutta* L.) and its feeding relationships with the salmon parr (*S. salar* L.) and the eel (*Anguilla anguilla* L.) in the river Teify, West Wales. *J. Anim. Ecol.*, **31**, 175–205.

Tusa I., 1969. On the feeding biology of the brown trout (*Salmo trutta* m. *fario* L.) in the course of day and night. *Zool. List.*, **18**, 275–284.

Utne F., 1979. Standards methods and terminology in finfish nutrition. In J. E. Halver and K. Tiews (Eds). *Proc. World Symp. on Finfish Nutrition and Fishfeed Technology*, Berlin, Heinemann, 438–443.

Vannote R. L., Minshall G. W., Cummins K. W., Sedell J. R., Cushing C. E., 1980. The river contuinum concept. *Can. J. Fish. Aquat. Sci.*, **37**, 130–137.

Vinyard G. L., O'Brien N. J., 1975. Dorsal light response as an index of prey preference in bluegill (*Lepomis macrochirus*). *J. Fish. Res. Board Can.*, **32**, 1860–1863.

Wankowski J. W. J., 1979. Morphological limitations, prey size selectivity and growth response of juvenile atlantic salmon, *Salmo salar. J. Fish Biol.*, **14**, 89–100.

Ware D. M., 1971. Predation of rainbow trout (*Salmo gairdneri*): the effect of experience. *J. Fish. Res. Board Can.*, **28**, 1847–1852.

Ware D. M., 1972. Predation by rainbow trout (*Salmo gairdneri*): the influence of hunger, prey density and prey size. *J. Fish. Res. Board Can.*, **29**, 1193–1201.

Ware D. M., 1973. Risk of epibenthic prey to predation by rainbow trout (*Salmo gairdneri*). *J. Fish. Res. Board Can.*, **30**, 787–797.

Warren C. E., Wales J. H., Davis G. E., Doudoroff P., 1964. Trout production in an experimental stream enriched with sucrose. *J. Wildl. Manage.*, **28**, 617–660.

Waters T. F., 1969. Invertebrate drift ecology and significance to stream fishes. In Northcote T. G., *Symposium on Salmon and Trout in Streams*, H. R. McMillan Lectures in Fish., Univ. Brit. Columbia, Van., 121–134.

Waters T. F., 1977. Secondary production in inland waters. *Adv. Ecol. Res.*, **10**, 91–104.

Weatherley A. H., Gill H. S., 1983. Protein lipid, water and caloric content of immature rainbow trout *Salmo gairdneri* Richardson, growing at different rates. *J. Fish Biol.*, **23**, 653–688.

Winberg G. G., 1956. Rate of metabolism and food requirements of fishes. *Transl. Russ. by Fish. Res. Board Can.*, **194**, 251 p.

Windell J. T., 1978a. Estimating food consumption rats of fish populations. In *Methods for assessment of fish production in freshwater*, IBP Handbook, vol. 3, 227–254, Blackwell Sci. Publ., London.

Windell J. T., 1978b. Digestion and the daily ration of fishes. In S. D. Gerking (Ed.), *Ecology of Freshwater Fish Production*, Blackwell Sci. Publ., London, 159–183.

Wissing T. E., Hasler A. D., 1968. Calorific values of some invertebrates in lake Mendota, Wisconsin. *J. Fish. Res. Board Can.*, **25**, 2515–2518.

Wissing T. E., Hasler A. D., 1971. Intraseasonal change in caloric content of some freshwater invertebrates. *Ecology*, **52**, 371–373.

Wootton R. J., 1985. Energetics of reproduction. In P. Tytler and P. Calow (Eds), *Fish Energetics. New Perspectives*, Croom Helm, London, 231–256.

Yurkowski M., Tabachek J. L., 1979. Proximate and aminoacid composition of some natural fish foods. In J. E. Halver and K. Tiews (Eds). *Proc. World Symp. on Finfish Nutrition and Fishfeed Technology*. Berlin, Heinemann, 436–445.

Zelinka M., 1974. Die Eintagsfliegen (Ephemeroptera) in Forellenbachen der Beskiden. III. Der Einfluss des verschiedenen Fischbestandes. *Vest. Cesk. Spol. Zool.*, **387**, 76–80.

4

Social organization and territoriality in brown trout juveniles during ontogeny

M. Héland

I. INTRODUCTION

The first studies of salmonid population dynamics have shown the determining role of behavioural factors, particularly competition for space and food in the limiting of natural populations and their distribution (Allen, 1969; McFadden, 1969; Le Cren, 1973). This role of behaviour and competition in the biology of populations and its consequences on fish production appears to be especially marked for the brown trout, *Salmo trutta* L., a strongly territorial salmonid species in freshwater lotic environments (Onodera, 1967; Noakes, 1978, Thorpe, 1982).

In freshwater steep streams, generally located in the upper part of rivers corresponding to the natural trout habitat, territorial competition is the rule (Frost and Brown, 1967). Hynes (1970) thought that settlement of territories by trout fry starts very early, as soon as the young fry emerge and are able to swim. It is also during the juvenile stage which is characterized by high mortality rates that the future population of the stream is established (Le Cren, 1961, 1973; Elliott, 1984, 1994). Le Cren (1961) put forward the hypothesis that territorial behaviour is responsible for the regulation of the density of juvenile fish in streams (Fig. 1).

With regard to territoriality in fishes, Greenberg (1947) was one of the first to demonstrate its existence, in the green sunfish, *Lepomis cyanellus* Rafinesque, according to the definition of Noble (1939) which included, essentially, the component of agonistic behaviour: 'the territory is any defended area'. In his work on the territorial behaviour of the stickleback, *Gasterosteus aculeatus* L., Assem (1967) noted that the concept of territory must contain two essential elements: spatial restriction of all or part of behaviour, and exclusion of conspecifics. He gives the following definition: the territory corresponds to any area where the presence and/or behaviour of a resident excludes or tends to exclude the simultaneous presence of conspecifics (Assem, 1970). The territory is distinguished from the home range of which it is a reduced portion. The home range is

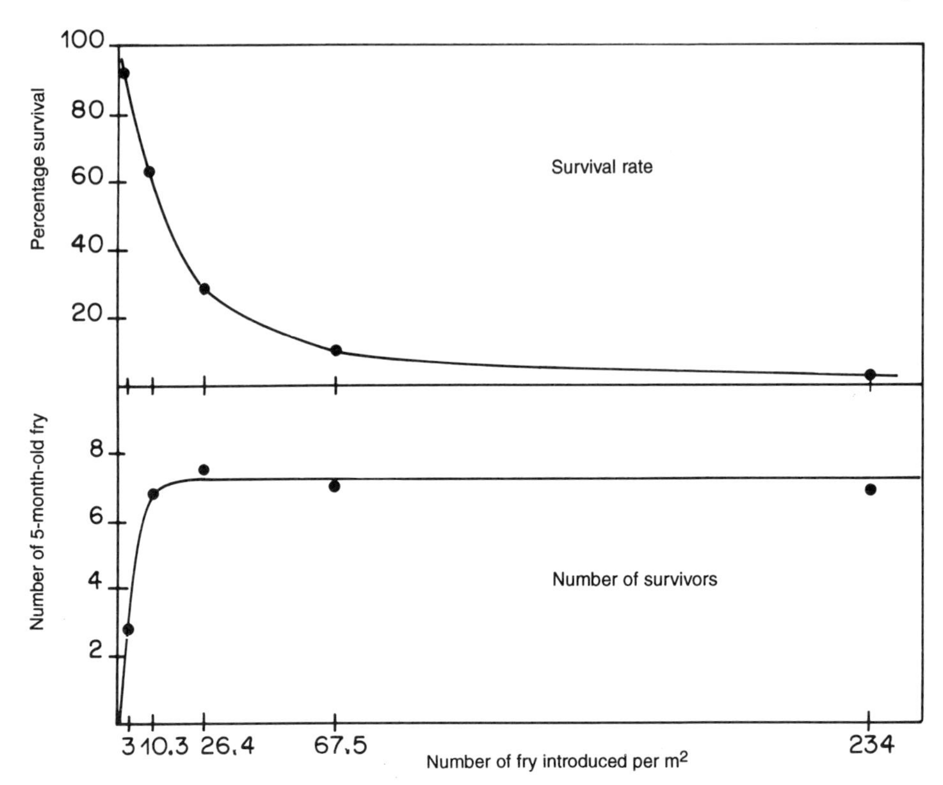

Fig. 1. Survival of trout fry used for stocking at different densities into sections of a natural stream. Upper graph = percentage survival from first feeding (start of May) until the beginning of September. Lower graph = number of individuals found per m^2 in September. (After Le Cren, 1961.)

defined for a given period as the assemblage of places frequented by an individual or group of individuals (Gerking, 1953; Richard, 1970).

Observations of territoriality outside the spawning season in the brown trout go back a long way, since Lindroth, in 1955, noted the behaviour in the river Indalsälven and established that the territory of each fry is approximately 1 m^2 at the end of June. Previously, Stuart (1953) had used the term 'territory', but suggested that fry, after emerging from the gravel, respected 'tolerated distances' of about 7.5 cm which increased with fry size. Kalleberg (1958) conducted the first study on territoriality in trout and Atlantic salmon, *Salmo salar* L., in an artificial stream, describing precisely the structure of the territory and agonistic behaviour linked to it and comparing it to the observations of other authors on the aggression of salmonids linked to spawning behaviour (Fabricius, 1951, 1953; Fabricius and Gustafson, 1954, 1955). Subsequently, numerous studies have dealt with, at least partially, territoriality in brown trout, but few of them have examined the appearance of this behaviour in the fry and its evolution during the course of ontogeny (Kennedy and FitzMaurice, 1968; Jenkins, 1969; Timmermans, 1972; Le Cren, 1973; Jones, 1975; Bachman, 1984; Grant and Kramer, 1990).

The objective of this chapter is to present an analysis of the social relationships and territorial competition in the brown trout during development throughout the immature juvenile phase, excluding migratory and reproductive phases. Numerous results illustrating the account are taken from experiments carried out at the Hydrobiological Station at Saint-Pée-sur-Nivelle, sometimes in the natural environment, but much more often in artificial streams reproducing the environment of a salmonid spawning area, the zone to which our study was limited.

II. DEVELOPMENT OF TERRITORIAL BEHAVIOUR IN THE FRY

1. Observations on early phases of social and territorial behaviour

Around the redds in both natural and artificial streams, trout fry appear at the time of emergence, in the interstices of the substrate. They are generally stationary on the bottom or move with weak swimming movements, remaining close to the bottom, fighting against the water current which tends to sweep them downstream. At this stage, the first feeding behaviour appears in the form of capturing larval invertebrates from the drift, which pass close by (Gaudin and Héland, 1995).

After several days, the behaviour of stationary swimming facing the current, also called 'static swimming' (Héland, 1975), becomes more secure and permanent (Héland, 1978). The fry therefore congregate in areas with moderate current (less than 20 cm/s) and search for a favourable swimming position, generally behind an obstacle in the substrate, such as a stone or large pebble, from where they capture their prey. This position is generally called 'holding station' or 'feeding station'. Note the development of two types of positioning in the fry: in relation to the water current (rheotaxis), against which the animal must resist and in relation to prey made up of invertebrate drift. In the very first days following emergence, the first aggressive behaviour between fry is expressed during confrontations while searching for holding positions or capturing prey (Gaudin and Héland, 1995).

During two consecutive identical experiments, in the same experimental stream, the change in numbers of fry holding their station and moving fry was noted from the end of emergence to approximately three weeks later (Héland, 1971a). The curves obtained for these two experiments are similar in appearance and demonstrate a general phenomenon observed in fry at emergence (Fig. 2). After yolk sac resorption and emergence from the gravel, the proportion of fry stationary swimming increases. This increase corresponds to the learning period of the static swimming behaviour facing into the current. At the same time, some 10–20% of fry change their holding position in order to look for better feeding stations.

In the following phase, around 5–6 days after hatching, the population of fry owning holding stations decreases greatly and then stabilizes 12–15 days after emergence (Fig. 2). During this phase, characterized by the appearance of much aggressive behaviour in the population, competition arises between the fry for possession of the best feeding positions. This competition ends with the elimination of some fry which are forced to hide under stones to avoid attacks. A social hierarchy is thus established in which dominant individuals defend a feeding station and its immediate surroundings, now called 'territory', while subordinate individuals are forced to subsist in hiding places

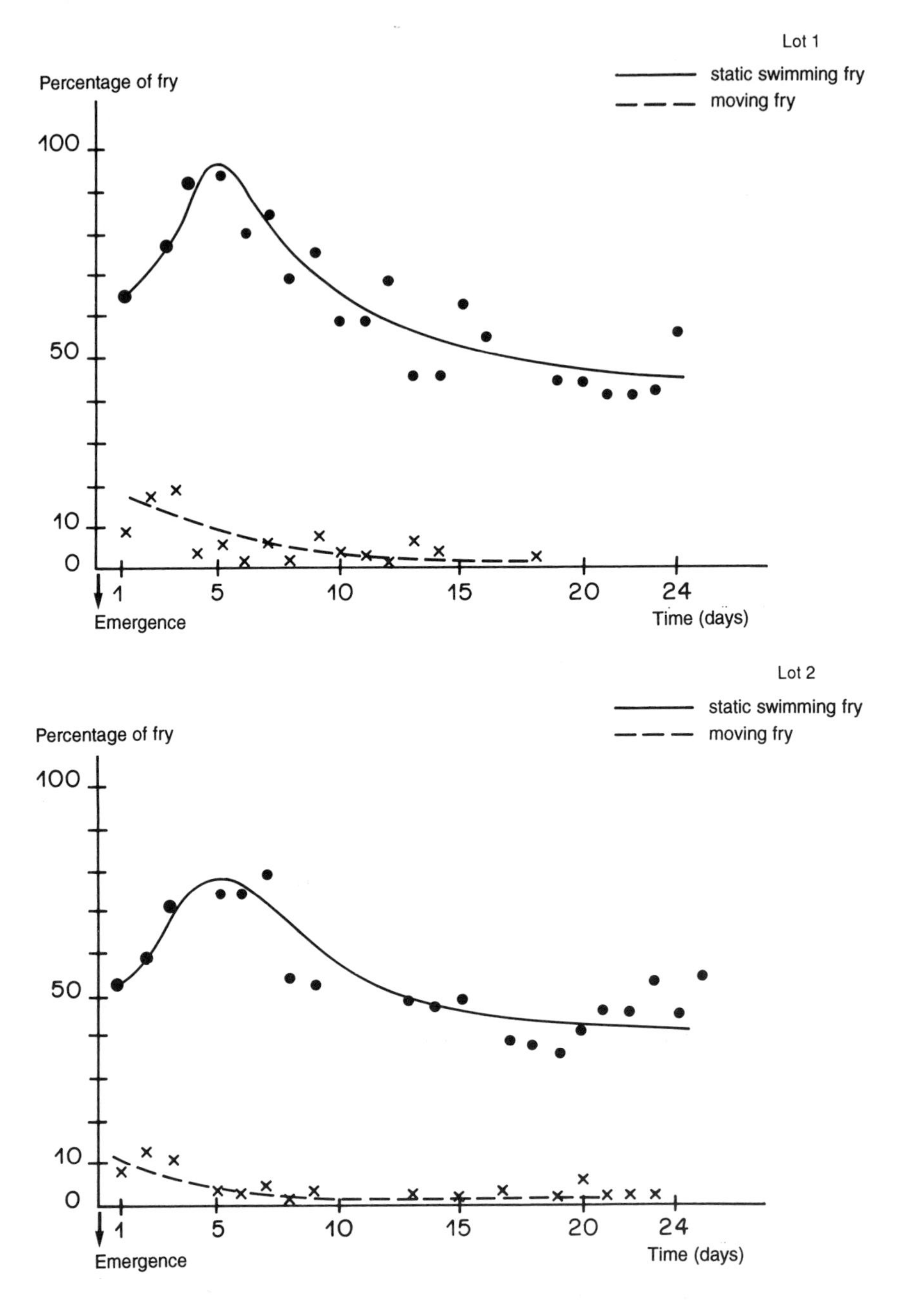

Fig. 2. Change in the proportions of static swimming fry and moving fry during two consecutive experiments in an experimental stream, during the first weeks after the end of emergence (after Héland, 1971a).

out of sight of the dominants or flee. A mosaic of contiguous and exclusive territories is created. It appears that visual perception of conspecifics is determinant in the establishment of territory borders: Stuart (1953) showed that positioning a vertical opaque screen amongst the trout fry after emergence decreased inter-individual distances on either side of the screen, while a transparent screen had no effect. Kalleberg (1958) showed that, when the substrate is diversified and allows visual isolation between fry, the number of territories and consequently the carrying capacity, increases (Fig. 3).

2. Life under gravel, emergence and learning to swim

The possible origin of certain behaviours in larvae expressed from emergence by the fry has been studied in artificial redds made of glass beads. The observations confirm those of Geiger and Roth (1962) and Roth and Geiger (1963): after hatching, the larvae, which exhibit negative phototaxis, bury themselves in the gravel and disperse. On the tenth day, approximately, post-hatching, this dispersion is increased, the larvae becoming more mobile. Then the ascent to the surface begins, followed by emergence which is spread over about 8 days for one batch. The fry frequently remain several days at the edge of the redds before starting to swim in the open water.

Interactions between larvae can explain their active spacing in the redd according to the hypothesis of Dill (1969) based on observations on Pacific salmon. However, no interactions between individuals of a social character or any indication of behaviour associated with capture of prey was seen inside the redds. This does not necessarily mean that the larvae do not have the capacity to express such behaviours. In open water, in a hatchery tank, for example, bites between larvae and capture of prey before complete yolk sac resorption can be observed.

At emergence, the learning of the behaviour of static swimming facing the current (rheotaxis) takes place. This very important period also corresponds to the transition between endogenous feeding from vitelline reserves to exogenous feeding from invertebrate drift (Héland, 1978). The precise observation of the establishment of holding station in fry at emergence shows that this activity develops progressively in increasingly longer but fewer sequences (Fig. 4). About 12 days after emergence, the fry have acquired the capacity for staying in one place by station holding in open water. Dill (1977) described the same phenomenon in the Atlantic salmon and rainbow trout (*Oncorhyncus mykiss* Walbaum). During this learning, the animal searches for the best place in the stream to hold station, sheltering from the current, while at the same time staying close to the main flow where the invertebrate drift is significant.

The establishment of holding station in the group of fry, emerging from the redd, is in preparation for the establishment of social relationships and territorial behaviour (Héland *et al.*, 1995).

3. Agonistic behaviour and site attachment

Competition for holding stations or access to invertebrate drift is accompanied by the exhibition of the whole range of agonistic behaviours from the most aggressive tendencies to submission. These behaviours are susceptible, for the most part, of being expressed very early on, from when the fry start to swim against the current and exploit the trophic environment. Their description has already been made by different authors

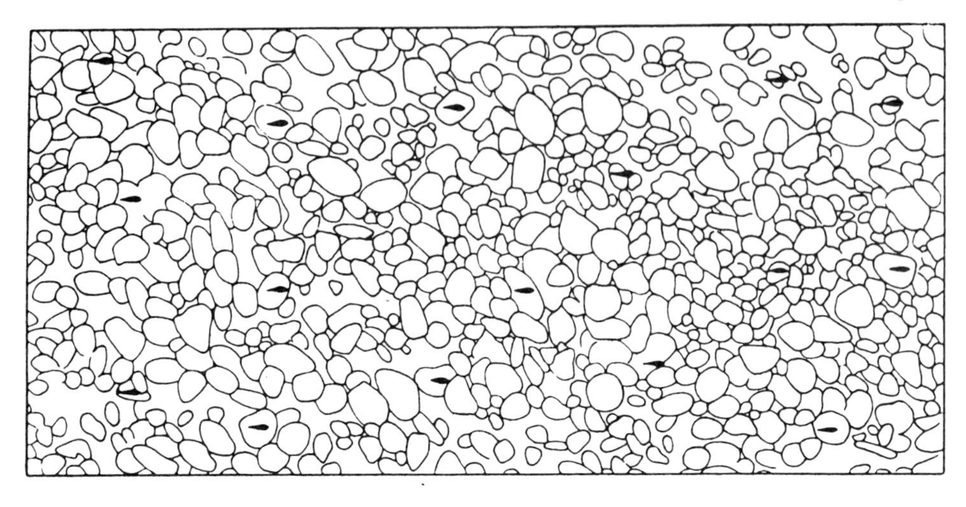

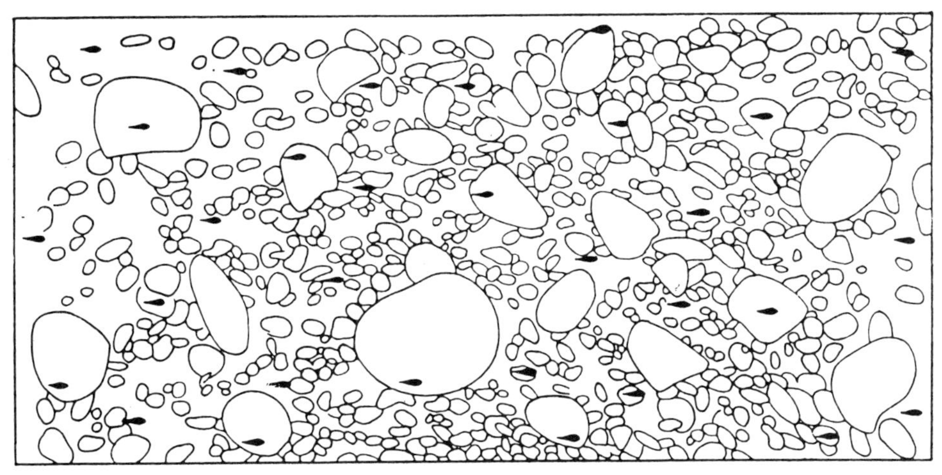

Fig. 3. 'Holding stations' used by Atlantic salmon fry within the territorial mosaics of two sections of artificial stream, with different substrates. Each section had an area of 170 × 80 cm (after Kalleberg, 1958).

since Stuart (1953) and Kalleberg (1958), although the terms chosen to describe them may vary according to author (Hartman, 1963; Jenkins, 1969).

(a) The charge

Executed by a fry while holding station, encountering another entering its immediate environment, it consists of direct and rapid swimming towards the intruder. It generally ends with a bite or an attempt to bite.

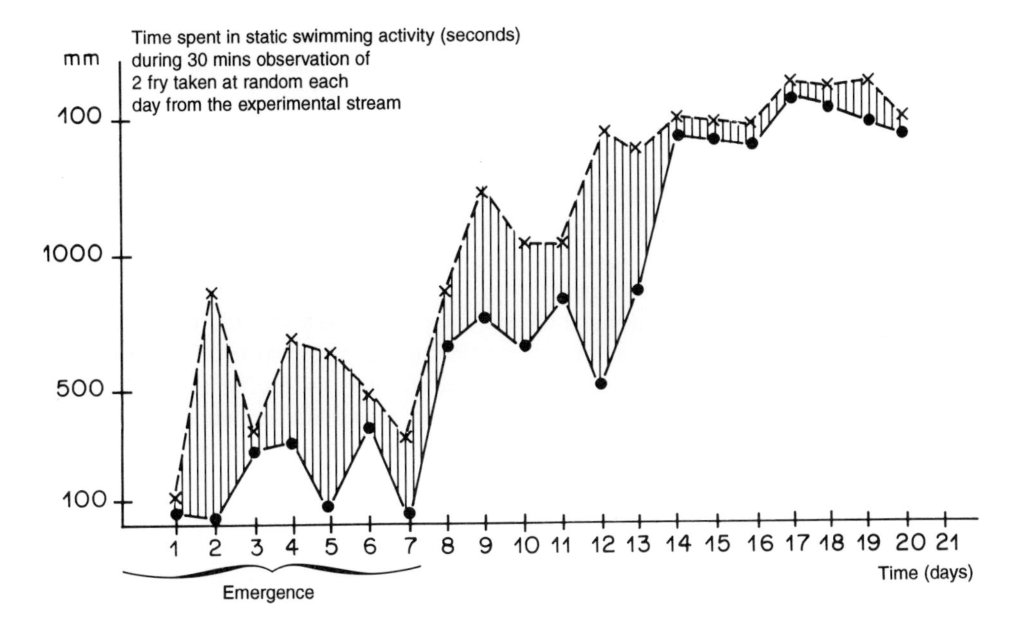

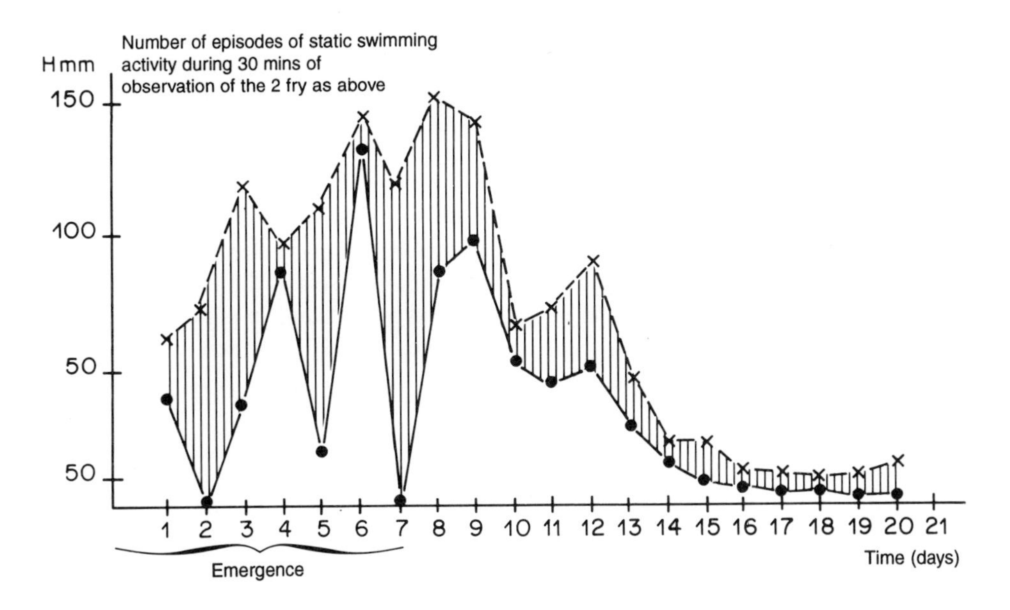

Fig. 4. Change, in two trout fry taken at random each day from an experimental stream, in the time spent in static swimming activity (upper figure) and the number of episodes of static swimming (lower figure) during daily observation periods of 30-min duration, from emergence to 3 weeks later. This change is represented by the shaded area between the upper and lower values noted, for the two fry (after Héland, 1978).

(b) The bite

This most aggressive behaviour arises after a charge or frontal threat display. The bite is not always effective as the attacked alevin sometimes manages to avoid contact.

(c) The chase

This results from repeated charges towards a fish which flees: it goes away a little further after each charge, to avoid the aggressor's bites.

(d) Frontal and lateral threat displays

The postures of frontal or lateral threat are stereotypically agonistic behaviours adopted by certain fish, amongst them the salmonids, to 'intimidate' an adversary. They consist of presenting an 'exaggerated' size, either face on (frontal) or profile (lateral) to the opponent, using contractions of particular muscles affecting the curvature of the body, erection of the fins, spreading of the opercula or lowering of the buccal floor (see the diagram in Kalleberg, 1958).

These posturings are agonistic behaviours which express conflict between the tendency to attack or flee in the animal. They generally appear between fry at the same social level, and consequently are rarely observed at emergence before the settlement of the hierarchy. They are frequent between dominant neighbour fry when one of the two enters the other's territory (McNicol and Noakes, 1981). A short time after emergence, the 'zone of influence' corresponds to a zone of intolerance to intruders around the resident, limited by the distances of attack or flight: the fish therefore respect the interindividual distances and can frequently change their position. Later, this zone of influence can progressively stabilize within the space and area defined as a territory, with temporary topographical limits, perpetually evolving as the fry grows.

(e) Submission posture

This is a posture shown exclusively by subordinates which many authors have observed without describing it as a separate behaviour. It manifests itself by the immobilization of the individual on the stream bed and is accompanied by a change in pigmentation of the fish which becomes colourless and pale, losing lateral markings which always show a high degree of contrast in dominant individuals.

(f) Flight

This behaviour, very frequent during aggressive encounters, is expressed by alevins subjected to attack; they flee very rapidly and immediately, usually towards shelter in a crevice in the substrate, most often under a stone or a pebble. Shelter is another component of the territory which allows the animal to escape from predators and to find, in certain cases, a protective cover from strong light (Butler and Hawthorne, 1968; Gibson and Power, 1975). A diverse substrate and a significant water depth increase the potential for shelter in a stream. In the brown trout, shelter is not always situated within the territory and can be beneath an undercut bank some distance away. However, the closeness of shelters play an important role in the selection of holding stations by the fry because of the necessity of avoiding competitors and predators and, also, to maximize food intake. In the brook charr, *Salvelinus fontinalis* Mitchell, the shelter-cover component plays an important role in the choice of territories with regard to social level (Caron, 1986).

While the concept of defence is essential to the notion of territory, that of site attachment associated with the restriction of activities to a particular area is equally so (Wickler, 1976). Testing site attachment (or territory attachment) allows the set-up of the link between the animal and its environment to be determined. These tests consist of noting the time for the return of a fry to its holding station, having left it after being shown a competitor model (Héland, 1971a). Using different competitor models to avoid habituation, the results are identical.

At the start, just after emergence, the tested fry move away and do not return. Thereafter, they return to their holding station more and more quickly in the following days (Fig. 5). Towards the end of the experiment, 2–3 weeks after emergence, certain fry adopt an aggressive attitude on encountering the competitor model: charge or showing lateral threat display. The longer the fry spend at one holding station, exploiting the invertebrate drift which passes by, the more attached they become to it and prevent access to conspecifics. Capture of prey in the drift comes in here as feeding reinforcement of attachment (Skinner, 1971; Fausch, 1984).

Such an attachment underlines how important it is for a fry to rapidly find a suitable station. Those which establish first will be able to 'appropriate' a station, which gives an important advantage to the fry which emerge precociously. Conversely, later-emerging

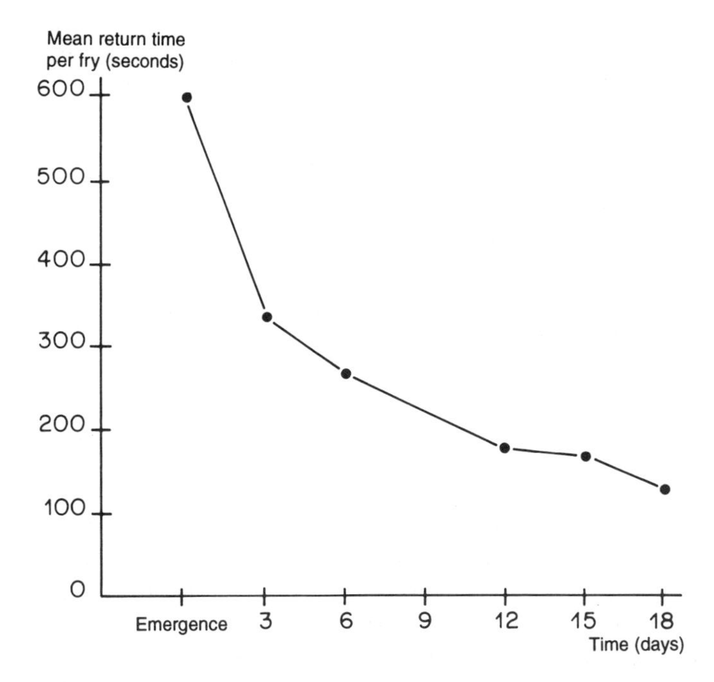

Fig. 5. Test of site attachment: change in time taken to return to holding station of trout fry after emergence, after being frightened by a predator model on the end of a glass rod (after Héland, 1971a).

fry will be excluded from stations already occupied and will have to seek out less favourable positions or move to other zones, exposing themselves to predation. This rule, showing the benefit of early occupation, was formulated by Jenkins (1969) and shows the reinforcement of the dominant position of residents relative to new arrivals (see also Johnson *et al.*, 1999). This phenomenon is quite widespread in fish (Zayan, 1975, 1976; Brown and Green, 1976). Egglishaw and Shackley (1973) experimentally reversed the dominance between two sympatric populations of trout and salmon in a natural stream, by introducing salmon fry which were more precocious than those of trout: contrary to the converse natural situation, the salmon were therefore 'dominant' to the small trout, which resulted in better growth and survival.

III. RELATIONSHIPS BETWEEN FEEDING, DENSITY AND TERRITORIALITY

The establishment of social behaviour and the territorial mosaic of the trout fry appears to be motivated by access to food in the form of invertebrate drift. It is of interest to know whether a change in numbers of fry or the amount of available food can lead to a modification of social and territorial organization.

1. Influence of density

According to Le Cren (1961), competition for acquisition of territories in small trout is responsible for the regulation of populations in streams (Fig. 1). This hypothesis was tested experimentally by examining whether this influence can be detected from the establishment of territories at the start of the fry's life in open water (Héland, 1971b).

In three parallel and identical, artificial streams, stocked with 100, 50 and 25 first feeding fry respectively, the change in static swimming fry numbers (and holding station with time) was followed daily during the first month post-emergence (Fig. 6). At the end of the experiment, these numbers tended towards the same values, and the number of territories counted at the end of the experiment (17, 15 and 13), did not differ significantly. This experiment suggests that for a given substrate, there is a maximum defined number of territories, according to the developmental stage of the fish, which confirms Le Cren's hypothesis.

After the effect of overpopulation was demonstrated, the effect of isolation was investigated: would a fry, isolated from the egg stage, be able to establish a territory? This experiment was carried out in 18 identical artificial streams, supplied from a single natural stream (Héland, 1982). The arrival of food in the form of invertebrate drift therefore followed a natural rhythm. The experiment lasted approximately 2 months, according to the rate of development of the larvae, and was repeated in two consecutive years. The alevins were introduced during yolk sac resorption in the form of one single alevin (isolated from egg stage) in six streams, two alevins in six other streams and a group of ten alevins (five in the second year of the experiment) in the remaining eight streams. The observations were carried out on the micro-separation of the alevins, as well as on their observed daily activities, whenever the turbidity allowed.

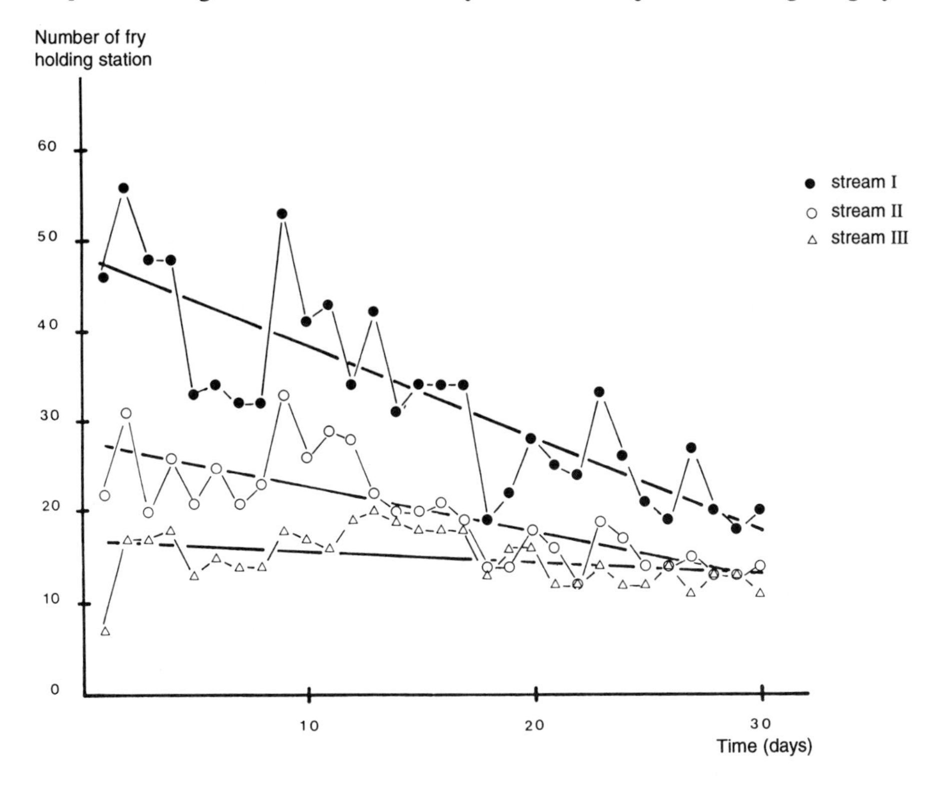

Fig. 6. Influence of initial population density on the settlement of territories in trout fry. Change over 1 month in the number of fry holding station in three artificial streams containing, at the start, 100(I), 50(II) and 25(III) fry at first feeding stage. The numbers of territories recorded at the end of the experiment did not differ significantly.

From emergence, the progressive occupation of the environment develops according to the same pattern in the three experimental situations: phase of instability during the establishment of swimming behaviour and the search for the most favourable position, then the stable position directed towards the taking of food. There is no difference between the activities of isolated fish and the others (except aggressive activities). Thus, several isolated fry remained hidden under a stone most of the time, exactly like certain subordinate fry. Tests of site attachment carried out on the three types of stream did not show any significant differences between the 'naive' (isolated) or 'experienced' (in a group) fry (Fig. 7).

Overall, there does not appear to be any notable difference between orientation in the environment and the behaviour of isolated fry compared to those in a group, which emphasizes the influential role of ecological factors, such as water velocity or invertebrate drift in the exploitation of the trophic environment in the form of territoriality. Independently of the presence of conspecifics, the fry attach themselves to a particular site in the stream bottom, from which they receive a feeding reinforcement in invertebrate drift form. Territorial behaviour is an adaptive response of the fry in the stream

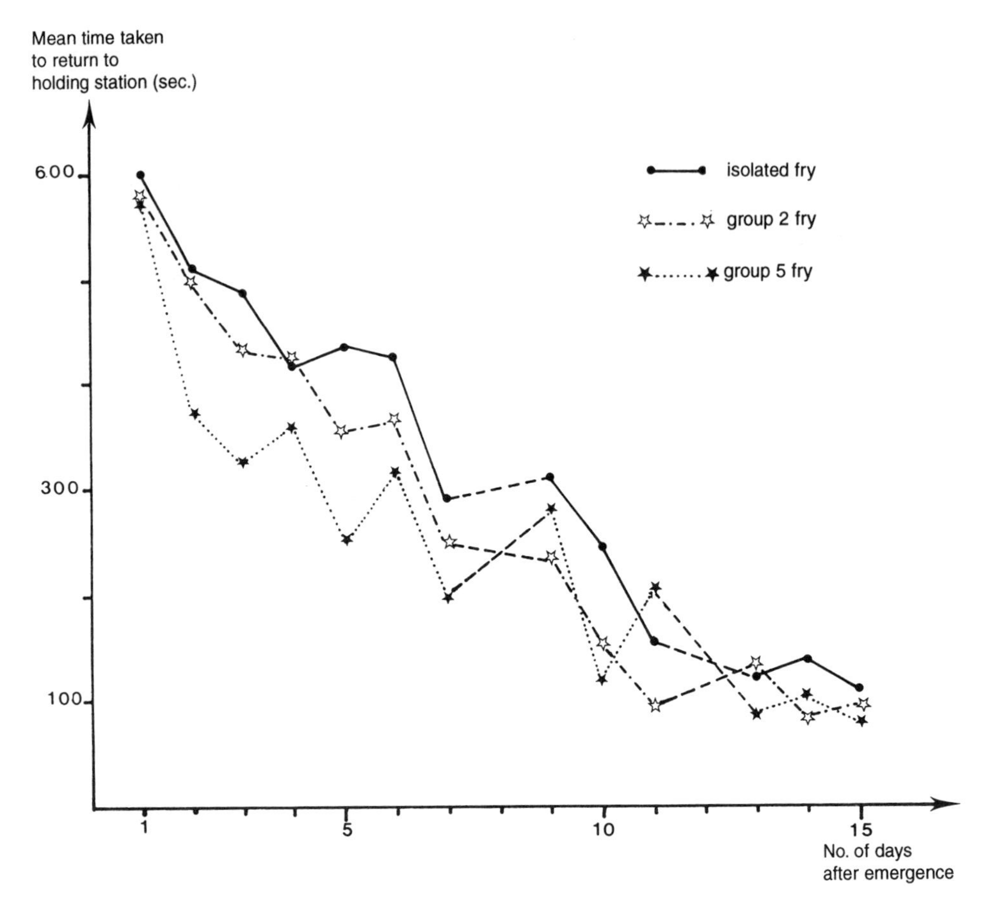

Fig. 7. Tests of site attachment in brown trout fry, either isolated or in a group, in 18 artificial, parallel, streams fed from a natural stream. The time taken for a fry to return to its holding station after being frightened by a predator model (after Héland, 1982).

environment, where prey are carried by the water current, the animal searching for the best balance between energy expenditure in swimming against the current and energy gain by food intake from the drift. This 'energetic' function of the territorial organization in juvenile salmonids in streams has been demonstrated in several species (Dill *et al.*, 1981; Hughes, 1990; Hughes and Dill, 1992; McNicol and Noakes, 1984; McNicol *et al.*, 1985; Puckett and Dill, 1985).

2. The influence of food

Chapman (1966) asserted that space is a regulator of population density in the salmonids, in as far as it is linked to food. From this statement, the density would be expected to adjust according to the abundance or scarcity of food. How does this relate to territoriality?

Slaney and Northcote (1974) showed that, for rainbow trout, it is possible to decrease the extent of territories by increasing prey density.

The brown trout shows some differences to the rainbow trout, both in terms of aggressive behaviours (Jenkins, 1969) and in terms of its occupied ecological niche (Dumas, 1976). In the brown trout at the juvenile stage in a flowing water environment close to the spawning zone, the hypothesis to be tested is that an increase in available food can result in an increase in the number of territories, while decreasing their area. This hypothesis was examined experimentally in fry at yolk sac resorption, then in 3-month-old fry.

In the *yolk sac resorption fry*, the experiment was carried out on two identical artificial streams, stocked with the same number of yolk sac fry (Héland, 1977). Each stream received a different ration quantity in the form of frozen plankton, distributed over a period of time at the start and the end of the day: 'normal' ration quantity in the first experimental stream, double ration quantity in the second stream. The number of fry holding station in the two streams was noted daily from the end of emergence until 2 weeks later (Fig. 8). At the start of the experiment, the static swimming fry appeared to be more numerous in experimental stream 2 (double feeding) but after territorial stabilization, the two curves were very close.

In *3-month-old fry*, a similar experiment was carried out in three streams with three food ration quantities: low, normal and double. As well as plankton, the food ration was

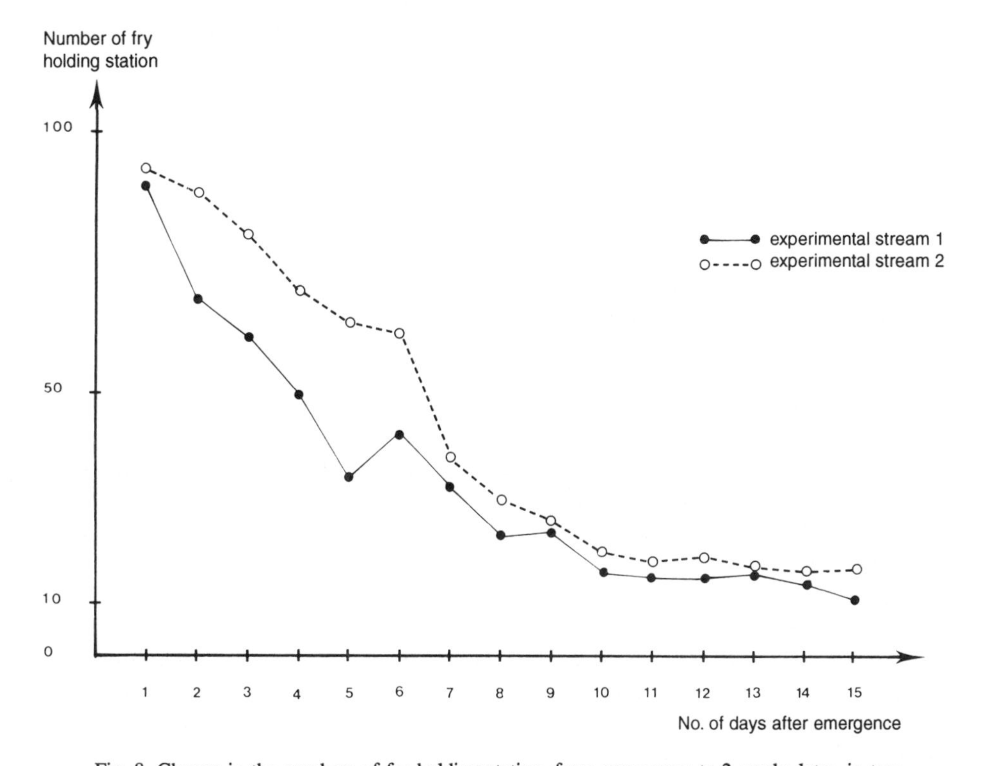

Fig. 8. Change in the numbers of fry holding station, from emergence to 2 weeks later, in two identical, parallel experimental streams fed by different ration quantities of frozen plankton: 'normal' ration in stream 1 and double ration in stream 2 (after Héland, 1977).

made up of weighed amounts of chironomid larvae (genus *Chironomus*). Despite very different amounts of distributed food, the pattern of change of fry numbers at the stationary swimming stage in the three streams is similar (Fig. 9). This result shows that the number, and consequently the size, of visible fry territories is not directly modified by food quantity. However, the number of dominant fry showing hiding behaviour, which remain in the stream at the end of the experiment is slightly greater when food is superabundant, as has been observed by Symons (1971) in the Atlantic salmon. On the other hand, when food is scarce (low food ration in the second experiment), the downstream displacement of fry is greater. In all these experiments in artificial streams, the fry can leave the stream in both an upstream and a downstream direction where they are caught in traps.

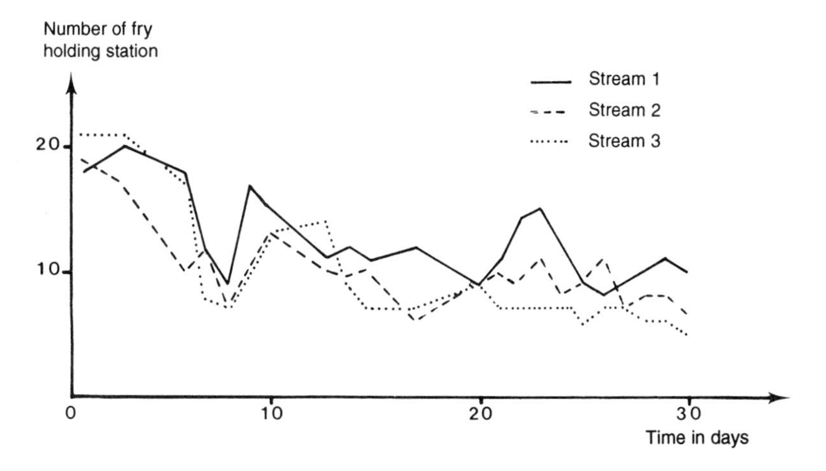

Fig. 9. Change in the numbers of 3–4-month-old fry holding station in three identical, parallel, experimental streams, under different feeding regimes (frozen plankton and chironomid larvae): normal ration in stream 1, double ration in stream 2, low ration in stream 3.

IV. EARLY DOWNSTREAM DISPLACEMENT OF FRY

The phenomenon of displacement of trout fry at the stage of emergence has long been noted by biologists (Huet, 1961; Elliott, 1966; Timmermans, 1966). In natural streams, it is common to catch fry in the spring in downstream traps. These downstream movements occur mainly nocturnally, with a peak corresponding to the period of emergence or a short time after. Some authors (Le Cren, 1961; Chapman, 1962) consider that the 'downstream moving' fry are excess numbers in the spawning zone, which will populate downstream sectors of the stream where recruitment is lower. Elliott (1987) noted that the majority of these fry are moribund and are lost from the ecosystem.

1. Observations in the natural environment

During the study of the brown trout fry population in the Lissuraga stream, a tributary of the Nivelle river, downstream movements were noted daily from a trap installed within a dam (Cuinat and Héland, 1979). This study was followed in three consecutive years. The results of the displacement of fry, as well as the concomitant variations in environmental factors, are shown for the year 1969, as an example (Fig. 10). The results obtained for the other two years are highly comparable.

Brown trout fry are displaced downstream in spring in March–April and May in the Lissuraga. The peak of downstream movement corresponds to the emergence period in the stream. The 'downstream moving' fry captured in the trap do not show particular signs of malformation or weakness. They are comparable in size to emerging fry sampled from the stream at the same time.

Fry are generally washed away during the night and more specifically in the early part of the night. Variations in environmental factors such as rainfall, water level or temperature do not appear to influence this phenomenon. However, as the efficiency of the trap decreases greatly in spates, the hypothesis of the influence of spates on displacement cannot be entirely dismissed.

During a major study of the downstream movement of sea trout fry in a natural stream, Elliott (1986) monitored upstream and downstream displacements of fry from five natural redds. In a first experiment, the traps were placed 1 m upstream and downstream of the redd and were observed over 20 days. In a second experiment, the traps were situated 10 m upstream and downstream and monitored over 28 days (Fig. 11). The results show that the fry are displaced downstream mainly at night, while they go upstream almost exclusively during the day. Elsewhere, traps situated very close to the redd, catch newly hatched fry, which still have vestiges of yolk sac. The shape of the displacement histogram, with its marked peak, is not dissimilar to that of the hatching histogram observed in Atlantic salmon (Marty and Beall, 1987). In traps situated 10 m from the redd, fry moving downstream are in poor physical condition with empty stomachs, while the fry captured in upstream traps are in good condition with full stomachs and of a size equal to those of the sedentary ones in the spawning area. Elliott (1986, 1994) opined that the downstream displaced fry are likely to be lost, especially in years where their density is high. Conversely, fry moving upstream are future residents which will colonize the upstream zone of the spawning ground.

2. Behaviour of downstream moving fry in artificial streams

The mechanisms of downstream movement in a fry population can be analysed in detail in an artificial stream. An experiment based on the principle of successive trapping allowed the characterization, isolation and observation of the 'downstream moving' fry compared to the sedentary ones (Héland, 1980a,b).

In an experimental stream, the 'downstream moving' fry were caught in traps and introduced to another stream, with no occupants. There, they could move downstream again; if this happened, they were marked by partial ablation of a pectoral fin and then replaced in the stream. The operation could be repeated with the partial ablation of the other pectoral fin. If they moved downstream again (fourth time), they were introduced into a third identical stream, divided into two sections, where they could not be displaced and their behaviour could be observed.

The change in the numbers of fry observed in static swimming activity in the three experimental streams (Fig. 12), showed a significant time interval in the process of

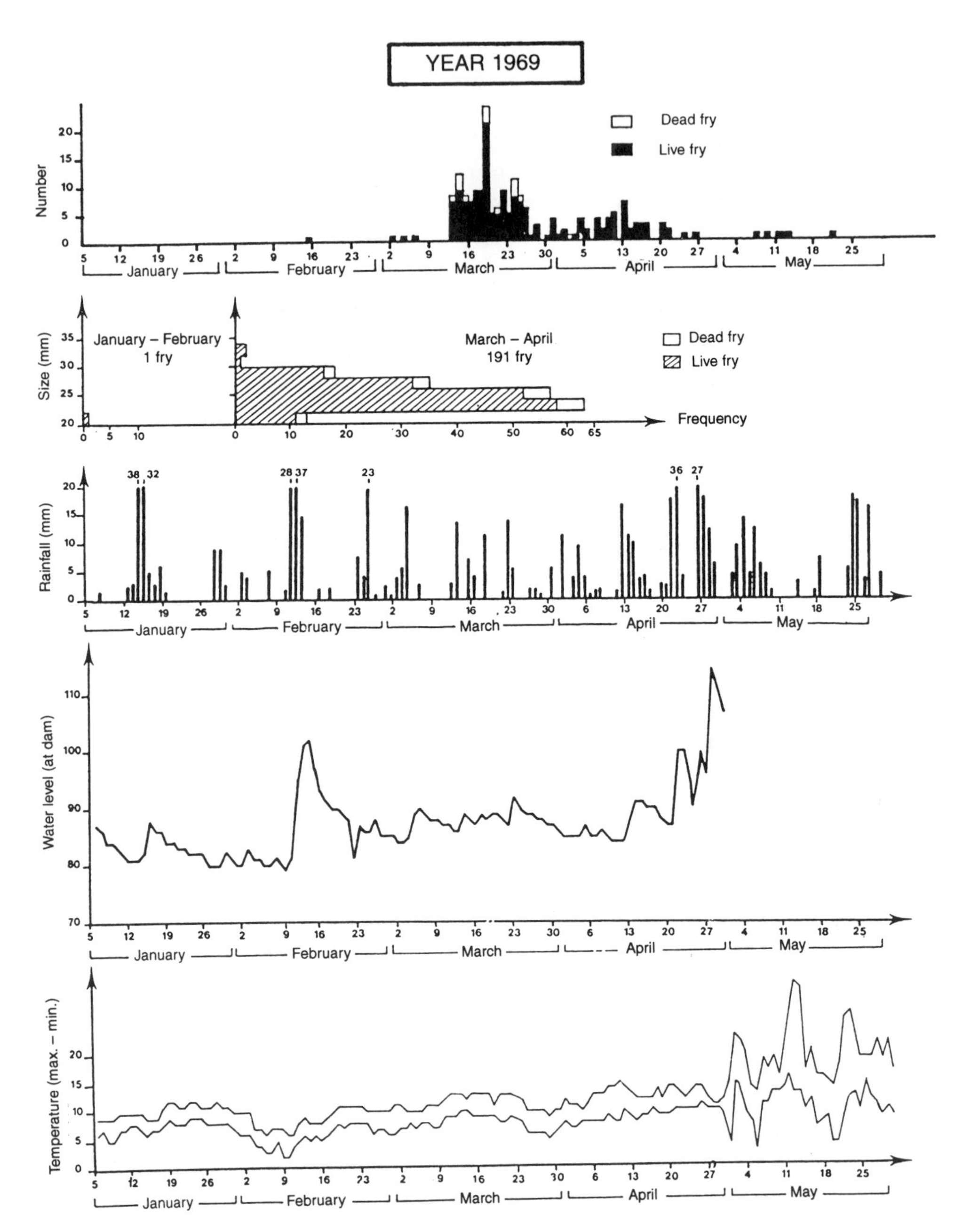

Fig. 10. Characteristics of the downstream movement of brown trout fry in the Lissuraga stream in 1969, monitored by a downstream trap. Variations in rainfall, water level and temperature are also given for the same period (after Cuinat and Héland, 1979).

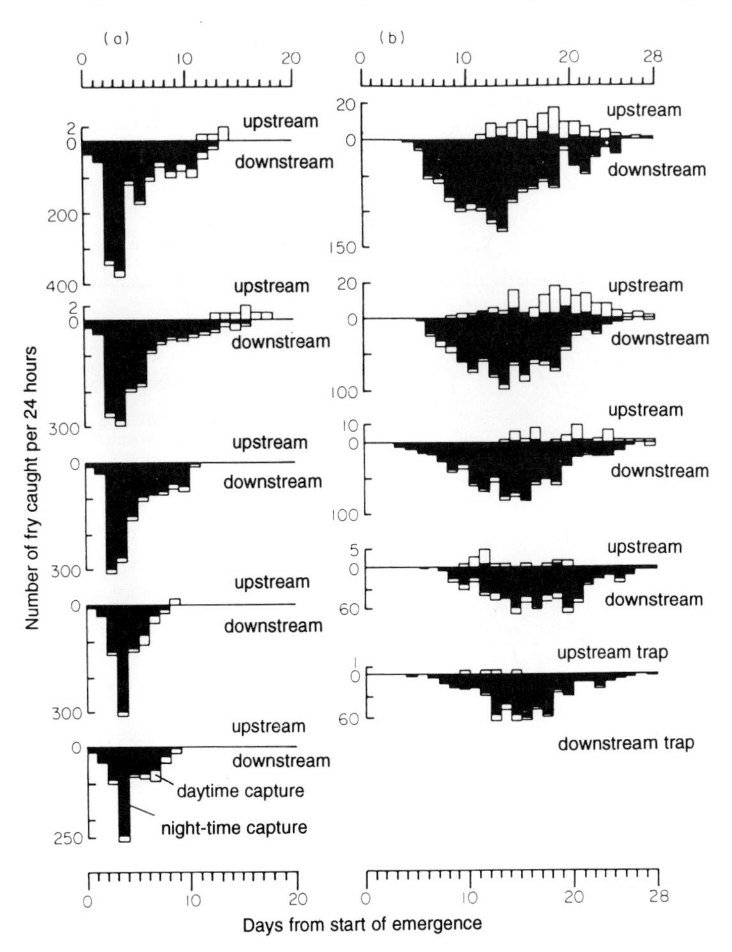

Fig. 11. Daily captures of 'upstream' and 'downstream' moving trout fry, after emergence from five natural spawning areas: (a) traps placed 1 m upstream and downstream of redds, (b) traps placed 10 m upstream and downstream of redds. Night-time captures are represented in black and daytime captures in white (after Elliott, 1986).

territorial stabilization, as defined at the start of this chapter (Fig. 2). In the first stream, where the yolk sac fry were introduced, an increase in the number of fish holding station occurred, which corresponds to the learning period of swimming in the sedentary fry and ended with the progressive and complete occupation of the whole environment. Then, with a hierachy in place and competition for the best positions, the number of fish holding station decreased until there was stabilization, corresponding to the number of dominant territorial fry.

In the second stream containing the downstream moving fry, the same phenomenon appeared but with a time interval of 4–5 days at least. Amongst the large number of fry holding station at the end of the experiment, some are probably fry still in the process of establishing their swimming behaviour, which explains their significant number.

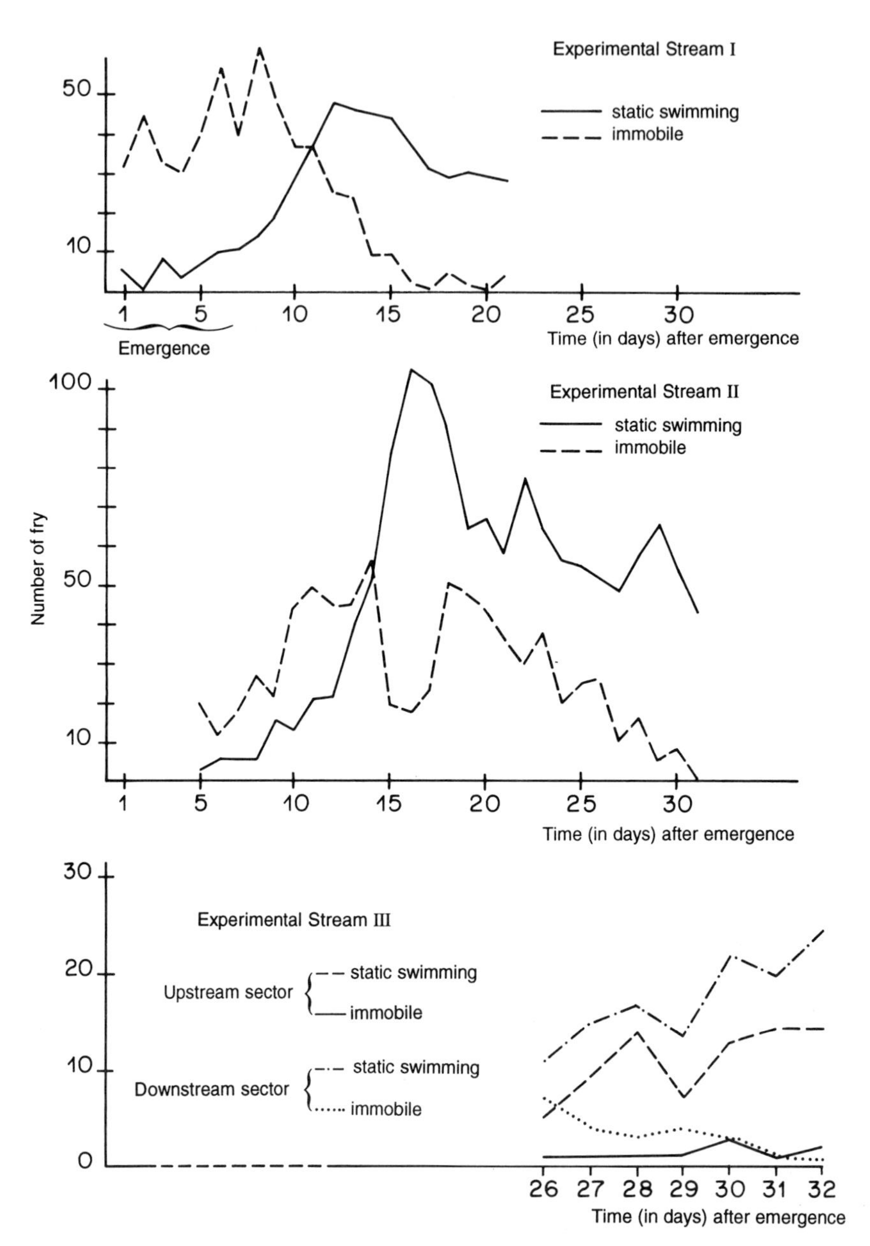

Fig. 12. Change in the number of fry observed in static swimming activity or immobile on the bottom in three experimental streams in the month following emergence. Experimental stream II received downstream moving fry from stream I and stream III, those which had moved downstream three times in stream II (monitored by marking) (after Héland, 1980a).

In the sections of experimental stream 3, containing fry which were displaced several times, stationary swimming and station-holding, the precursor to territorial stabilization, started to occur. This experiment showed that downstream moving fry prove to be capable of establishing a territory, although later, to a greater or lesser degree, than sedentary fry.

The aforesaid experiments show that downstream movement appears to be a regulatory mechanism of natural fry populations, allowing late-emerging fry to hold feeding stations in areas where competition is reduced. Further, data from recent studies show that downstream movement can also be an avoidance reaction to other species, potentially predators (Gaudin and Héland, 1984; Beall *et al.*, 1994; Bardonnet and Héland, 1994; Héland and Beall, 1997).

V. EVOLUTION OF TERRITORIALITY DURING ONTOGENY

The results presented until now have concerned trout fry from emergence until their first summer, that is during the initial recruitment phase where determinism is of major importance in the regulation of natural populations.

However, social relations change during the course of ontogeny together with the ecological conditions necessary for the development of the individuals. In fact, the mosaic of contiguous and exclusive, well-defined territories, arising after emergence, is likely to be transformed in response to new needs of the fish during growth. Variations in environmental conditions, notably cyclical (circadian or seasonal) can lead to necessary behavioural adaptations in the juvenile trout.

1. From the territorial mosaic of fry to the partial territories of small trout

In artificial streams, exact data about the boundaries of territories from observation of feeding stations and occupied shelters allows a map of the territories to be constructed. For the fry, these maps constitute a mosaic (see Part II, Chapter 1). During growth, the fry increase the extent of their territories by occupying new feeding stations. The hierarchy tends to get complicated by the frequent appearance of three-sided relationships, while the general aggression decreases at the population level, due to the development of habituation reactions between neighbouring individuals who avoid conflict while still respecting the hierarchy order. From this, each individual benefits from access to food in a smaller or larger territory, according to their position in the hierarchy. In surface area terms, this represents, for 3-month-old fry, 0.1–0.2 m^2 for a subordinate individual, to several m^2 for those higher up the social scale.

This process of change ends in a significant superimposition of territories, which increases with time. Between the third and fourth month after emergence, the total surface of the occupied territories can increase by 50% in one stream (Héland, 1977). The vertical dimension of the territories partially explains this increasing overlap. In fact, this depends mainly on the facultative occupation of feeding stations between neighbouring individuals; the small trout exploit the trophic resources according to a system where feeding stations are partitioned in space and time. Jenkins (1969) observed the same phenomenon in wild trout of 20–30 cm in length in a river and uses the term 'partial territories', a term defined by Greenberg (1947), or 'rotating territories', using the definition of Newman (1956). The choice of feeding stations always occurs in relation to invertebrate drift and water velocity, according to the principle of 'energy conservation' (Hughes, 1992; and compare Part III, Chapter 1).

With partial territories, Jenkins constructs a system of local hierarchies which he explains by the stability of the social group and the sedentary nature of the great majority of fishes (Fig. 13). These observations are confirmed by Bachman (1981) who was able to show the stable hierarchical organization of wild trout followed individually in a zone of a river, over three consecutive years.

At a later stage, when the adult trout reaches a certain size, its dietary regime tends to change and evolve towards a piscivorous diet. These fish usually confine themselves to the deeper areas and feed very little from the drift, but mainly around eddies, within their home range. According to the observations of Jenkins (1969), these large trout appear to be ignored by younger trout at their feeding stations in the current, as the habitats they occupy are different.

On the subject of competition between cohorts, the most complex case is the confrontation between young fry and young trout of the preceding year, which occupy the same areas. This confrontation is minimized in streams where the two age classes use and exploit different microhabitats: shallower, pebbly areas for the fry and deeper areas with more diverse substrate for the small trout (Bohlin, 1977; Kennedy and Strange, 1982).

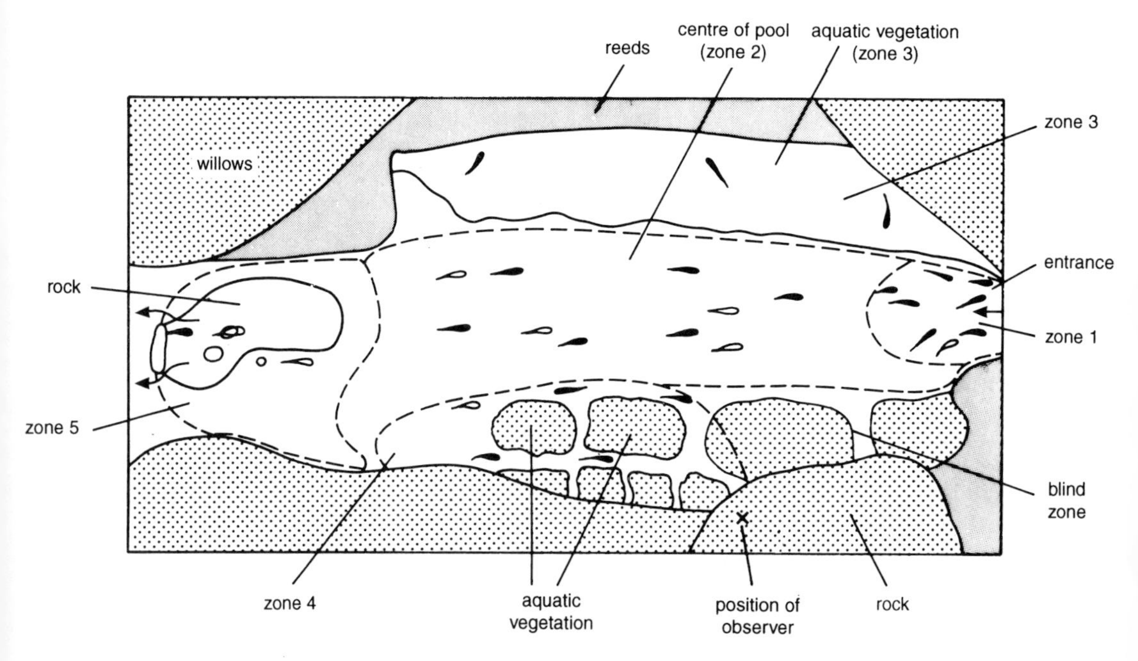

Fig. 13. Example of distribution of trout in an observation area in a 15-m long deep zone of the Owens River (Sierra Nevada, USA). Fish shown in black and white represent individuals at the surface and in deep water, respectively. The broken lines represent the subdivisions into local hierarchies of trout exploiting the zone (after Jenkins, 1969).

The presence of older conspecifics, as it is with other species, all potential predators, leads to an avoidance reaction of fry by increasing downstream movements (see before) or by a segregative interaction process through different microhabitat-choices (Hearn, 1987; Bardonnet and Héland, 1994; Héland and Beall, 1997; Roussel and Bardonnet, in press).

2. Dynamics of territoriality during daily and seasonal rhythms

Social organization, as for all animal activity, is subjected to the influence of environmental periodicity, mainly in the form of circadian or seasonal rhythms. The fry or young trout are no exception to this rule and their overall activity will vary according to daily or seasonal rhythms.

Perez (1986) showed the existence of a circadian rhythm of activity in the brown trout fry. We have seen before (Part IV) that the day/night pattern directly influences the downstream movement of fry which takes place during the night. Elliott (1973) has shown that trout fry feed mainly during twilight hours at the beginning and end of the night; this was confirmed by Neveu and Thibault (1977) and implies significant activity during these twilight and nocturnal hours. Chaston (1968) demonstrated mainly nocturnal activity in groups of wild trout by video recordings in artificial tanks.

The question needing to be answered is whether the partitioning of trophic resources, according to a hierarchic and/or territorial order that applies during the hours of light, is also applicable during the dark phase. Observations in natural or artificial streams with the aid of photographic or video recording, suggest a reduction of intra-population aggression during the dark phase. Reduced visual capacities can explain this decrease, which is similar to the effects of increased turbidity (Berg and Northcote, 1985). During the dark phase, the subordinate fry may have better access to prey in the drift. Thus, the partitioning in time and space of the environment's resources—space (feeding stations) and feeding (invertebrate drift)—can be suspended at night, ending with a partial redistribution of trophic exploitation of the environment between dominant and subordinate individuals. This hypothesis, based on fragmented observations, remains to be confirmed.

In an experimental stream, Roussel and Bardonnet (1995) showed that one year-old trout were active daily on feeding stations mainly in riffles but were concentrated in resting position in pools during the night without agonistic behaviour exchanges. Concerning the fry, a quite similar phenomenon was observed but the 'refuge' zone for resting (and also predator avoidance) corresponded to the stream-margin habitat (Roussel and Bardonnet, in press).

With regard to seasonal rhythms, the great majority of work published up until now on territoriality in salmonids refers to studies carried out in spring or summer. This state of affairs is the consequence of the much greater ease of sampling at this time of year, but has given rise to the apparent non-existence of social activities in salmonids during the cold period. Some studies carried out in winter have shown a decrease in aggression and territoriality among gregarious fish gathered in pools, as has been shown in the brook charr by Cunjack and Power (1987). During the cold season, invertebrate drift is very low and the metabolic rate of trout decreases when the temperature is falling. Heggenes *et al.* (1993) observed a decreasing territorial defence activity of brown trout juveniles in winter in relation to a sheltering behaviour or aggregating in pools for minimizing energy expenditure.

In the trout fry at the end of yolk sac resorption, an experiment to artificially simulate winter conditions (short photoperiod and cold temperature) showed a delay in emergence and in the establishment of swimming behaviour preceding territorial behaviour (Calleja Boveda, 1987). This experiment emphasizes the adaptive role of territoriality in mountain ecosystems, within the limits: a mild winter favours precocious emergence and a cold spring slows the melting of the snows and can lead to the annihilation of a cohort of fry, which explains the frequently unbalanced structure of these populations (Gayou and Simonet, 1978; Abad, 1982).

In terms of seasonal variations, it appears that territoriality manifests itself mainly during the fine weather, at the time when invertebrate drift and the arrival of the new generation of fry in the environment require adjustments to be made. Autumn and winter are often times of reorganization of the population within the ecosystem, with downstream migrations of juveniles prior, for the anadromous salmonids, to the downstream migration as smolts (Bjornn, 1971; Thorpe, 1974; Solomon and Templeton, 1976; Godin, 1982; Solomon, 1982; Elliott, 1985, 1986, 1988). It should also be noted that spawning migrations occur at the end of autumn and start of winter, with the arrival of spawners in the redd zones. They will establish reproductive territories within these zones, to which they will deny access to conspecifics, by the expression of the repertoire of aggressive behaviours from the juvenile phase (Jones and Ball, 1954).

VI. CONCLUSION

The social organization of juvenile brown trout changes throughout development, from a strict territoriality in the fry to partial territoriality in the juvenile. This evolution leads to a partitioning in space and time of access to trophic resources for individuals in the same social group. This access is advantageous to a greater or lesser extent, in energetic terms, in relation to social standing. Other fry do not remain in the spawning zone, but drift down to the lower zones of the stream, which they may be able to colonize.

In summary, territoriality represents a behavioural adaptation to the exploitation of invertebrate drift in the stream and tends towards an adjustment of the density of fish to the resources of the environment. This adjustment can be achieved more completely through a social organization which, after the initial dispersal, favoured by territorial competition, allows the subordinates to access the food source and a fraction of nomadic individuals to exploit zones downstream of the spawning reaches. The partial territories of older fry and small trout, in the same way as different microhabitats preferred by fry or juveniles of >1 year, contribute to the organization of a dynamic partitioning of resources of space and food.

In addition to this, territorial behaviour is not constant in time and appears to vary as a function of time of day or season of the year. Similarly, when the diet of the trout changes—to the piscivorous diet of large trout for example—territoriality appears to be abandoned. Elsewhere, when the current which transports the invertebrates is no longer present, either artificially or naturally as in lakes or the sea, territorial behaviour is abandoned in favour of a school structure with some hierarchy (Kalleberg, 1958).

There appears to be a certain flexibility in social behaviour which depends on individual characteristics and the influence of the environment. It is clear that this socio-plasticity starts with the fry which develop their capacities for colonizing and exploiting the trophic environment. This phenomenon of precocious exploration of new ways as a means of constructing the relationship between an animal and its environment is common in the vertebrates (Gauher, 1982). These processes of ontogeny explain in part the multiplicity of adult phenotypes observed in a species such as the brown trout which can occupy all aquatic environments. Conversely, other salmonid species such as certain Pacific salmon (*Oncorhynchus* spp., e.g. *O. nerka, O. keta* and *O. gorbuscha* Walbaum) which are closely dependent on particular environments such as estuary or lake, for example, seem less flexible in their ontogenetic development, particularly with regard to their social structure.

BIBLIOGRAPHY

Abad N., 1982. Ecologie et dynamique des populations de truites commune (*Salmo trutta fario* L.) dans le bassin du Tarn. Thèse 3e cycle, Sci. Techn. Prod. Anim., Inst. Nat. Polytech. Toulouse, 221 pp.

Allen K. R., 1969. Limitations on production in salmonid populations in streams. In T. G. Northcote (Ed.), *Salmon and Trout in Streams*, 3–20, H. R. McMillan Lect. Fisheries, Univ. Br. Columbia, Vancouver.

Assem J. Van Den, 1967. Territory in the three-spined stickleback (*Gasterosteus aculeatus* L.), an experimental study in intraspecific competition. *Behaviour*, Suppl. 16, 1–164.

Assem J. Van Den, 1970. Less problèmes du territorialism chez les poissons. In G. Richard, *Territoire et Domaine Vital*, 21–34. Masson, Paris.

Bachman R. A., 1984. Foraging behavior of free-ranging wild and hatchery brown trout in a stream. *Trans. Am. Fish. Soc.*, **113** (1), 1–32.

Bardonnet A., Héland M., 1994. The influence of potential predators on the habitat preferenda of emerging brown trout. *J. Fish Biol.*, **45**, suppl. A, 131–142.

Beall E., Héland M., Marty C., 1989. Interspecific relationships between emerging Atlantic salmon, *Salmo salar*, and coho salmon, *Oncorhynchus kisutch*, juveniles. *J. Fish Biol.*, **35**, suppl. A, 285–293.

Berg L., Northcote T. G., 1985. Changes in territorial, gill-flaring, and feeding behaviour in juvenile coho salmon (*Oncorhynchus kisutch*) following short-term pulses of suspended sediment. *Can. J. Fish. Aquat. Sci.*, **42**, 1410–1417.

Bjornn T. C., 1971. Trout and salmon movements in two Idaho streams as related to temperature food stream flow cover and population density. *Trans. Am. Fish. Soc.*, **100** (3), 423–438.

Bohlin T., 1977. Habitat selection and intercohort competition of juvenile sea trout, *Salmo trutta. Oikos*, **29**, 112–117.

Brown J., Green J. M., 1976. Territoriality, habitat selection and prior residency in underyearling *Stichaeus punctatus* (Pisces: Stichaeidae). *Can. J. Zool.*, **54**, 1904–1907.

Butler R. L., Hawthorne V. M., 1968. The reactions of dominant trout to changes in overhead artificial cover. *Trans. Am. Fish. Soc.*, **97**, 37–41.

Calleja Boveda P., 1987. Influence de la température et de la photopériode sur le développement des activités chez l'alevin de truite commune, *Salmo trutta* L., en aquarium. D.E.A. Ecol. exp., Univ. Pau et Pays Adour, 43 pp.

Daron J., 1986. L'organisation sociale et l'utilisation de l'espace chez l'omble de fontaine (*Salvelinus fontinalis*): les effets de la compétition pour la nourriture, l'abri et le couvert en ruisseau artificiel et en dehors de la période de reproduction. Thèse Doct., Univ. Québec, Montréal, 188 pp.

Chapman D. W., 1962. Aggressive behaviour of juvenile coho salmon as a cause of emigration. *J. Fish. Res. Board Can.*, **19**, 6, 1047–1080.

Chapman D. W., 1966. Food and space as regulators of salmonid populations in streams. *Am. Nat.*, **100**, 913, 345–358.

Chaston I., 1968. Influence of light on activity of brown trout (*Salmo trutta*). *J. Fish. Res. Board Can.*, **25** (6), 1285–1289.

Cuinat R., Heland M., 1979. Observations sur la dévalaison d'alevins de truite commune (*Salmo trutta* L.) dans le Lisssuraga. *Bull. Fr. Piscic.*, **274**, 1–17.

Cunjack R. A., Power G., 1987. The feeding and energetics of stream-resident trout in winter. *J. Fish Biol.*, **31**, 493–511.

Dill L. M., 1969. The sub-gravel behaviour of Pacific salmon larvae. In T. G. Northcote (Ed.), *Salmon and Trout in Streams*, 89–99, H. R. McMillan Lect. Fisheries, Univ. Br. Columbia, Vancouver.

Dill L. M., Ydenberg R. C., Fraser A. H. G., 1981. Food abundance and territory size in juvenile coho salmon (*Oncorhynchus kisutch*). *Can. J. Zool.*, **59**, 1801–1809.

Dill P. A., 1977. Development of behaviour in alevins of Atlantic salmon, *Salmo salar*, and rainbow trout, *S. gairdneri. Anim. Behav.*, **25**, 116–121.

Dumas J., 1976. Dynamique et sédentarité d'une population naturalisée de truites arc-en-ciel (*Salmo gairdneri* Richardson) dans un ruisseau de montagne, l'Estibère (Hautes-Pyrénées). *Ann. Hydrobiol.*, **7** (2), 115–139.

Egglishaw H. J., Shackley P. E., 1973. An experiment on faster growth of salmon (*Salmo salar* L.) in a Scottish stream. *J. Fish Biol.*, **5**, 197–204.

Elliott J. M., 1966. Downstream movements of trout fry (*Salmo trutta*) in a Dartmoor stream. *J. Fish. Res. Board Can.*, **23**, 1, 157–159.

Elliott J. M., 1973. The food of brown and rainbow trout (*Salmo trutta* and *S. gairdneri*) in relation to the abundance of drifting invertebrate in a mountain stream. *Oecologia*, **12**, 329–347.

Elliott J. M., 1984. Numerical changes and population regulation in young migratory trout, *Salmo trutta*, in a lake district stream, 1966–83. *J. Anim. Ecol.*, **53**, 327–350.

Elliot J. M., 1985. Population regulation for different life-stages of migratory trout, *Salmo trutta*, in a lake district stream, 1966–83. *J. Anim. Ecol.*, **54**, 617–638.

Elliott J. M., 1986. Spatial distribution and behavioural movements of migratory trout, *Salmo trutta*, in a lake district stream. *J. Anim. Ecol.*, **55**, 907–922.

Elliott J. M., 1987. The distances travelled by downstream moving trout fry, *Salmo trutta*, in a lake district stream. *Freshwater Biol.*, ,**17**, 491–499.

Elliott J. M., 1988. Growth, size, biomass and production in contrasting populations of trout, *Salmo trutta*, in two Lake District streams. *J. Anim. Ecol.*, **57**, 49–60.

Elliott J. M., 1994. *Quantative Ecology and the Brown Trout.* Oxford University Presss, Oxford, 286 pp.

Fabricius E., 1951. The topography of the spawning bottom as a factor influencing the size of territory in some species of fish. *Rep. Inst. Freshwater Res. Drottningholm*, **32**, 43–49.

Fabricius E., 1953. Aquarium observations on the spawning behaviour of the char, *Salvelinus alpinus* L. *Rep. Inst. Freshwater Res., Drottningholm*, **34**, 14–48.

Fabricius E., Gustafson K. J., 1954. Further aquarium observations on the spawning behaviour of the char, *Salmo alpinus* L. *Rep. Inst. Freshwater Res., Drottningholm*, **35**, 58–104.

Fabricius E., Gustafson K. J., 1955. Observations on the spawning behaviour on the grayling, *Thymallus thymallus* L. *Rep. Inst. Freshwater Res., Drottningholm*, **36**, 75–103.

Fauch K. D., 1984. Profitable stream positions for salmonids: relating specific growth rate to net energy gain. *Can. J. Zool.*, **62**, 441–451.

Frost W. E., Brown M. E., 1967. *The Trout.* Collins, London, 286 pp.

Gaudin P., Héland M., 1984. Influence d'adults de chabots (*Cottus gobio* L.) sur des alevins de truite commune (*Salmo trutta* L.): étude expérimentale en milieux semi-naturels. *Acta Oecologica, Oecol. Applic.*, **5**, 71–83.

Gaudin P., Héland M., 1995. Stratégie d'utilisation de l'habitat par les alevins post-émergents de truite commune (*Salmo trutta*) et de saumon atlantique (*Salmo salar*). *Bull. Fr. Pêche Piscic.*, **337/338/339**.

Gautier J. Y., 1982. Socioécologie. L'animal social et son univers. Privat, Toulouse, 267 pp.

Gayou F., Simonet F., 1978. Dynamique des populations de truites (*Salmo trutta fario* L.). Aménagements piscicoles en haute vallée d'Aure. Thèse 3e cycle, Sci. Tech. Prod. anim., Inst. Nat. Polytech., Toulouse, 244 pp.

Geiger W., Roth H., 1962. Observations on artificial trout redds. *Schweiz. Z. Hydrol.*, **24**, 76–89.

Gerking S. D., 1953. Evidence for the concept of home range and territory in stream fishes. *Ecology*, **34**, 347–365.

Gibson R. J., Power G., 1975. Selection by brook trout (*Salvelinus fontinalis*) and juvenile Atlantic salmon (*Salmo salar*) of shade related to water depth. *J. Fish. Res. Board Can.*, **32** (9), 1652–1656.

Godin J.-G. J., 1982. Migrations of salmonid fishes during early life history phases: daily and annual timing. In. E. L. Brannon and E. O. Salo (Eds): *Salmo and Trout Migratory Behavior Symposium*, 22–50, School of Fisheries, Univ. of Washington, Seattle, USA.

Grant J. W. A., Kramer D. L., 1990. Territory size as a predictor of the upper limit to population density of juvenile salmonids in streams. *Can. J. Fish. Aquat. Sci.*, **47**, 1724–1737.

Greenberg B., 1947. Some relations between territory, social hierarchy and leadership in the Green Sunfish (*Lepomis cyanellus*). *Physiol. Zool.*, **20**, 267–299.

Hartman G. F., 1963. Observation on behavior of juvenile brown trout in a stream aquarium during winter and spring. *J. Fish. Res. Board Can.*, **20** (3), 769–787.

Hearn W. E., 1987. Interspecific competition and habitat segregation among stream-dwelling trout and salmon: a review. *Fisheries*, **12**, 24–31.

Heggenes J., Krog O. M., Lindas O. R., Dokk J. G., Bremnes T., 1993. Homeostatic behavioural responses in a changing environment: brown trout (*Salmo trutta* L.) become nocturnal during winter. *J. Anim. Ecol.*, **62**, 295–308.

Héland M., 1971a. Observations sur les premières phases du comportement agonistique et territorial de la truite commune, *Salmo trutta* L., en ruisseau artificiel. *Ann. Hydrobiol.*, **2** (1), 33–46.

Héland M., 1971b. Influence de la densité du peuplement initial sur l'acquisition des territoiries chez la truite commune, *Salmo trutta* L., en ruisseau artificiel. *Ann. Hydrobiol.*, **2** (1), 25–32.

Héland M., 1977. Recherches sur l'ontogenèse du comportement territorial chez l'alevin de truite commune, *Salmo trutta* L. Thèse 3e cycle, Biol. Anim., Fac. Sci., Univ. Rennes, 239 pp.

Héland M., 1978. Observations sur l'établissement du comportement de nage face au courant chez l'alevin de truite, *Salmo trutta* L., en ruisseau artificiel. *Ann. Limnol.*, **14** (3), 273–280.

Héland M., 1980a. La dévalaison des alevins de truite commune, *Salmo trutta* L. I. Caractérisation en milieu artificiel. *Ann. Limnol.*, **16** (3), 233–245.

Héland M., 1980b. La dévalaison des alevins de truite commune, *Salmo trutta* L. II. Activité des alevins 'dévalants' comparés aux sédentaires. *Ann. Limnol.*, **16** (3), 247–254.

Héland M., 1982. Influence de l'isolement sur l'établissement du comportement territorial chez l'avevin de truite commune, *Salmo trutta* L. Commun. Coll. SFECA, Tours, March 1982. *Bull. Int. SFECA, Rennes*, **2**, 49–61.

Hélandf M., Beall E., 1997. Etude expérimentale de la compétition interspécifique entre juvéniles de saumon coho, *Oncorhynchus kisutch*, et de saumon atlantique, *Salmo salar*, en eau douce. *Bull. Fr. Pêche Piscic.*, **344–345**, 241–252.

Héland M., Gaudin P., Bardonnet A., 1995. Mise en place des premiers comportements et utilisation de l'habitat après l'émergence chez les Salmonidés d'eau courante. *Bull. Fr. Pêche Piscic.*, **337/338/339**, 191–197.

Huet M., 1961. Reproduction et migration de la truite commune dans un ruisselet salmonicole de l'Ardenne belge. *Verh. Int. Verein Theor. Angew. Limnol.*, **14**, 757–762.

Hughes N. F., 1992. Selection of positions by drift-feeding salmonids in dominance hierarchie: model and test for Arctic grayling (*Thymallus arcticus*) in subarctic mountain streams, interior Alaska. *Can. J. Fish. Aquat. Sci.*, **49**, 1999–2008.

Hughes N. F., Dill L. M., 1990. Position choice by drift-feeding salmonids: model and test for Arctic grayling (*Thymallus articus*) in subarctic mountain streams, interior Alaska. *Can. J. Fish. Aquat. Sci.*, **47**, 2039–2048.

Hynes H. B. N., 1970. *The Ecology of Running Waters*. Liverpool Univ. Press, Liverpool, 555 pp.

Jenkins T. M. Jr, 1969. Social structure, position choice and microdistribution of two trout species (*Salmo trutta* and *Salmo gairdneri*) resident in mountain streams. *Anim. Behav. Monog.*, **2** (2), 55–123.

Johnsson J. I., Nöbbelin F., Bohlin T., 1999. Territorial competition among wild brown trout fry: effects of ownership and body size. *J. Fish Biol.*, **54**, 469–472.

Jones A. N., 1975. A preliminary study of fish segregation in salmon spawning streams. *J. Fish Biol.*, **7** (1), 95–104.

Jones J. W., Ball J. N., 1954. The spawning behaviour of brown trout and salmon. *Br. J Anim. Behav.*, **2**, 103–114.

Kalleberg H., 1958. Observations in a stream tank of territoriality and competition in juvenile salmon and trout (*Salmo salar* L. and *S. trutta* L.). *Rep. Inst. Freshwater Res., Drottningholm*, **39**, 55–88.

Kennedy G. J. A., Strange C. D., 1982. The distribution of salmonids in upland streams in relation to depth and gradient. *J. Fish Biol.*, **20**, 579–591.

Kennedy M., Fitzmaurice P., 1968. The early life of brown trout (*Salmo trutta* L.). *Ir. Fish. Invest.*, ser. A, 4, 31 pp.

Le Cren E. D., 1961. How many fish survive? *Yb. River Bds Ass.*, 57–64.

Le Cren E. D., 1973. The population dynamics of young trout (*Salmo trutta*) in relation to density and territorial behaviour. *Rapp. P.V. Reun. Cons. int. Explor. Mer*, 164, 241–246.

Lindroth A., 1955. Distribution territorial behaviour and movements of sea trout fry in the River Indalsäven. *Rep. Inst. Freshwater Res., Drottningholm*, **36**, 104–119.

Marty C., Beall E., 1987. Rythmes journaliers et saisonniers de dévalaison d'alevins de saumon atlantique à l'émergence. In M. Thibault and R. Billard (Eds) *La Restauration des Rivières à Saumons*, 283–290, INRA, Paris.

McFadden J. T., 1969. Dynamics and regulation of salmonid populations in streams. In T. G. Northcote (Ed.), *Salmon and Trout in Streams*, 313–322, H. R. McMillan Lect. Fisheries, Univ. Br. Columbia, Vancouver.

McNicol R. E., Noakes D. L. G., 1981. Territories and territorial defense in juvenile brook charr, *Salvelinus fontinalis* (Pisces: Salmonidae). *Can. J. Zool.*, **59**, 22–28.

McNicol R. E., Noakes D. L. G., 1984. Environmental influences on territoriality of juvenile brook charr, *Salvelinus fontinalis*, in a stream environment. *Environ. Biol. Fishes*, **10** (1/2), 29–42.

McNicol R. E., Scherer E., Murkin E. J., 1985. Quantitative field investigations of feeding and territorial behaviour of young-of-the-year brook charr, *Salvelinus fontinalis. Environ. Biol. Fishes*, **12** (3), 219–229.

Neveu A., Thibault M., 1977. Comportement alimentaire d'une population sauvage de truites fario (*Salmo trutta* L.) dans un ruisseau des Pyrénées-atlantiques, le Lissuraga. *Ann. Hydrobiol.*, **17** (2), 111–128.

Newman M. A., 1956. Social behavior and interspecific competition in two trout species. *Physiol. Zool.*, **29**, 64–81.

Noakes D. L. G., 1978. Social behavior as it influences fish production. In S. D. Gerking (Ed.), *Ecology of Freshwater Fish Production*, 360–386, Blackwell Sci. Publ., London.

Noble G. K., 1939. The role of dominance in the social life of birds. *Auk*, **56**, 263–273.

Onodera K., 1967. Some aspects of behaviour influencing production. In S. D. Gerking

(Ed.), *The Biological Basis of Freshwater Fish Production*, 345–355, Blackwell Sci. Publ., Oxford.

Perez E., 1986. Rôle de facteurs externes et internes dans la mise en place due rythme circadien d'activité au cours de l'ontogenèse de la truite (*Salmo trutta* L.). Thèse Biol. Anim., Fac. Sci. Tech. St-Etienne, 295 pp.

Puckett K. J., Dill L. M., 1985. The energetics of feeding territoriality in juvenile coho salmon (*Oncorhynchus kisutch*). *Behaviour*, **92** (1–2), 97–111.

Richard P. B., 1970. Le comportement territorial chez les vertébres. In G. Richard, *Territoire et Domaine Vital*, 1–9, Masson, Paris.

Roth H., Geiger W., 1963. Experimental studies on brown trout fry in the gravel. *Schweiz. Z. Hydrol.*, **25**, 202–218.

Roussel J. M., Bardonnet A., 1995. Activité nycthémérale et utilisation de la séquence radier-profond par les truitelles d'un an (*Salmo trutta* L.). *Bull. Fr. Pêche Piscic.*, **337/338/339**, 221–230.

Roussel J. M., Bardonnet A., (in press) Ontogeny of diet pattern of stream-margin habitat use by emerging brown trout (*Salmo trutta*) in experimental channels: influence of food and predator presence. *Env. Biol. Fish.*

Skinner B. F., 1971. *L'analyse expérimentale du comportement. Un essai théorique*. 2nd edn, Dessart & Mardaga, Brussels, 408 pp.

Slaney P. A., Northcote T. G., 1974. Effects of prey abundance on density and territorial behaviour of young rainbow trout (*Salmo gairdneri*) in laboratory stream channels. *J. Fish. Res. Board Can.*, **31** (7), 1201–1209.

Solomon D. J., 1982. Migration and dispersion of juvenile brown and sea trout. In E. L. Brannon and E. O. Salo (Eds): *Salmon and Trout Migratory Behavior Symposium*, 136–145, School of Fisheries, Univ. of Washington, Seattle, USA.

Solomon D. J., Templeton R. G., 1976. Movements of brown trout, *Salmo trutta* L., in a chalk stream. *J. Fish Biol.*, **9**, 411–423.

Stuart T. A., 1953. Spawning, migration, reproduction and young stages of loch trout (*Salmo trutta* L.). *Freshwater Salm. Fish. Res.*, **5**, 39.

Symons P. E. K., 1971. Behavioural adjustment of population density to available food by juvenile Atlantic salmon. *J. Anim. Ecol.*, **40**, 569–587.

Thorpe J. E., 1974. The movements of brown trout, *Salmo trutta* L., in Loch Leven Kinross, Scotland. *J. Fish Biol.*, **6** (2), 153–180.

Thorpe J. E., 1982. Migration in salmonids, with special reference to juvenile movements in freshwater. In E. L. Brannon and E. O. Salo (Eds): *Salmon and Trout Migratory Behavior Symposium*, 86–97, School of Fisheries, Univ. of Washington, Seattle, USA.

Timmermans J. A., 1966. Etude d'une population de truites (*Salmo trutta fario* L.) dans une petite rivière de l'Ardenne belge. *Verh. Int. Theor. Angew. Limnol.*, **16** (2), 1204–1211.

Timmermans J. A., 1972. La territorialité de la truite fario. *Trav. Stn. Rech. Eaux & For. Groenendaal-Hœilaart*, D, **42**, 7–15.

Wickler W., 1976. The ethological analysis of attachment. *Z. Tierpsychol.*, **42**, 12–28.

Zayan R. C., 1975. Modification des effets liés à la priorité de résidence chez *Xiphophorus* (Pisces, Poeciliidae): le rôle des manipulations expérimentales. *Z. Tierphysiol.*, **39**, 463–491.

Zayan R. C., 1976. Modification des effets liés à la priorité de résidence chez *Xiphophorus* (Pisces, Poeciliide): de rôle de l'isolement et des différences de taille. *Z. Tierpsychol.*, **41**, 142–190.

Part II

Ecological plasticity and genetic diversity in trout

1

Main characteristics of the biology of the trout (*Salmo trutta* L.) in Lake Léman (Lake Geneva) and some of its tributaries

A. Champigneulle, B. Buttiker, P. Durand and M. Melhaoui

I. INTRODUCTION

Divided between France (41%) and Switzerland (59%), Lake Léman is the largest lake in western Europe, with a surface area of 58 240 ha, (Fig. 1). The main physical, chemical and biological characteristics of this subalpine, mesoeutrophic lake, have been detailed in a report (C.I.P.E.L., 1984). The most important physical data concerning the basin flowing into the lake are shown in Fig. 1. Twenty-three species of fish have been reported (Laurent, 1972) but the fishing (Gerdeaux, 1988) is aimed principally at seven species: perch (*Perca fluviatilis*), whitefish (*Coregonus lavaretus*), Arctic charr (*Salvelinus alpinus*), brown trout (*Salmo trutta*), burbot (*Lotta lotta*), roach (*Rutilus rutilus*) and pike (*Esox lucius*). Catches of trout represent between 2% and 5% of the total annual tonnage of fish captured in Lake Léman. The main characteristics of the fishing and release of trout in Lake Léman in the course of the last 12 years are shown in Table 1. The lake trout is mainly caught within the lake (Fig. 2), either by amateurs using dragline from motorboats, or with nets by professional fishermen. Since 1950, the annual catch of trout declared by professional fishermen has varied between 9 and 32 tonnes/year (Fig. 2). The amateurs have recently (since 1986) been required to declare their catches: the latter were around three quarters of those of professionals (Fig. 2). Occasional catches of adult lake-dwelling trout are made by rod and line in the river, before or after spawning, but they remain poorly known, as they do not have to be declared.

Apart from Forel (1904), studies published on the biology of trout in Lake Léman have been relatively recent (Melhaoui, 1985; Buttiker and Matthey, 1986; Buttiker *et al.*, 1987; Champigneulle, 1987; Champigneulle *et al.*, 1988a, 1990a,b; Durand and Pilotto, 1989). The genotypic structure of the type of populations of trout of Lake Léman and its basin has been studied and compared to those of reference strains described in France by Chevassus and Guyomard (1983) and Krieg (1984). The genetic study (Guyomard,

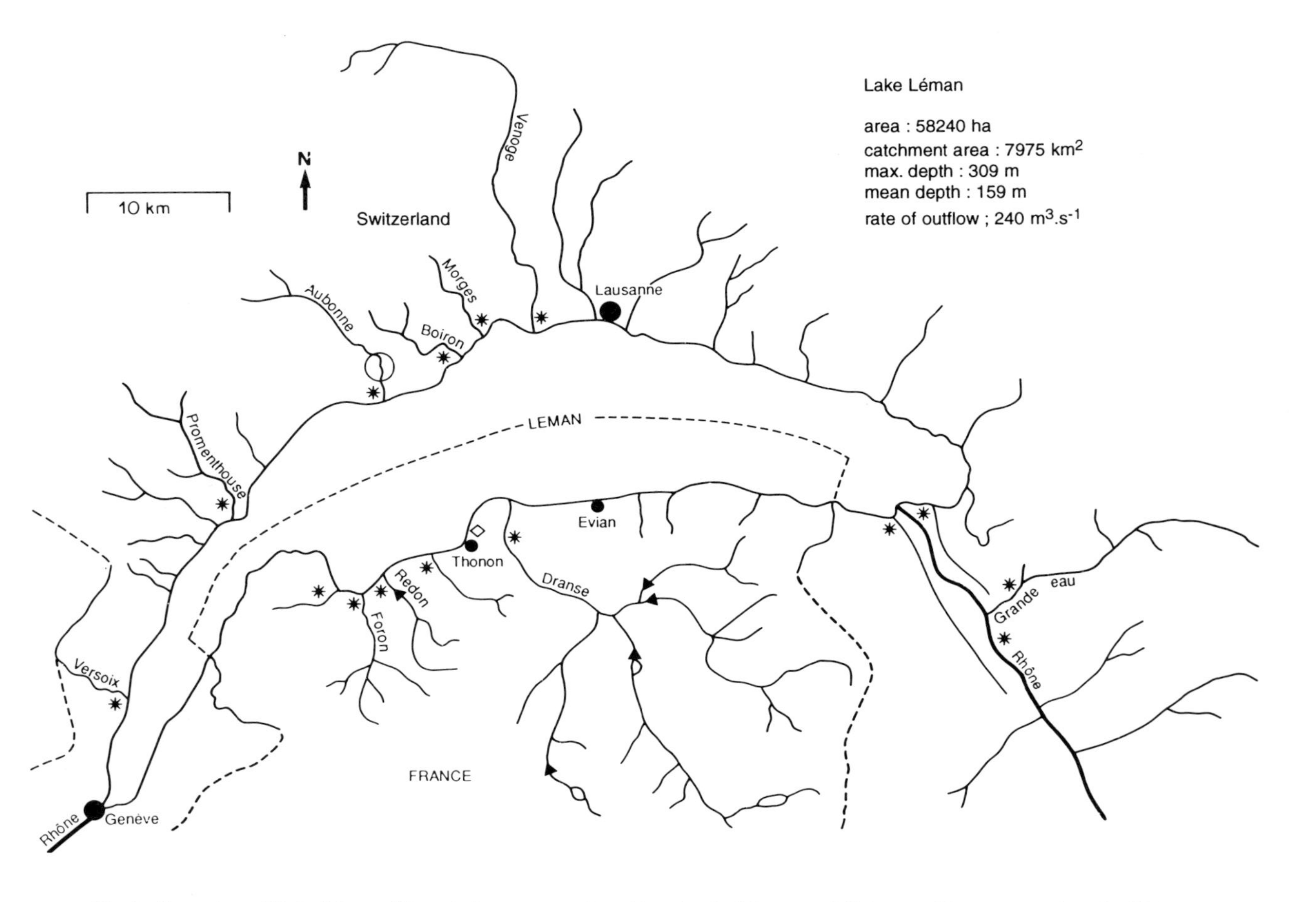

Fig. 1. Ecosystem of Lake Léman. (*), main known spawning tributaries for lake trout; (◀), impassable upstream obstacle; (○), trout trap; (◇), fish farm producing fish for restocking.

Table 1. Main features of the fishery and releases of brown trout in Lake Léman from 1986 to 1997

Lake fishery (1986–97)			
	Open season: 15 January–5 October		
	Legal capture size: 35 cm		
	Methods		
		Professional	
			Number of permits: 145
			Nets with minimal mesh size of 48 mm
			No quota for catches
		Amateur	
			Number of permits: 2465
			Draglines with a maximum of 20 lures/boat
			Quota: 8 trout/day and 250 trout/year/permit
	Mean annual catches		
		Professional:	16 tons
		Amateur:	12 tons
Restocking into Lake Léman			
	Stage at restocking:		fed fry (5–10 cm)
	Origin of parents:		(LT): Wild or captive stock from Lake Léman trout
			(DT): Domesticated brown trout
	Mean annual number stocked:		
		1986–93	LT: 250 000 DT: 750 000
		1994–97	LT: 1 000 000

1989a) of a sample of trout produced from mature lake trout ascending the River Aubonne in Switzerland showed for a small part of them a genetic proximity to populations of brown trout of the rivers flowing into the French Mediterranean, and, for the largest part, genetic proximity with the Atlantic form of brown trout. The examination of the same sample also revealed a phenomenon of inbreeding between Atlantic and Mediterranean forms which could very likely be partly linked to restocking operations with Atlantic domestic stocks. In practice, the juveniles, released in large numbers over several decades, are derived not only from lake trout broodstock captured in various tributaries of

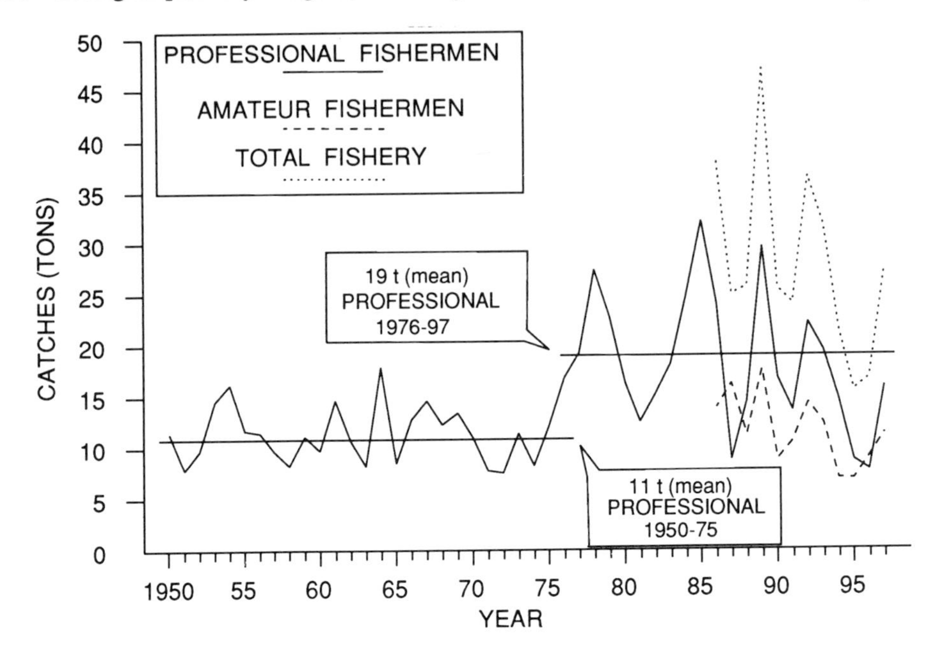

Fig. 2. Catches (in tons) of lake-dwelling trout in Lake Léman (data from DDA for France and Conservation de la Faune for Switzerland).

Lake Léman, but also from spawners from aquaculture stocks. The releases carried out directly into the lake are well recorded (Table 1) but those carried out throughout the tributaries are no longer integrated into the planned hatchery programme for the Léman system. In this chapter, the term 'lake trout' will be employed for trout of the species *Salmo trutta* captured in the lake or identified as having had a growth phase in the lake. The term will also be used to designate juveniles, wild or stocked having a lake trout parental origin. Because of the many possible modes of recruitment (natural or artificial) and the lack of data on their respective efficiencies, it is difficult, apart from where trout are marked, to determine the origin of lake trout actually captured in Lake Léman. However, in order to illustrate the possible life history pathways, the main lines of the biological cycle of the lake trout are described in Fig. 3.

II. JUVENILE PHASE IN THE TRIBUTARIES

1. The populations

At the time when they are living in the river, no single external characteristic can be used to distinguish potential lake trout from potential sedentary trout (trout having their entire life cycle in the river). Elsewhere, apart from the case where marked fish can be checked, it is difficult to appreciate the respective contributions of natural production and that of very diverse restocking methods (Fig. 3) and their quantitative importance in the lake (Table 1) and the tributaries. Consequently, it is only possible to make an overall statement of the characteristics of juveniles in certain tributaries of Lake Léman (Fig. 1;

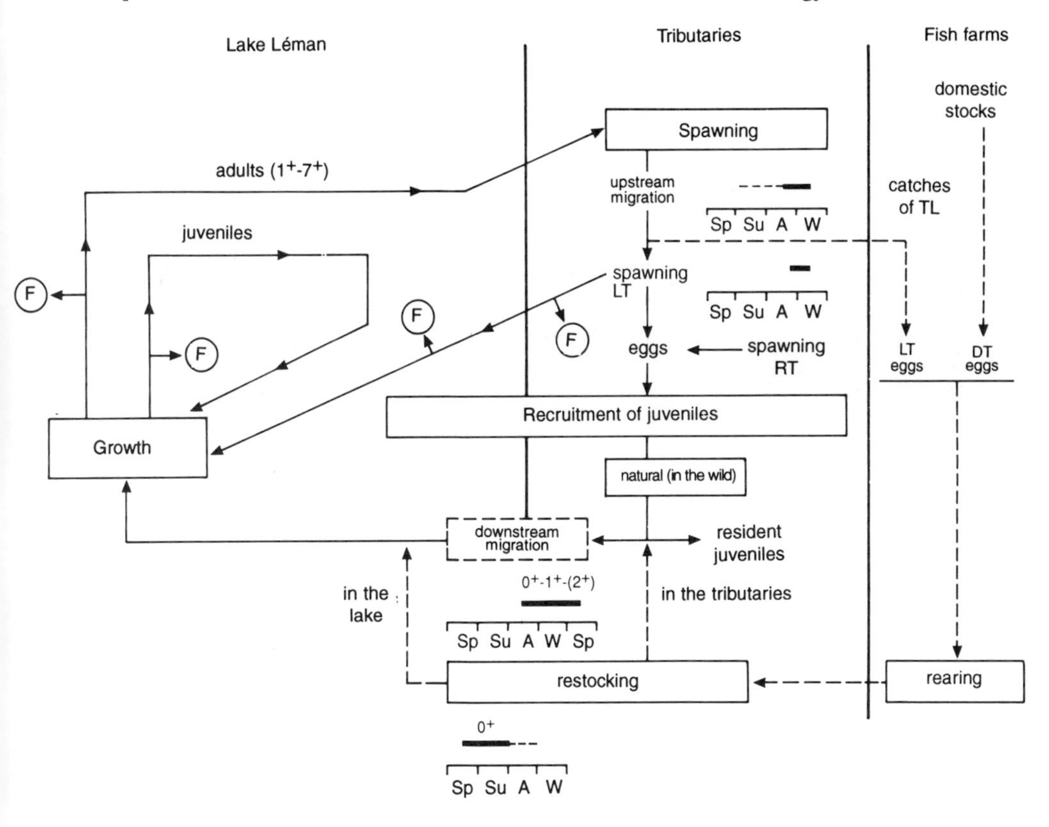

Fig. 3. Main features of the life cycle of the lake trout in Lake Léman. (LT), lake trout; (RT) resident trout; (DT), domestic trout; (→F), fishing; (Sp), spring; (S), summer; (A), autumn; (W), winter.

Table 2) and in particular in four spawning tributaries: the Redon in France, the Versoix, the Aubonne and the Promenthouse in Switzerland. The tributaries, because of their steep slopes, are dominated by 'rapids' and 'riffles', which are favourable to trout. The data on the upstream part of the tributaries, usually open to fishing, are missing at present.

In the Redon, inventories carried out at the end of October, from 1983 to 1987 (Melhaoui, 1985; Champigneulle, 1987; Champigneulle *et al.*, 1988a, 1990a) show, in areas used for spawning by lake trout that are open to fishing, the density of trout aged ⩾2+ is low (<2 ind./100 m^2) (Table 2). In contrast, the structure of the population is characterized by the dominance of juveniles, 0$^+$ (11.7 to 72.5 ind./100 m^2) and 1$^+$ (1.8 to 20.9 ind./100 m^2) (Table 2). However, the autumnal density of 0$^+$ trout resulting from natural recruitment remains lower at 35 ind./100 m^2 in the zone open to fishing (Champigneulle, 1987; Champigneulle *et al.*, 1988a). In areas where fry are released, natural recruitment represents, according to year and area, between 7% and 72% of 0$^+$ fish present in the autumn (Champigneulle *et al.*, 1990a). Juvenile 0$^+$ and 1$^+$, resulting from restocking at the fed fry stage can represent, depending on sector and year, between 28%

Table 2. Characteristics of populations in the main flow of some tributaries of Lake Léman. (*W*): width of sector in m; (*D*): density as number of individuals per 100 m^2, estimated using method of Lury (2–3 samples). (*LT*): total mean length in mm. (*): zone frequented by spawning lake trout. (**): zone with restocking.

Authors	Tributary	*W* (m)	Fishing pressure	Time period	0^+ *D* (n/100 m^2)	0^+ *LT* (mm)	1^+ *D* (n/100 m^2)	1^+ *LT* (mm)	$\geqslant 2^+$ *D* (n/100 m^2)
Buttiker, 1984	Greny	5	none	mid-May 1982	2.1	87	3.3	153	2.4
Champigneulle *et al.* (1988a)	Redon	4	none	end Oct. 1983 to 1987	15.6–43.8	90–105	3.8–17.9	167–209	3.1–5.6
				end Feb. 1984–85	24.1–28.1	101–116	4.6–5.1	203–207	1.0–2.8
	Redon (*) (**)	4–4.5	high	end Oct. 1983 to 1987	11.7–72.5	83–117	1.8–20.9	164–193	0–1.6
				end Feb. 1984–85	10.6–35.9	100–119	2.0–5.8	158–193	0–1.1
Durand (unpubl. data)	Versoix (*)	10	high	mid-July to end-July 1984–87	1.5–14	51–59	1.3–7.7	148–163	0.1–1.1
	(**)			end Nov. 1985	26.9			214	0.25

and 93% (7–65 ind./100 m^2) of 0^+ fish present in the autumn and between 39 and 91% (4–17 ind./100 m^2) of 1^+ fish (Champigneulle *et al.*, 1990a). The results were obtained from reared fry (TL = 3–5 cm), produced from eggs collected from spawning lake trout or domestic trout, dispersed (40–50 ind./100 m^2) at the end of spring – start of summer in the downstream part of the Redon.

Studies carried out on the Swiss tributaries open to fishing confirm the tendencies observed in the Redon. In fact, on the Versoix, low densities of trout aged 2 or more years were noted (0.1–1.1 ind./100 m^2) (Table 2) and also small trout 0^+ (1.5–14 ind./100 m^2), resulting from natural recruitment. Elsewhere, the autumnal releases of marked fish-farm-reared juveniles (5–8 cm), produced from eggs taken from lake trout were followed in experimental sectors of the Versoix, the Promenthouse and the Aubonne. The study (Durand and Pilotto, 1989) shows that, for a range of stocking rates (13–53 ind./100 m^2), the impact of the release increases with the density of the input and manifests itself mainly in the sectors where the restocking occurs. The released fry represent, depending on sector and tributary, from 10% to 56% of the juvenile 1^+ fish in place, 8 to 10 months after the release.

2. Growth

Growth of juveniles in the lower part of Lake Léman tributaries can be compared to that observed in other French rivers (Cuinat, 1960; Prouzet *et al.*, 1977; Baglinière *et al.*, 1989). In fact, during the second fortnight in July, the mean size (total length = TL) is already 51–59 mm for the 0^+ fish and 148–163 mm for the 1^+ fish (Table 2). The mean autumnal size of 0^+ and 1^+ fish in the Redon varies between 83–117 mm and 164–209 mm respectively (Table 2) depending on sector and year.

The multiplicity of possible origins of recruits (Fig. 3) renders accurate studies on growth difficult, unless the origin of the samples is well known, usually due to marking. Durand and Pilotto (1989) followed the growth of small trout hatched from lake trout eggs, reared to 65 mm on a fish farm and then released in November 1987 into three Swiss tributaries of Lake Léman. The mean monthly growth in the river, evaluated for the first 8–10 months after release, varied between 6 and 10 mm/month depending on the tributary. According to the authors, growth was equally great whether the tributary was productive or non-productive in terms of young trout, already resident at the time of release. Champigneulle *et al.* (1990a) studied the growth of fry (from spawning lake trout eggs), reared to 30–40 mm on a fish farm, marked and then released, in the Redon at the end of spring – start of summer. For each of the years studied, the mean autumnal size at stage 0^+ fish was smaller, by 1–3 cm, for the released trout than for those resulting from natural reproduction of lake trout in the Redon. The authors attribute this retardation in part to the fact that, at the time of release, the size of the released fry was already smaller than those of the resident fry. In contrast, the difference in size no longer existed at stage 1^+ in the autumn of the following year. The same study shows that, as a result of their fast growth, small trout of lacustrine origin can contribute to the fishery in the Redon in their third year (age 2^+), since some of them are still present on the river and have already attained the legal size (21–23 cm in French tributaries)..

3. Mortality and movements

Champigneulle *et al.* (1990a) demonstrated, on the Redon, a high rate of disappearance in the spawning zone of lake trout (>98%) between egg stage and the 0^+ stage in mid-autumn. The respective parts played by natural mortality, movements within the tributary or possible displacements to the lake at a very early stage are still not known. Counts carried out on the Redon at the end of the winter of 1983–84 and 1984–85 showed that there is a reasonably sharp fall (about 50% on average) in the densities of 0^+ and 1^+ fish between the end of October and the beginning of March (Melhaoui, 1985; Champigneulle, 1987; Champigneulle *et al.*, 1990a). Part of this decrease in numbers can be attributed to downstream movements. Marking experiments (Champigneulle *et al.*, 1988a) have shown that the period at the end of autumn – winter is characterized by instability of the juvenile populations, with movement downstream. Elsewhere, an experiment of partial trapping of fish moving downstream, carried out in the downstream current (Melhaoui, 1985; Champigneulle *et al.*, 1988a) in autumn–winter and the start of spring demonstrated the existence of displacements towards the lake of young trout of 1 or 2 years of age. Contrary to the case of small trout displacing in autumn, these ones migrating in winter and spring show traits of smolts and presmolts. These results agree with studies carried out (Stuart, 1957; Hunt and Jones, 1972; Thorpe, 1974; Arawomo, 1982; Craig, 1982) in lakes in Great Britain, where downstream migration appears after 1 to 3 seasons of growth in the river and during the period from mid-autumn to the start of spring. A study on the Redon was done to compare, as a function of size (two subsamples differently marked), the destiny of fry (from eggs from spawning lake trout) released before summer. Early results (Champigneulle, unpublished data) suggest that migration to the lake at stage 0^+ can occur before mid-autumn and the latter concerns the fry which were largest at the time of release. These studies are still too limited to allow quantitative information to be obtained on the patterns of displacement of juvenile trout in Lake Léman. Durand (unpublished data) also showed the existence of downstream migratory movements of immature lake trout leaving Lake Léman by its outflow: the downstream Rhône. These movements were observed in spring and the size of migrating trout varied between 15 and 50 cm. Because of dams preventing them returning upstream, these displacements constitute a loss to the system.

In the absence of any salinity barrier, the movements of juveniles between the flowing and non-flowing zones can take many different forms in the case of the lake trout similar to that of the sea trout. For example, displacement can involve smaller trout and extends over a longer period of time than the migration of sea trout smolts (Piggins, 1975). Autumnal displacements can occur as well as the spring migration of smolts. Elsewhere, some authors (Stuart, 1957; Jonsson, 1985) noted a springtime recolonization of lake tributaries by juveniles which had been displaced into the lake the previous autumn.

III. CHARACTERISTICS OF ADULTS

1. Age determination by scale reading

In all studies of the biology of the trout in Lake Léman, scales have been used for age determination and growth studies (Melhaoui, 1985; Buttiker *et al.*, 1987; Champigneulle *et al.*, 1988a). For the trout in Lake Léman, most studies of age and growth have been carried out on adults during the spawning season. While the determination of age does not pose any particular problems for resident trout, the interpretation of structures is more complex for the lake trout (Buttiker *et al.*, 1987). The use of reference scales, from trout whose age and history are known from being marked, has allowed the criteria for scale interpretation to be elucidated (Buttiker *et al.*, 1987; Champigneulle *et al.*, 1990b).

The scales of Lake Léman trout generally show the existence of one or two (rarely three) years of slow initial growth followed by a faster growth phase, indicated by an increase in spacing, thickness and sometimes the annual number of circuli (Fig. 4). These initial years of slow growth are generally interpreted (Allen, 1938; Holcik and Bastl, 1970; Craig, 1982) in other lakes as the period of life spent in the river. In Lake Léman, while this interpretation is no doubt valid for many lake trout, it is not necessarily applicable everywhere. In fact, lake trout produced from 0^+ juveniles released directly into the lake (Champigneulle and Durand, unpublished) show the same type of initial scale structure. The change in structure could therefore equally correspond to an acceleration of growth linked to a transfer to an ichthyophagous diet, which would generally

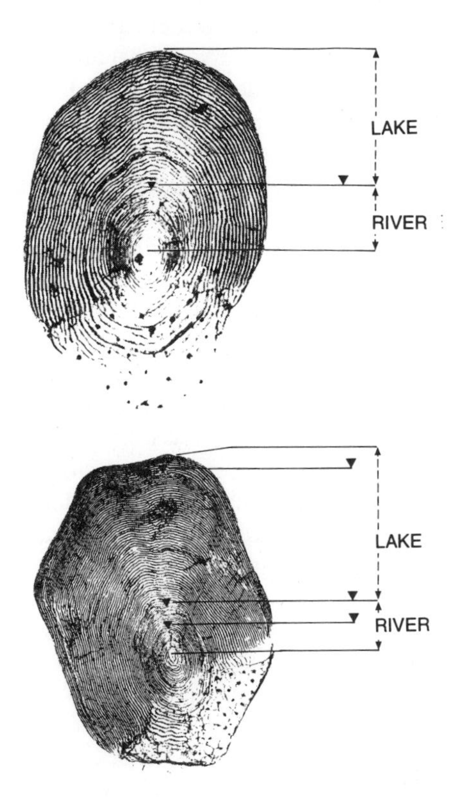

Fig. 4. Scales from lake trout from Lake Léman, caught in winter. 4a, 2-year-old spawners (1.0^+): 1 year of slow, initial 'river type' (type 1) growth and one season of fast growth in the lake. TL = 330 mm. (▼), winter. 4b, 4-year-old spawners ($2.I^+$): 2 years of slow, initial 'river type' (type 2) growth and two seasons of fast growth in the lake. TL = 585 mm. (▼), winter.

correspond to a transfer into the lacustrine environment by small trout produced in the tributaries.

In the following text, we will talk, therefore, about the number of years of initial slow growth as 'river type' and the number of years fast growth as 'lake type'. The terminology 'trout of type 1 or 2' will indicate trout having 1 or 2 years' initial slow 'river type' growth respectively, or also trout in which the 'juvenile' phase has lasted 1 or 2 years. For the adults, studies of age and growth have been carried out on trout caught at the end or very beginning of the year. This is why in the following text, the terms trout of 2, 3, 4 years and so on will be equivalent to trout at the end of stage 1^+, 2^+, 3^+ respectively.

2. Age, sexual maturity and sex-ratio

(a) Comparative data for stream-resident trout (from the Redon)

A study carried out in the autumn on the Redon (Champigneulle *et al.*, 1990a) allowed the sexing of spawning resident trout. Ripe males were sexed by obtaining milt on squeezing the abdomen and the ripe females by serodiagnostic methods as described by Le Bail *et al.* (1981). The study shows that sexual maturity can be observed from the age of 2 years in both sexes, but in a higher percentage of males (16–27%) than females (3–4%). However, it is only at the age of 3 years that almost all resident trout are mature (Fig. 5A). According to Champigneulle *et al.* (1988a), in the zone open to fishing, spawning stream-resident trout are very scarce in autumn (density of males < 3 ind./100 m^2 and density of females < 1 ind./100 m^2) with sex ratio (males/females) greatly in favour of the males (4.0:1) and very few trout aged 3 years and over (Fig. 5a). In contrast, in the protected zone, the density is increased overall (density of females: 3.5–5.8 ind./100 m^2; density of males: 4.3–6.4 ind./100 m^2) with a more even sex ratio and with the presence of trout 3 years and over (Fig. 5A).

(b) Lake trout

Breeding lake trout were captured in the Swiss tributaries of Lake Léman between 1964 and 1974; of those whose age could be determined, all the females captured were more than 2 years old, while some males were 2 years old (Buttiker *et al.*, 1987). Because of the nature of the objectives (egg collecting), these catches were very selective (females and large individuals were caught preferentially) and did not permit the sex-ratio or the age structure of the breeding fish to be determined. In contrast, during work on the Redon (Champigneulle *et al.*, 1988a, 1990b), the Versoix and the Brassu (Durand, unpublished data), the authors attempted to reduce the sampling bias by fishing frequently and examining all the breeding fish caught. This approach has shown, in the case of the Redon followed from 1983 to 1988, the presence of breeding lake trout aged between 2 and 7 years (Table 3). Although 2-year-old females (7% of females) were captured, it was the males which showed a significant percentage (41%), but very variable according to year (15–72%) for 2-year-old males (Champigneulle *et al.*, 1990b). Over all five years sampled, while the percentage of individuals of 3 years is close in the two sexes (Table 3), the percentages of breeding fish aged between 4 and 7 years is significantly higher for the females (67%) than for the males (27%). The percentage of breeding fish of 1, 2 or 3 years of initial 'river type' growth was 68, 31 and 1% respectively for the males and 38,

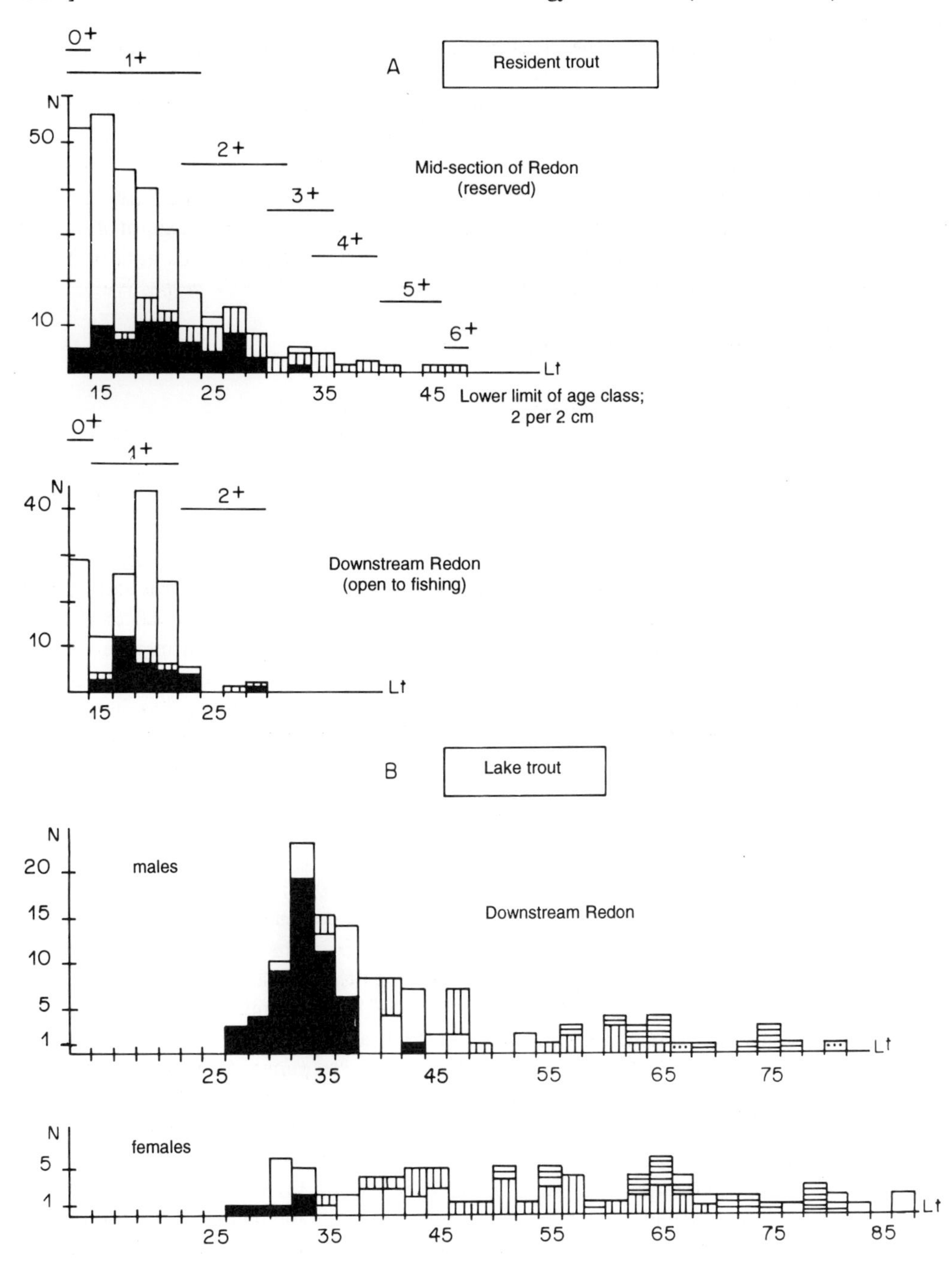

Fig. 5. Size structure (TL in mm) of spawning lake trout and resident trout caught in the Redon. 5a: Resident trout at the end of autumn in the reserve zone and zone open to fishing. (□) juveniles; (■) spermiating males; (□)| mature females. 5b: Spawning lake trout. Age class: (■) 1⁺; (□) 2⁺; (◫) 3⁺; (⊟) 4⁺; (⊡). 5⁺–6⁺.

61 and 1% for the females. The difference is essentially due to the presence of many young 2-year-old males. For 3-year-old breeding fish, the percentage of individuals of type 2 is more increased overall for females (82%) than for males (40%) while there is little difference for the older spawners (Champigneulle *et al.*, 1988a).

Table 3. Age structure of male and female spawning lake trout caught in the Redon between 1983 and 1988. Data are expressed as a percentage of the total age (2 years old = 1^+; 3 years old = 2^+ etc.)

Total age / Sex / N = number	2 years	3	4	5	6–7
Male ($N = 128$)	41.4%	31.2%	16.4%	9.4%	1.6%
Female ($N = 86$)	7.0%	25.6%	37.2%	22.1%	8.1%

The sex-ratio (males/females) of the total (226) catch was about 1.5:1 and varied, depending on year, between 1 and 2. However, the value of the sex-ratio may have been overestimated in favour of the males by using the technique of electro-fishing for capture as staying time in the river appears to be shorter for the females than for the males (Champigneulle *et al.*, 1988a).

In general, in a given environment and for a given population, faster-growing trout reach sexual maturity at an earlier age (Alm, 1959; Baglinière *et al.*, 1981; Craig, 1982; Burger, 1985; Dellefors and Faremo, 1988; Jonsson, 1989). However, maturation rate can vary depending on stocks or environment (Alm, 1952; Quillet *et al.*, 1986; Dellefors and Faremo, 1988). Elsewhere, the relationship between ‘fast growth and sexual precocity’ cannot be applied systematically (Stuart, 1953; Jonsson, 1982) and can be inverted when comparing the migratory fraction of a population (to lake or sea) to that remaining in the river (Craig, 1982). Data collected from the Redon (Fig. 5; Champigneulle *et al.*, 1988a, 1990b) show, despite their fast growth, the existence of a later sexual maturation in lake trout than in sedentary trout. Melhaoui (1985) and Champigneulle *et al.* (1990b) noted the presence of numerous large immature lake trout aged between 3 and 6 years, while Champigneulle *et al.* (1988a) indicated that the majority (minimum of 90%) of sedentary trout aged 3 years or older are mature. The later maturation of lake trout was also noted in Lake Constance where, according to Ruhlé (1983), 8% of lake trout are mature at 3 years old, 22% at 4 years, 75% at 5 years and 100% at 6 years. However, age at first maturity is not always late in lake trout. In fact Maisse (1985) noted, in a sample of 86 lake trout from Lake Léman with two years ‘river type’ slow growth, the presence of a precocious spawning mark at the time of the second winter. The author indicates that this precocious spawning can be observed as often in males as in females and involves 20% of the sample examined.

In order to make a comparison with the broodstock of lake trout going back up the Redon, a sample of 126 immature lake trout captured at the same time in the pelagic zone of Lake Léman were examined (Champigneulle *et al.*, 1990b). In this sample which contained immature trout aged between 2 and 6 years, the sex-ratio (males/females) was about 0.8:1; in contrast, in the case of the Redon spawners, the age structure and division of trout into types 1, 2 and 3 (44, 52 and 4% respectively for the whole sample) varied little with sex.

3. Growth

Buttiker *et al.* (1987) determined the age and studied the growth of lake trout spawners captured between 1964 and 1974 in the Swiss tributaries of Lake Léman. For this sample of spawning trout, composed essentially of trout aged 3 years or more, the medians of total length for the different age classes are shown in Table 4. The authors noted a very clear acceleration in growth between 2 and 3 years. At the same age, it is possible to find significant differences in mean length between females from different tributaries. Elsewhere, no immediate modification of growth appears to have taken place, despite eutrophication, in cohorts born between 1958 and 1970. The authors indicate possible bias linked to under-sampling of small spawning fish and to the fact that only scales which were easy to read were included. Scale readings have also been carried out on parents of lake trout captured more recently, since 1983 (Melhaoui, 1985; Champigneulle *et al.*, 1988a; Champigneulle *et al.*, 1990b; Durand, unpublished data). The mean size of captured spawning fish are shown in Table 4. Champigneulle *et al.* (1990b) studied the age of growth of the broodfish of Redon trout from 1983 to 1987, and distinguished between spawners with 1 (type 1) or 2 (type 2) years initial slow 'river type' growth. To do back-calculations, the authors used an adjusted linear model on a sample composed of both adult lake trout and juveniles from the Redon. The distinction between the two types of spawners allowed it to be shown that, for a given type of broodfish (1 or 2) there is little difference in growth between the sexes. In contrast, for equal total ages, size is significantly greater in type 1 broodfish than for type 2 (Fig. 6). This more rapid growth occurs at an early stage since the back-calculated length at 1 years is significantly higher (129 mm) for type 1 spawners than those of type 2 (100 mm). This difference could be partly explained by passage of juveniles into the lake at 1 or 2 years respectively for the type 1 and 2 spawners. Several authors (Allen, 1938; Stuart, 1957; Treasurer, 1976; Arawomo, 1982; Jonsson, 1982) indicate, for other lakes, that the staying time in the river is shorter for juveniles which have grown faster. According to Treasurer (1976), it seems preferable to describe the growth of lake trout using a distinction between groups depending on their number of years' initial, slow 'river type' growth.

Champigneulle *et al.* (1990b) demonstrated the presence in Lake Léman of immature trout of equal age but larger size at the start of winter, than spawning lake trout. Examining their growth by back-calculation showed that, although they had slightly slower growth during the first years, during the last year, the immature fish caught up and exceeded the size of the spawners. This catching up can partly explain the fact that sexual maturation slows growth, a well-known phenomenon in aquaculture (Burger, 1985; Quillet *et al.*, 1986). Elsewhere the examination of stomach contents showed that the immature fish continue to feed actively at the end of autumn – start of winter, while the mature lake trout stop feeding in the river, during the spawning period.

Table 4. Total length (TL in mm) of spawning (male + female) lake trout caught in various tributaries of Lake Léman, in relation to total age (2 to 7 years). (m): mean length; (md): median length.

Authors	Tributaries	Period	Tl in mm (n; σ)					
			2	3	4	5	6	7
Buttiker *et al.* (1987)	Aubonne	1964–1974 (md)		388 (1294)	500 (1494)	588 (1332)	666 (793)	730 (282)
Champigneulle *et al.* (1989b)	Redon	1983–87 (m)	341 (58; 30)	394 (62; 55)	532 (51–91)	685 (32; 81)		
Durand (unpublished data)	Versoix	1985–87 (m)	355 (4;50)	375 (12; 59)	525 (6; 89)	623 (21; 65)	710 (11; 97)	749 (9; 101)
	Brassu	1985–87 (m)	345 (2;0)	391 (10; 63)	438 (6; 86)	519 (4; 47)	618 (8; 17)	650 (1; 0)

Compared to other lakes, the growth of trout appears to be rapid in Lake Léman (Melhaoui, 1985; Buttiker *et al.*, 1987); it is even comparable to that of sea trout (Euzenat and Fournel, 1979; Richard, 1981). At the same age, trout size is clearly higher in Lake Léman than in the river (Fig. 6), although growth in the downstream part of the tributaries can be considered as being fast (Melhoui, 1985; Champigneulle *et al.*, 1988a). The thermal regime of Lake Léman is overall more favourable to growth than that of the tributaries. Thermal inertia limits cooling down in winter. Elsewhere, in summer, although the temperature is usually higher than 15°C in a tributary, the pelagic zone of the lake, as a result of thermal stratification, has temperatures of 12–14°C (optimum for growth of the species according to Elliott (1982)) close to the thermocline. Nettles *et al.* (1987) demonstrated, in Lake Ontario, the preferential localization of lake trout close to the thermocline in summer. Apart from the thermal factor, the better growth in the lake can be explained by the difference in other factors: lack of current, food abundance and lower trout density. According to Héland (1977), these changes lead to a release from territorial behaviour in favour of shoaling behaviour in small groups which are better adapted for capturing prey in the pelagic environment. The richness of fish prey items is one of the main explanatory factors of the high growth of trout usually observed in eutrophic or mesoeutrophic lakes such as Lake Léman. However, large interannual fluctuations in abundance of prey (juvenile roach and perch) could partly explain the variations in trout growth and catches observed in Lake Léman (Champigneulle *et al.*, 1990b). Studies by Jensen (1977) show that, in the lacustrine environment, an increase in trout density can be accompanied by a decrease in growth rate. The development of acoustic techniques (echolocation, wide band sonar) would allow a better understanding of fluctuations in abundance of fish prey items in the pelagic zone of lakes (Dziedzic and Gerdeaux, personal communication).

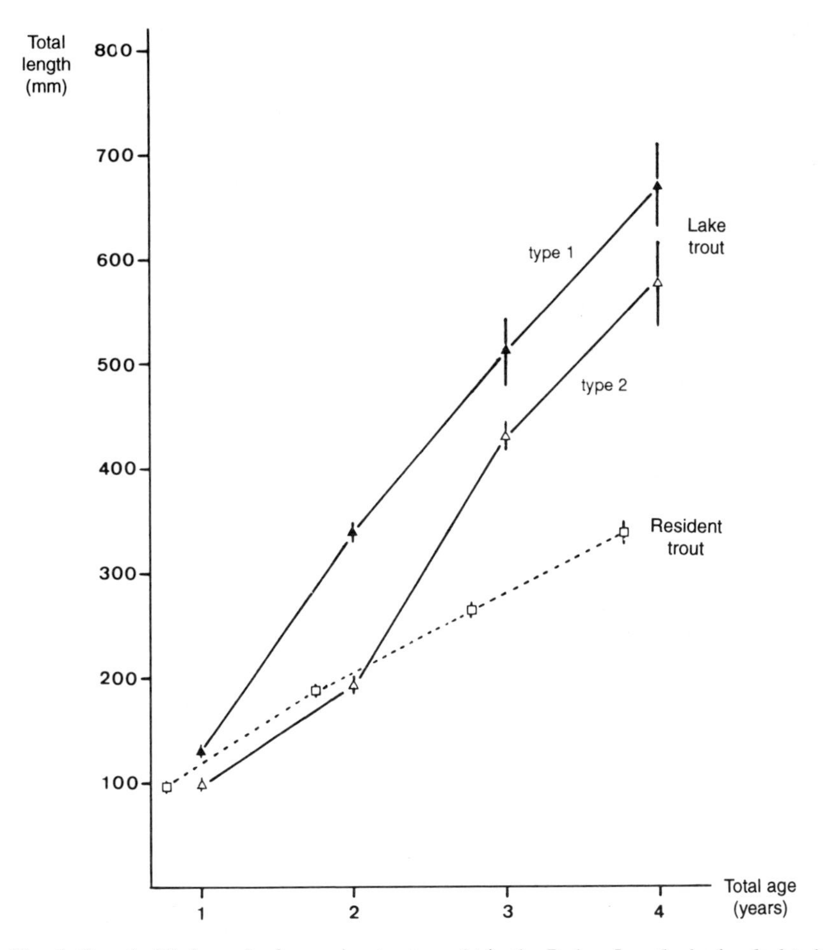

Fig. 6. Growth (TL in mm) of spawning trout caught in the Redon. Lengths back-calculated (linear model) for lake trout, 1983–1987, taking into account the number of years slow 'river type' growth: (▲), type 1; (Δ) 2, type 2. (□) resident trout, mean length measured in mid-autumn, 1984–85 (after Champigneulle *et al.*, 1988a). Vertical bars indicate 95% confidence interval.

IV. FEEDING IN LAKE TROUT

Some of the characteristics associated with the feeding of trout in Lake Léman between December 1983 and January 1985 have been demonstrated by Melhaoui (1985). He examined the stomach contents of 208 large trout (35–81 cm) caught in the pelagic zone. The examination of this sample showed a very strong tendency towards ichthyophagy (mostly roach at the time of sampling) in trout caught in spring, summer and autumn. In contrast, trout caught in winter were ichthyozooplankton eaters. In this season, trout can be seen close to the surface. Elsewhere, Champigneulle *et al.* (1986) showed, from two trials in 1983 and 1986, that winter catches per unit effort were on average 5 to 9 times greater when the nets (3 m deep) were positioned on the surface instead of 2 m below the surface (Table 5). Examination of immature trout caught in this way shows that they continue to feed actively despite the low water temperature (<6°C). Examination of a sample of 76 smaller trout (6–32 cm) caught in winter in the littoral zone showed them to be ichthyobenthophages (Fig. 7; Melhaoui, 1985). The tendency towards ichthyophagy has also been noted (Wojno, 1961; Holcik and Bastl, 1970; Antioniazza and Pedroli, 1983; Aass, 1984; Brandt, 1986) in some other lakes with different 'prey' species: roach, perch, bleak, whitefish, trout, shad.

Devaux and Monod (1987) evaluated the conversion rate (weight gain/weight ingested) of roach by Lake Léman trout while studying the accumulation of PCB and DDT (organochloride pollutants) as a function of age, annual growth and the contamination of roach. The conversion rate was about 0.4 between 3 and 4 years and about 0.17 between 4 and 5 years.

Several studies (Nilsson, 1963; Nilsson and Pejler, 1973; Svardson, 1976) show, for Scandinavian lakes, the existence of interactions between the Arctic charr and brown trout resulting in significant suppression of the trout and profiting the charr, for trophic reasons. This phenomenon has not been observed in Lake Léman. The large size of the lake, the abundance of zooplankton, the generally deep localization of the charr and above all the abundance of fish prey serve to limit the feeding competition between the charr and the trout in Lake Léman and in addition, according to Svardson (1976), the trout starts to be piscivorous earlier than the charr. Nevertheless, interactions are likely to

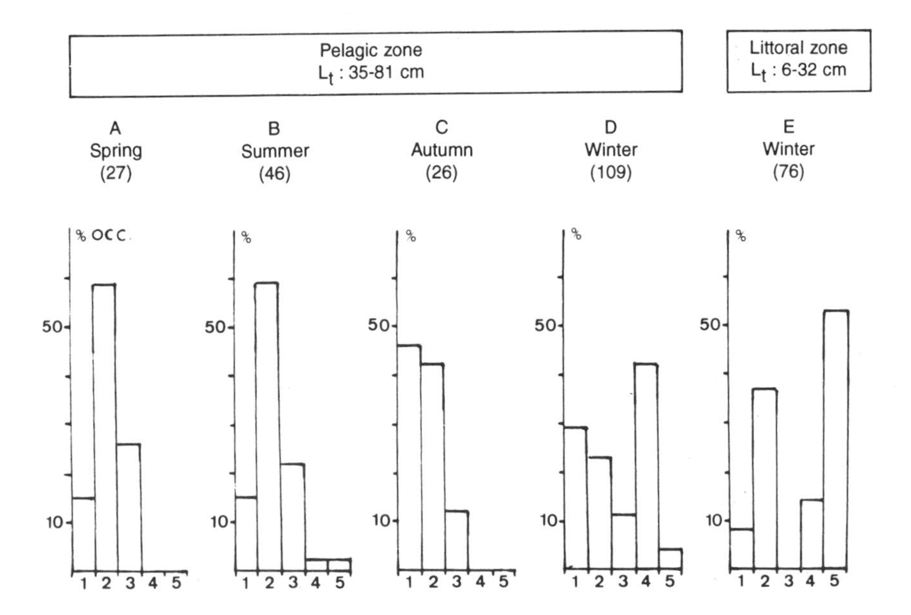

Fig. 7. Feeding of lake trout in Lake Léman from the end of 1983 until the beginning of 1985 (after Melhaoui, 1985). Large trout (L_t = total length: 35–81 cm) in the pelagic zone: (A) spring; (B) summer; (C) autumn; (D) winter. Small (7–32 cm) trout in the littoral zone in winter: (E) data expressed as percentage occurrence: (1) empty, (2) roach, (3) other fish (bleak, perch) and unidentified species, (4) zooplankton, (5) benthic macroinvertebrates. Values in parentheses: number of trout examined.

evolve from the fact that restocking of charr is increasing (Champigneulle *et al.*, 1988b) and because there are large fluctuations in the abundance of fish prey items.

Table 5. Comparison of two capture methods for trout per unit effort at the start of 1983 and 1986; setting of a net either at the surface or 2 m below the surface (after Champigneulle *et al.*, 1986).

		1983 (13/02 to 11/03)	1986 (17/01 to 23/02)
Set at the surface	Fishing effort (length of net set, m)	7 840	114 300
	Mean catch (kg/100 m of net)	1.25	1.8
	Mean catch (no./100 m of net)	1.0	1.6
Set 2 m below the surface	... See above ...	21 075	3 330
		0.22	0.2
		0.20	0.2
Relationship between mean catches: (surface)/(−2 m)	By weight	5.7	9.0
	By number	5.0	8.0

V. REPRODUCTION

1. Upstream migration and spawning sites

It appears from many studies, reviewed by Melhaoui (1985), that the lake trout usually spawns in tributaries. However, in certain Scandinavian lakes, spawning can take place in the outflow rivers (Runnström, 1949, 1952, 1957). Elsewhere, Stuart (1953) and Frost and Brown (1972) have observed trout spawning at the edge of lakes, both in Ireland and in Scotland. In Europe, spawning migration generally takes place from summer to the start of winter and appears to be earlier, as is spawning, in the higher latitudes (Allen, 1938; Runnstrom, 1949, 1952 and 1957; Stuart, 1957; Sokowicz, 1961; Holick and Bastl, 1970; Jensen, 1977; Rippmann, 1983).

In Lake Léman, the main identified spawning tributaries of lake trout are shown in Fig. 1. For many of them the best zones (e.g. the Dranse) for reproduction and production of juveniles have been rendered inaccessible by the creation of dams. Data on upstream migration and characteristics of the breeding fish have been collected from tributaries on the Swiss bank, mainly from fisheries for spawners (electro-fishing, and partial trapping of the Aubonne; Fig. 1), in order to provide eggs for fish farms producing fish for restocking (Buttiker and Matthey, 1986). On the French bank, data on trout spawning (lake and sedentary) have been collected in the Redon in association with autumnal inventories and sexing of the spawners, then from several samples by electro-fishing and by following the spawners during the reproductive period (Champigneulle *et al.*, 1988a). These studies have shown the existence of upstream migrations, essentially from November to the end of January with, as in the Redon (Fig. 8), peaks of catches varying from one year to another, but usually observed during and just after the strongest spates. In certain large rivers, however, such as the Dranse (Fig. 1), catches of lake trout by amateur fishermen indicate that certain spawners are already present in summer and at the beginning of autumn. Nevertheless, these early upstream migrations are not of the same scale observed in the sea trout (Piggins, 1975; Euzenat and Fournel, 1979; Richard, 1986).

2. Spawning sites and spawning activity

The characteristics of spawning sites and spawning activity have been described for lake and stream-resident trout in the main Redon river (Champigneulle *et al.*, 1988a). In this tributary, spawning is on average later in lake trout than in stream-resident trout. On the date when 50% of spawning lake trout redds were built, the cumulative percentage of built resident trout redds was already close to 90% (Fig. 8). For the resident trout, the peak spawning period occurs between the start of November and mid-December, whereas for lake trout, it occurs between the beginning of December and the beginning of January (Fig. 8).

Fig. 9 shows the morphometric characteristics of redds (length, water depth, substrate) of lake trout in the Redon. In this tributary, in relation to the larger size of the migratory spawners, the redds of lake trout are usually larger, deeper and with coarser substrate than those of sedentary trout (Melhaoui, 1985; Champigneulle *et al.*, 1988a). On this tributary, the opening up of some redds in 1984 (Melhaoui, 1985) showed a good survival rate (85%) of eggs at the eyed stage. However, at the time of unfavourable winter water conditions (e.g. end of autumn 1985 without spates, then strong spates at the start of 1986; Fig. 8), some redds can be totally destroyed (Champigneulle *et al.*, 1988a).

3. Fecundity and the phenomenon of multiple spawnings

Table 6 shows the length–weight and length–fecundity relationships established by Melhaoui (1985) from samples of lake trout from Lake Léman. The mean relative fecundity calculated by Melhaoui (1985) from a sample of 26 trout of between 42 and 71 cm is 2460 ± 220 oocytes/kg female. The relative fecundity fluctuates between 1300 and 3700 oocytes/kg depending on the females, confirming the wide variability of this parameter, as already noted in France for other populations of river or sea trout (Euzenat and Fournel, 1976, 1979).

In Lake Léman, multiple spawnings were demonstrated by the examination of spawning marks on the scales or using marking procedures on individual spawners (Melhaoui, 1985; Buttiker and Matthey, 1986; Champigneulle *et al.*, 1990b). Table 7 shows that individuals can have four successive spawnings while indicating the rarity of individuals which have spawned three–four times (two–three recaptures), compared to those spawning twice. In other European lakes, several authors (Arvidson, 1935; Alm, 1949;

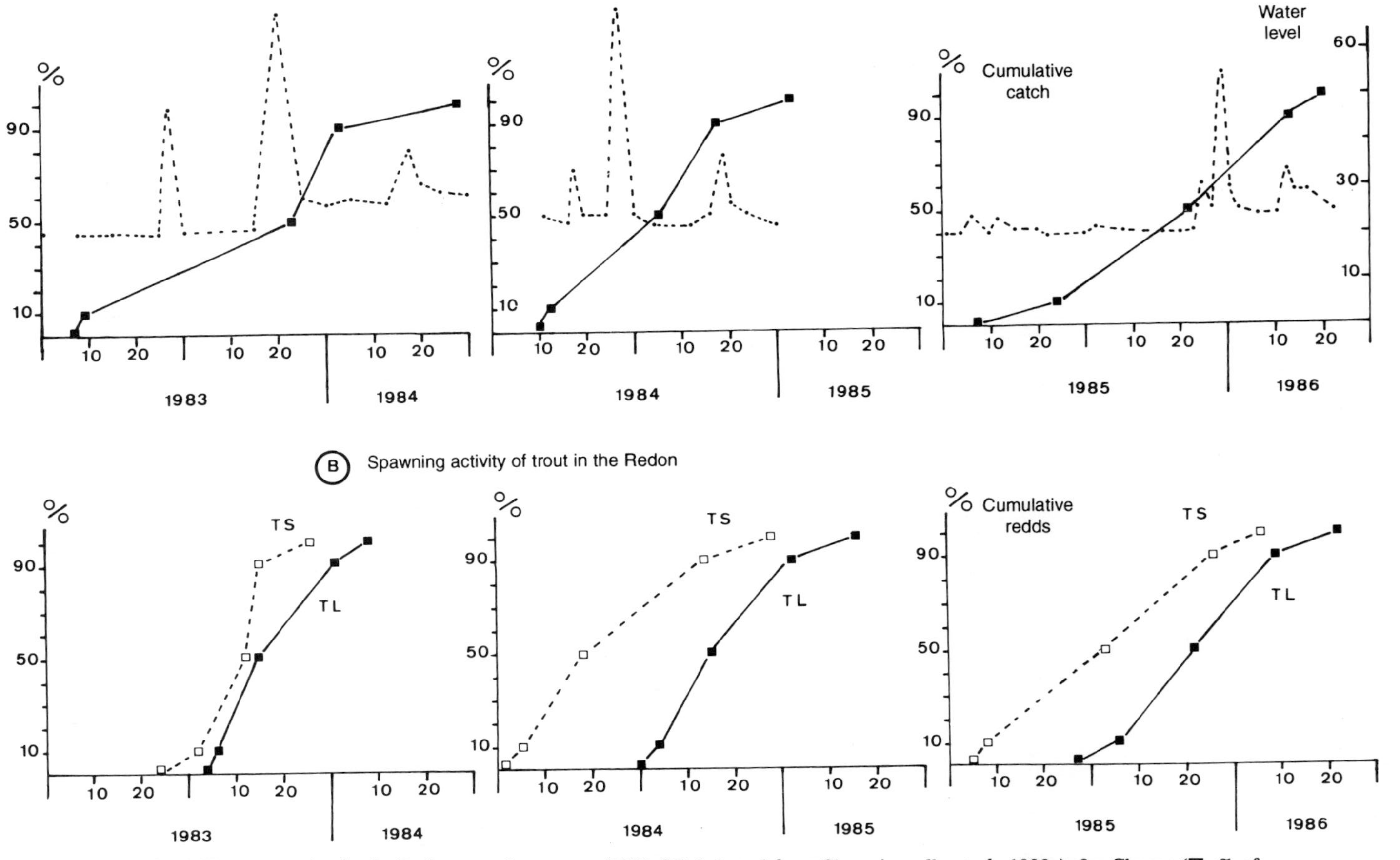

Fig. 8. Trout spawning in the Redon over three years (1983–85) (adapted from Champigneulle *et al.*, 1988a). 8a: Change (■: % of total caught) in catches of spawning lake trout in relation to period and water level (●—●). 8b: Change in cumulative number (%) of lake trout redds (■) and resident trout redds (■) in the downstream section of the Redon.

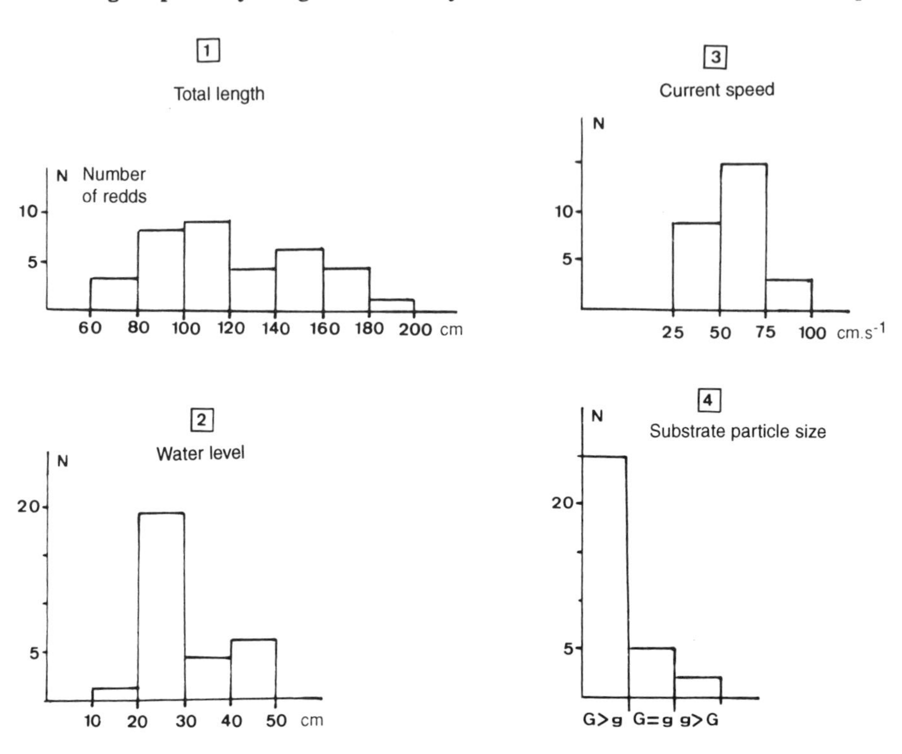

Fig. 9. Physical characteristics of lake trout redds during winter spawning 1984–85 in the Redon. (1) total length; (2) water level; (3) current speed; (4) substrate particle size of the stream bed; mixture of gravels (t: 2 mm–2 cm) and pebbles (G: 2 cm–10 cm) with (G > g) pebbles dominant; (g > G) gravel dominant; (G = g) similar proportions of gravel and pebbles.

Table 6. Relationships between total length (TL in mm) and weight (g), and between total length and fecundity for lake trout in Lake Léman (*n*): number of individuals. (Data from Melhaoui, 1985).

n		TL in mm	Relationship formula	*r*
105	49 males + 56 females immature	200–800	Length (TL in mm)–Weight (in g) log P = 3.115 log LT – 12.101	0.995
26	female mature	420–710	Length (TL in mm)–absolute fecundity (Fa) log Fa = 3.0159 log LT – 10.549	0.8044
			Relative mean fecundity: Fr = 2460 ov/kg ± 220	

Table 7. Return to site of initial capture by spawning lake trout during fishing for breeding fish (interval between two captures >200 days). (A): recaptured in initial river, (B): recaptured in another river. (Data from Buttiker & Matthey, 1986.)

	River of origin			
	Aubonne		Promenthouse	
Number of recaptures	A	B	A	B
1	89	1	40	3
2	2	0	5	0
3	0	0	2	0
Total	91	1	47	3

Runnstrom, 1952; Stuart, 1953; Sakowicz, 1961) indicated that some breeding fish (mainly 2-year-olds) may cease reproduction between two spawnings. This phenomenon was noted by Champigneulle *et al.* (1990b) for some trout in Lake Léman. The number of spawning trout exhibiting multiple spawnings is partly limited by the significant mortality due to fishing. In fact, campaigns for marking spawning fish and for the declaration of recaptures by fishermen have resulted in rates of declared recaptures of 10% (Buttiker and Matthey, 1986) to 30% (Champigneulle *et al.*, 1988a).

4. Homing

The bibliographical review carried out by Saglio (1986) indicated that homing, defined as the return of the adult, for reproduction, to the river where it was produced as a juvenile, has been particularly well demonstrated in the Atlantic salmon (*Salmo salar*) and Pacific salmon (genus *Oncorhynchus*). Although less well known, this phenomenon has also been shown in the brown trout in the lake-tributary ecosystem and a process of memory, with olfactory support, seems to be involved (Stuart, 1957; Tilzey, 1977; Scholz *et al.*, 1978).

In Lake Léman, trout fry produced either from breeding lake trout caught in the Swiss tributaries of Lake Léman or from fish farm stock, were raised in lake water until they were 3–5 cm in length. After marking, they were released at the end of spring – start of summer into the Redon, a tributary on the French bank (Melhaoui, 1985; Champigneulle *et al.*, 1990a). Following the spawning fish in the Redon allowed the return of breeding lake trout, resulting from marked fry (Champigneulle *et al.*, 1990b) to be shown for both types of fish but with higher returns for the lake trout origin. The latter authors discovered that a major part (46% of females and 22% of males) of the spawning cohort in 1983, returning to spawn in the Redon, was produced from a small release of 6000 fry reared in aquaculture (origin: lake trout; length: 4 cm) released into the Redon at the start of August 1983.

Loyalty to the same river over successive spawnings, shown by Stuart (1957) and Tilzey (1977), has also been demonstrated in Swiss tributaries of Lake Léman by Buttiker and Matthey (1986). Between 1964 and 1974, 2588 breeding lake trout were caught (mainly in two tributaries 14 km apart: 1871 in the Aubonne and 577 in the Promenthouse), marked and released. This study shows that most breeding fish which are marked and recaptured at the time of several successive spawnings, were found to be in the river in which they were first caught (Table 7). The cases of 'wanderings' have probably been underestimated, since all Lake Léman tributaries were not sampled equally. However, out of a total of 145 recaptures achieved in the Aubonne or the Promenthouse, one single trout originating from the Aubonne was recaptured in the Promenthouse and only two spawning fish originating from the Promenthouse were recaptured in the Aubonne. Examination of recaptures in Lake Léman of marked spawning trout in Swiss tributaries shows that they have a tendency to stay in one large (90–300 km^2), but well-defined, part of the lake, surrounding the mouths of these same tributaries (Buttiker and Matthey, 1986). Phenomena of attachment to one particular zone have also been demonstrated in some other lakes (Gustafson *et al.*, 1969; Jensen, 1977).

VI. RESEARCH PERSPECTIVES—CONCLUSIONS FOR MANAGEMENT

The current study has allowed the definition of the principal characteristics of the biology of the trout in Lake Léman and some of its tributaries. New research is necessary to better define, as has been done in other environments (Stuart, 1953, 1957; Jonsson, 1989; Baglinière *et al.*, 1989; Elliott, 1989), the strategies employed by the trout to colonize the Lake Léman ecosystem.

The existence of possible genetic differentiation of trout populations in Lake Léman and its tributaries remains to be studied, since a preliminary study (Guyomard, personal communication), suggests that the population does not appear to be homogeneous in this ecosystem. Crozier and Ferguson (1989) and Ferguson (1989) demonstrated the existence of genetic differences in trout present in the same lake. In Lough Melvin, which is small (2200 ha) compared to Lake Léman, Ferguson (1989) indicated the presence of three sympatric populations, with different spawning grounds, food preferences and growth rates. One of the main failings in the research carried out on Lake Léman is that the possible biological variability of the lake trout has only been partially explored. In fact, studies have almost all been carried out on the growth, reproduction and population dynamics in downstream parts of tributaries of average size (3 to 10 m wide). In contrast, the characteristics and biology of trout frequenting the two largest tributaries (upstream Rhône and the Dranse), the upstream parts of the tributaries and the littoral zone of Lake Léman are very poorly understood. In addition, the existence of a heterogeneity created by fry-rearing programmes cannot be excluded, varying until now from one country to the next and between one river and another.

It is necessary to develop a better understanding of the fluctuations, determinism of natural recruitment of juveniles into the lake and the relationships between resident and migratory trout. This research necessitates choosing and following the functioning of one or several reference tributaries equipped with trapping systems (upstream and down stream). The carrying capacity of 0^+ trout in the tributaries is relatively easy to evaluate by quantifying the habitat characteristics and density in reference sectors (Jones, 1976; Baglinière and Champigneulle, 1982). In contrast, the capacity of the lake for juvenile production and ongrowing is more difficult to evaluate. The species (mainly roach and perch) used as prey items at a young stage, are characterized by very larger interannual fluctuations in recruitment, which are difficult to control and predict as they are partly linked to the weather. It would nevertheless be useful to have an index of abundance which would take into account these interannual fluctuations in this important component of the carrying capacity in the lake. In parallel, it would be interesting to study the spatial distribution (in both littoral and pelagic zones), movements, growth and feeding of trout in the lake, in relation to life-cycle stages, seasons and different levels of abundance of fish prey items.

The management of the trout in Lake Léman has become more complex as a result of the increased pressure from fishing and the change of environments and populations. In the tributaries, there is a decrease in the areas available for reproduction and natural production of juveniles (dams, reservoirs, river widening, pollution, water abstraction). Improvement of natural production therefore begins with the protection of the tributaries. From a practical point of view, the operations (passes, pollution control) which are likely to give quantitative results remain to be identified and prioritized.

Restocking constitutes a potentially important means of management (Champigneulle, 1985), on the condition that the state of natural production, the holding capacity of the whole ecosystem (lake–tributaries) and the possible impacts on the entire population (competition, predation in relation to other exploited or restocked species) are taken into account. Research is still necessary to optimize releases in relation to production by trout in the lake. It will be important to follow up the existing work to evaluate and compare the efficiency of various methods of restocking (size, timing, place: lake–tributary) for different identified stocks of trout (or their crosses) or used in practical restocking. Based on the current state of knowledge, it appears to be possible to improve the production of juveniles from the tributaries by releasing lake-reared fry there, up to saturation point of the carrying capacity, at the end of spring – beginning of summer. This method of fry rearing, orientated towards the production of lake trout, could use mainly juveniles produced from eggs collected from spawning lake trout caught preferentially in the tributaries concerned. It is nevertheless important to determine whether this method of release deteriorates or ameliorates the natural production of juveniles and that of sedentary trout exploitable by river fishing. The setting up and maintenance in aquaculture of a captive stock made up of breeding lake trout caught in Lake Léman tributaries little affected by restocking could be an interesting avenue to test for the restocking carried out in the Lake Léman area (Guyomard, 1989b). The practice of fishing for breeding trout and monitoring traps will be of great value in determining if the state of the exploited stocks can produce sufficient spawners to saturate the holding capacity of the tributaries and allow the collection of eggs for restocking operations, the efficiency of which is known and optimized in relation to management objectives (lake ranching). The optimization of release directly into the lake would very likely require the relatively flexible production of juveniles (number, size, timing). In fact, releases carried out directly into the lake would have to be organized to fill in a deficit in recruitment of trout from the tributaries, allow additional recruitment of juveniles to the littoral zone or to better exploit (or even regulate) the strong age classes of fish prey species.

The optimization of exploitation is difficult where the size and rate of exploitation of a stock are not known. The recent improvement in fishery statistics collection (obligatory daily declaration for amateur and professional fishermen) constitutes a first improvement which must be continued in order to determine fishing effort. This has shown that the mean weight of trout caught by professional fishermen is about twice as high (1100–1300 g) as that for the amateurs (600–750 g), which suggests different exploitation of the stock depending on fishing method. In addition, it will be important to take samples which will allow the selectivity of different fishing methods and characteristics of the catch (size, age, sex, maturity) to be determined.

The optimization of trout fishery management in Lake Léman will inevitably come from an improvement of the knowledge of the biology and ecology of the species in this ecosystem.

ACKNOWLEDGEMENTS

The studies on the trout of Lake Léman were carried out, on the French side, as part of a CSSP-INRA contract (83-625 from 6/12/83) and an ATP-INRA contract on the functioning of lacustrine ecosystems. On the Swiss side, the work benefited from help and financial support of the Service des Forêts, de la Faune et de la Protection de la Nature (Geneva district) and of the Service de Conservation de la Faune (Vaud district).

BIBLIOGRAPHY

Aass P., 1984. Brown trout stocking in Norway. EIFAC symposium on stock enhancement on the management of freshwater fisheries. *EIFAC Tech. Pap./Doc. Tech. CECPI*, 42, Supp. 1, 123–138.

Allen K. R., 1938. Some observations on the biology of the trout (*Salmo trutta*) in Windermere. *J. Anim. Ecol.*, **7**, 333–349.

Alm G., 1949. Influence of heredity and environment on various forms of trout. *Rep. Inst. Freshwater Res. Drottningholm*, 29, 29–34.

Alm G., 1959. Connection between maturity, size and age fishes. *Rep. Inst. Freshwater Res. Drottningholm*, 40, 5–145.

Antoniazza Y., Pedroli J. C., 1983. Contribution à l'étude de la biologie et de la pêche de la truite de lac (*Salmo trutta*) dans le lac de Neuchâtel. Rapp. Comm. Pêche lac Neuchâtel, May 1983, 48 pp.

Arawomo G. A. O., 1982. The age and growth of juvenile brown trout in Loch Leven, Kinross, Scotland. *Arch. Hydrobiol.*, **93**, 466–483.

Ardvidson, G., 1935. Marking av Lxoring i Vattern. *Mednk. Lantbr Styr.n.s.*, **4**, 16.

Baglinière J. L., Champigneulle A., 1982. Densité des populations de truite commune (*Salmo trutta* L.) et de juvéniles de saumon atlantique (*Salmo salar* L.) sur le cours principal du Scorff (Bretagne): preferendums physiques et variations annuelles (1976–1980). *Acta Oecol., Oecol. Appl.*, **3**, 241–256.

Baglinière J. L., Le Bail P. Y., Maisse G., 1981. Détection des femelles en vitellogénèse. II. Un exemple d'application: recensement dans la population de truite commune (*Salmo trutta*) d'une rivière de Bretagne sud (le Scorff). *Bull. Fr. Piscic.*, **283**, 89–95.

Baglinière J. L., Maisse G., Lebail P. Y., Nihouarn A., 1989. Population dynamics of brown trout (*Salmo trutta* L.) in a tributary in Brittany (France): spawning and juveniles. *J. Fish Biol.*, **34**, 97–110.

Brandt S. B., 1986. Food of trout and salmon in Lake Ontario. *J. Great Lakes Res.*, **12**, 200–205.

Burger G., 1985. *Relations entre génotype, croissance et âge à la première maturation sexualle chez la truite arc-en-ciel* (Salmo gairdneri). Mémoire DES, Univ. Pierre et Marie Curie, Paris VI, 26 pp.

Buttiker B., 1984. Inventaire et estimation du rendement piscicole d'un ruisseau à truites: le Greny. *Bull. Soc. Vaud. Sci. Nat.*, **366**, 119–134.

Buttiker B., Matthey G., 1986. Migration de la truite lacustre (*Salmo trutta lacustris* L.) dans le Léman et ses affluents. *Schweiz. Z Hydrol.*, **48**, 153–160.

Buttiker B., Matthey G., Bel J., Durand P., 1987. Age et croissance de la truite lacustre (*Salmo trutta lacustris* L.) du Léman. *Schweiz. Z. Hydrol.*, **49**, 316–328.

Champigneulle A., 1985. Analyse bibliographique des problèmes de repeuplement en omble chevalier (*Salvelinus alpinus*), truite fario (*Salmo trutta*) et corégones (*Coregonus* sp.) dans les grands plans d'eau. In R. Billard and D. Gerdeaux (Eds, *Gestion Piscicole des Lacs et Retenues Artificielles*, INRA, Paris, 187–217.

Champigneulle A., Pattay D., Buttiker B., 1986. Pêches hivernales de truites: comparaison de la pêche avec des filets tendus en surface et à 2 m de profondeur. *Rapport du groupe Recherches Piscicoles. Rapport interne de la Commission Consultative de la Pêche au Léman*, June 1986, 5 pp.

Champigneulle A., 1987. Etude sur la truite de lac et mises au point techniques. *Rapp. Inst. Limnol.*, Thonon, 1 vol., 10 p.

Champigneulle A., Melhaoui M., Maisse G., Baglinière J. L., Gillet C., Gerdeaux D., 1988a. Premières observations sur la population de truite (*Salmo trutta* L.) dans le Redon, un petit affluent-frayère du lac Léman. *Bull. Fr. Pêche Piscic.*, **310**, 59–76.

Champigneulle AA., Michoud M., Gerdeaux D., Gillet C., Guillard J., Rojs-Beltran R., 1988b. Suivi des pêches de géniteurs d'omble chevalier (*Salvelinus alpinus* L.) sur la partie française du lac Léman de 1982 à 1987. Premières données sur le pacage lacustre de l'omble. *Bull. Fr. Pêche Piscic.*, **310**, 85–100.

Champigneulle A., Melhaoui M., Gerdeaux D., Guillard J., Rojas-Beltran R., Gillet C., 1990a. La truite commune (*Salmo trutta* L.) dans le Redon, un petit affluent du lac Léman. I. Caractéristiqes de la population en place (1983–88) et premières données sur l'impact des relachers d'alevins prégrossis. *Bull. Fr. Pêche Piscic.*, **319**, 181–196.

Champigneulle A., Melhaoui M., Gerdeaux D., Rojas-Beltran R., Gillet C., Guillard J., Moille J. P., 1990b. La truite commune (*Salmo trutta* L.) dans le Redon, un petit affluent du lac Léman. II. Caractéristiques des géniteurs de truite de lac (1983–88) et premières données sur l'impact des relachers d'alevins prégrossis. *Bull. Fr. Pêche Piscic.*, **319**, 197–212.

Chevassus B., Guyomard R., 1983. Recherches sur la génétique de la truite fario et du saumon atlantic. *Rapport final de la convention INRA-CSP 1980–82*, 1 vol., 58 pp.

Cuinat R., 1960. Croissance et taille légale de la truite fario dans quelques rivières françaises. *Ann. Stn Cent. Hydrobiol.*, **8**, 225–261.

Cipel, 1984. *Le Léman Synthèses des travaux de la CIPEL de 1957 à 1982*. Lausanne, 1 vol., 650 pp.

Craig J. F., 1982. A note on growth and mortality of trout (*Salmo trutta* L.) in afferent streams of Windermere. *J. Fish Biol.*, **20**, 423–429.

Crozier W. W., Ferguson A., 1986. Electrophoretic examination of the population structure of brown trout (*Salmo trutta*) from the Lough Neagh catchment, Northern Ireland. *J. Fish Biol.*, **28**, 459–477.

Dellefors C., Faremo, 1988. Early sexual maturation in males of wild sea trout (*Salmo trutta*) inhibits smoltification. *J. Fish Biol.*, **33**, 741–749.

Devaux A., Monod G., 1987. PCB and p,p'-DDE in Lake Geneva brown trout (*Salmo trutta* L.) and their use as bioenergetic indicators. *Environ. Monit. Assess.*, **9**, 105–114.

Durand P., Pilotto J. D., 1989. Projet de recherche sur la truite lacustre. Etude du repeuplement effectué dans quelques affluents du Léman. Rapport intermédiaire January 1989. Ecotec, 18 pp.

Elliott J. M., 1982. The effects of temperature and ratio size on the growth and energetics of salmonids in capacity. *Comp. Biochem. Physiol.*, **73**(B), 81–91.

Elliott J. M., 1989. The natural regulation of number and growth in contrasting populations of brown trout (*Salmo trutta*) in two Lake District streams. *Freshwater Biol.*, **21**, 7–19.

Euzenat G., Fournel F., 1976. *Recherches sur la truite commune* (Salmo trutta) *dans une rivière de Bretagne: le Scorff.* Thèse Doct. *3e* cycle Biol. Anim., Fac. Sci. Univ. Rennes, 243 pp.

Euzenat G., Fournel F., 1979. Etude sur les Salmonidés migrateurs du Bassin de l'Arques (Seine maritime). *Bull. Inf. du C.S.P.*, **115**, 67–90.

Fergusson A., 1989. Genetic differences among brown trout (*Salmo trutta*) stocks and their importance for the conservation and management of the species. *Freshwater Biol.*, **21**, 35–46.

Forel F. A., 1904. *Le Léman. Monographie limnologique*, vol. 3, F. Rouge, Lausanne. 715 pp.

Frist W. E., Brown N. E., 1972. *The Trout*. Collins, London, 286 pp.

Gerdeaux D., 1988. La gestion piscicole d'un lac international: le lac Léman. EIFAC Symposium on Management Schemes for Inland Fisheries. Göteborg (Suède)—31 May to 3 June 1988.

Gustafson K. J., Lindstrom T., Fagerstrom A., 1969. Distribution of trout and char within a small Swedish high mountain lake. *Rep. Inst. Freshwater Res. Drottningholm*, 49, 63–75.

Guyomard R., 1989a. Diversité génétique de la truite commune. *Bull. Fr. Pêche Piscic.*, **314**, 118–135.

Guyomard R., 1989b. Gestion génétique des populations naturelles: l'exemple de la truite commune. *Bull. Fr. Pêche Piscic.*, **314**, 136–145.

Heland M., 1977. Recherches sur l'ontogenèse du comportement territorial chez l'alevin de truite commune (*Salmo trutta*). Thèse Doct. 3e cycle Biol. Anim. Fac. Sci. Univ. Rennes, 239 pp.

Holcik J., Bastl I., 1970. Notes on the biology and origin of the trout (*Salmo trutta* m. *lacustris*) in the Orava valley reservoir (Northern Slovakia). *Zool. List.*, **19**, 71–85.

Hunt P. C., Jones J. W., 1972. Trout in Llyn Alaw, Anglesey, North Wales. II. Growth. *J. Fish Biol.*,**4**, 409–424.

Jensen K. W., 1977. On the dynamics and exploitation of the population of brown trout (*Salmo trutta* L.) in Lake øvre Heimdalsvatn, Southern Norway. *Rep. Inst. Freshwater Res. Drottningholm*, 56, 18–69.

Jones A. N., 1975. A preliminary study of fish segregation in salmon spawning streams. *J. Fish Biol.*, **7**, 95–104.

Jonsson B., 1982. Life History patterns of freshwater resident and sea-run migrant brown trout in Norway. *Trans. Am. Fish. Soc.*, **114**, 182–194.

Jonsson B., 1989. Life history and habitat use of Norwegian brown trout (*Salmo trutta*). *Freshwater Biol.*, **21**, 78–86.

Krieg F., 1984. Recherche sur la différenciation génétique entre populations de *Salmo trutta*. Thèse 3e cycle, Univ. Paris Sud Orsay, 1 vol., 92 pp.

Laurent P. J., 1972. Lake Leman: effects of exploitation, eutrophication and introductions, on the salmonid community. *J. Fish. Res. Board Canada*, **29**, 867–875.

Le Bail P. Y., Maisse G., Breton B., 1981. Détection des femelles de Salmonidés en vittelogénèse. 1. Description de la méthode et mise en œuvre pratique. *Bull. Fr. Piscic.*, **283**, 79–88.

Maisse G., 1985. ATP INRA no. 4355. Connaissance et gestion des écosystèmes lacustres subalpins. Compte-rendu des travaux du Laboratoire de Physiologie et d'Ecologie des Poissons, INRA Rennes, 20 pp.

Melhaoui M., 1985. Eléments d'écologie de la truite de lac (*Salmo trutta* L.) du Léman dans le système lac-affluents. Thèse Doct. 3e cycle, Univ. Pierre et Marie Curie, Paris VI, 127 pp.

Nettles D. C., Haynes J. M., Olson R. A., Winter J. D., 1987. Seasonal movements and habitats of brown trout (*Salmo trutta*) in south central Lake Ontario. *Great Lakes Res.*, **13**, 168–177.

Nilsson, N. A., 1963. Interaction between trout and char. *Trans. Am. Fish. Soc.*, **92**, 276–285.

Nilsson N. A., Pejler B., 1973. On the relation between fish fauna and zooplankton composition in north Swedish lakes. *Rep. Inst. Freshwater Res. Drottningholm*, 53, 51–77.

Piggins D. J., 1975. Stock production, survival rates and life history of sea trout of the Burrishoole River system. *Annu. Rep. Salm. Res. Trust Ireland Inc.*, **XX**, 45–57.

Prouzet P., Harache Y., Danel P., Branellec J., 1977. Etude de la croissance de la truite commune (*Salmo trutta fario* L.) dans deux rivières du Finistère. *Bull. Fr. Piscic.*, **267**, 62–84.

Quillet E., Chevassus B., Krieg F., Burger G., 1986. Données actuelles sur l'élevage en mer de la truite commune (*Salmo trutta*). *Pisc. Fr.*, **86**, 48–56.

Richard A., 1981. Observations préliminaires sur les populations de truite de mer (*Salmo trutta*) en Basse-Normandie. *Bull. Fr. Piscic.*, **283**, 114–124.

Richard A., 1986. Recherches sur la truite de mer (*Salmo trutta*) en Basse-Normandie: scalimétrie, sexage, caractéristiques biométriques, démographiques et migratoires. Thèse 3e cycle, Fac. Sci. Univ. Rennes, 1 vol., 54 pp.

Rippmnn U., 1983. La pêche de la truite dans le lac des Quatre-Cantons. Off. Féd. de la Protection de l'Environnement (Berne). *Cah. Pêche*, **41**, 119–135.

Ruhle C., 1983. Wachstumsverhältnisse und Reifeent wicklung bie der Seeforelle (*Salmo trutta lacustris* L.) des Bodensees. *Osterreichs Fisch.*, **36**, 196–201.

Runnstrom S., 1949. Control of trout migration by a fish ladder. *Rep. Inst. Freshwater res. Drottningholm*, 29, 85–107.

Runnstrom S., 1952. The population of trout (*Salmo trutta* L.) in regulated lakes. *Rep. Inst. Freshwater Res. Drottningholm*, 33, 179–198.

Runnstrom S., 1957. Migration, age, growth of the brown trout (*Salmo trutta* L.) in lake Rensjön. *Rep. Inst. Freshwater Res. Drottningholm*, 38, 194–246.

Saglio P., 1986. Considérations sur le mécanisme chimiosensoriel de la migration reproductrice chez les Salmonidés. *Bull. Fr. Pêche Piscic.*, **301**, 35–55.

Sakowicz SS., 1961. Reproduction of the lake trout (*Salmo trutta lacustris* L.) from Wdzydze Lake. A biologico-managing monograph of the lake trout from Wdzydze Lake. *Rocz. Nak. Roln.*, **93**(D), 501–556.

Scholz A. T., Cooper J. C., Horrall R. M., and Hasler A. D., 1978. Homing of morpholine-imprinted brown trout (*Salmo trutta* L.). *Fish Bull.*, **76**, 293–295.

Stuart T. A., 1953. Spawning migration, reproduction and young stages of loch trout (*Salmo trutta* L.). *Sci. Invest. Freshwater Salm. Fish. Res. Scotl.*, 5, 39 pp.

St uart T. A., 1957. The migrations and homing behaviour of brown trout (*Salmo trutta* L.). *Freshwater Salm. Fish. Res.*, **18**, 1–27.

Svardson G., 1976. Interspecific population dominance in fish communities of Scandinavian lakes. *Rep. Inst. Freshwater Res. Drottningholm*, 55, 144–171.

Thorpe J. E., 1974. The movements of brown trout (*Salmo trutta* L.) in Loch Leven, Kinross, Scotlnd. *J. Fish Biol.*, **6**, 153–160.

Tilzey R. D. J., 1977. Repeat homing of brown trout (*Salmo trutta* L.) in Lake Eucumbene, New South Wales Australia. *J. Fish. Res. Board Can.*, **34**, 1085–1094.

Treasurer J. W., 1976. Age, growth and length–weight relationshupo of brown trout (*Salmo trutta* L.) in the lock of Stathberg, Aberdeenshire. *J. Fish Biol.*, **8**, 241–253.

Wojno T., 1961. The feeding of trout (*Salmo trutta*) from Wdzydze Lake. *Rocz. Nauk. Roln.*, **93**(D), 681–702.

2

Sea trout (*Salmo trutta* L.) in Normandy and Picardy

G. Euzenat, Françoise Fournel and A. Richard, with the technical collaboration of J. L. Fagard

I. INTRODUCTION

Sea trout, the anadromous form of the brown trout, *Salmo trutta*, is present in most water courses running into the European side of the Atlantic.

Although, in recent years, it has not benefited from the same level of interest as the Atlantic salmon, the sea trout has been the subject of numerous studies since the start of the century in various European countries. The first descriptions were of Scottish (Nall, 1930) and Scandinavian (Jarvi and Menzies, 1936) stocks, then those, mainly Polish, of the Baltic Sea (Backiel and Sych, 1958; Skrochowska, 1969; Bartel, 1977).

Studies aiming to gain a better knowledge of the functioning of populations are mainly those started at the end of the 1960s, often linked to studies of salmon stocks (Went, 1962, 1967; Jensen, 1968; Campbell, 1972; Paterson, 1973; Piggins, 1976; Pratten and Shearer, 1983; Elliott, 1984, 1985).

In France, the first investigations were started in Upper-Normandy (Arrignon, 1967, 1968a,b; Demars, 1976) but outside the regional plane—the work centred mainly around the reintroduction of Atlantic salmon, in the main part of the first national programme for restoration of Atlantic salmon—little is known about the sea trout in French national terms, in the absence of structured monitoring effort. An inquiry (Euzenat, Fournel, unpublished) carried out under the auspices of the Délégations Régionales du Conseil Supérieur de la Pêche and the Fédérations Départementales des Asssociations de Pêche revealed in summary:

- that the sea trout is present in many water courses on the Atlantic side of the English Channel (small coastal rivers and the downstream parts of larger rivers) and, in the opinion of the experts, of the Mediterranean, although it is not possible to produce facts on the state, qualitative or quantitative, of the populations;

- that the origin of returning adults is poorly understood: they may be relics of indigenous stock, Polish sea trout introduced in the 1960s or recolonized trout.

In the 1980s, under the initiative and finance from the Minister of the Environment, within the framework of the Salmon Plan, work was expanded to other major migratory fish and programmes of specific research were instigated: firstly in Upper-Normandy/ Picardy, in 1979, then in Lower-Normandy in 1981. It is in these two regions where 95% of the total catches of sea-trout in France (voluntary declaration) were made in 1997. Other regions concerned to a much smaller extent with sea-trout are Brittany and the South-West (Fournel, 1998).

This chapter presents a summary of the results of the programmes led, in these two regions within the Délégations Régionales No. 1 (Compiègne) and 2 (Rennes) of the Conseil Supérieur de la Pêche, by two small units based respectively in Eu (Seine-Maritime) and Caen (Calvados).

Some of the results shown have been published before (Fournel and Euzenat, 1979, 1982, 1987; Fournel *et al.*, 1986, 1987; Richard, 1981, 1982, 1986); they are complemented here by original data which have been acquired more recently.

II. OBJECTIVES

In Upper-Normandy/Picardy, as in Lower-Normandy, the studies carried out aim mainly to establish the biological basis for rational management on the catchment scale, taking into the account regional peculiarities.

The investigations mainly concern the migratory phases: smolts, spawning adults and kelts. They include, in the two regions, reasonably comparable methods of approach, with as large as possible a spread of basic data (characteristics, biological cycle); after this, the study lines were clearly differentiated as:

- in Upper-Normandy there was a predominance of the orientation on the descriptive aspects: characterization of populations in several coastal waters and definition of methodological tools (scale reading, sexing);
- in Upper-Normandy/Picardy, direction was towards a predominance of the functional aspects: study of the sea trout population dynamics in a single river.

Most of the data collected have not until now undergone in-depth statistical analysis, a basic processing being generally sufficient for use in management of the stocks.

III. SITES AND STUDY METHODS

The data shown come mainly from four river catchments: Bresle and Arques in Upper-Normandy/Picardy, Orne and Touques in Lower-Normandy (Fig. 1); their physical characteristics are summarized in Table 1.

Concerning the characterization of populations, the results are obtained mainly from trapping running fish on their way up or downstream, the methods employed varying from site to site:

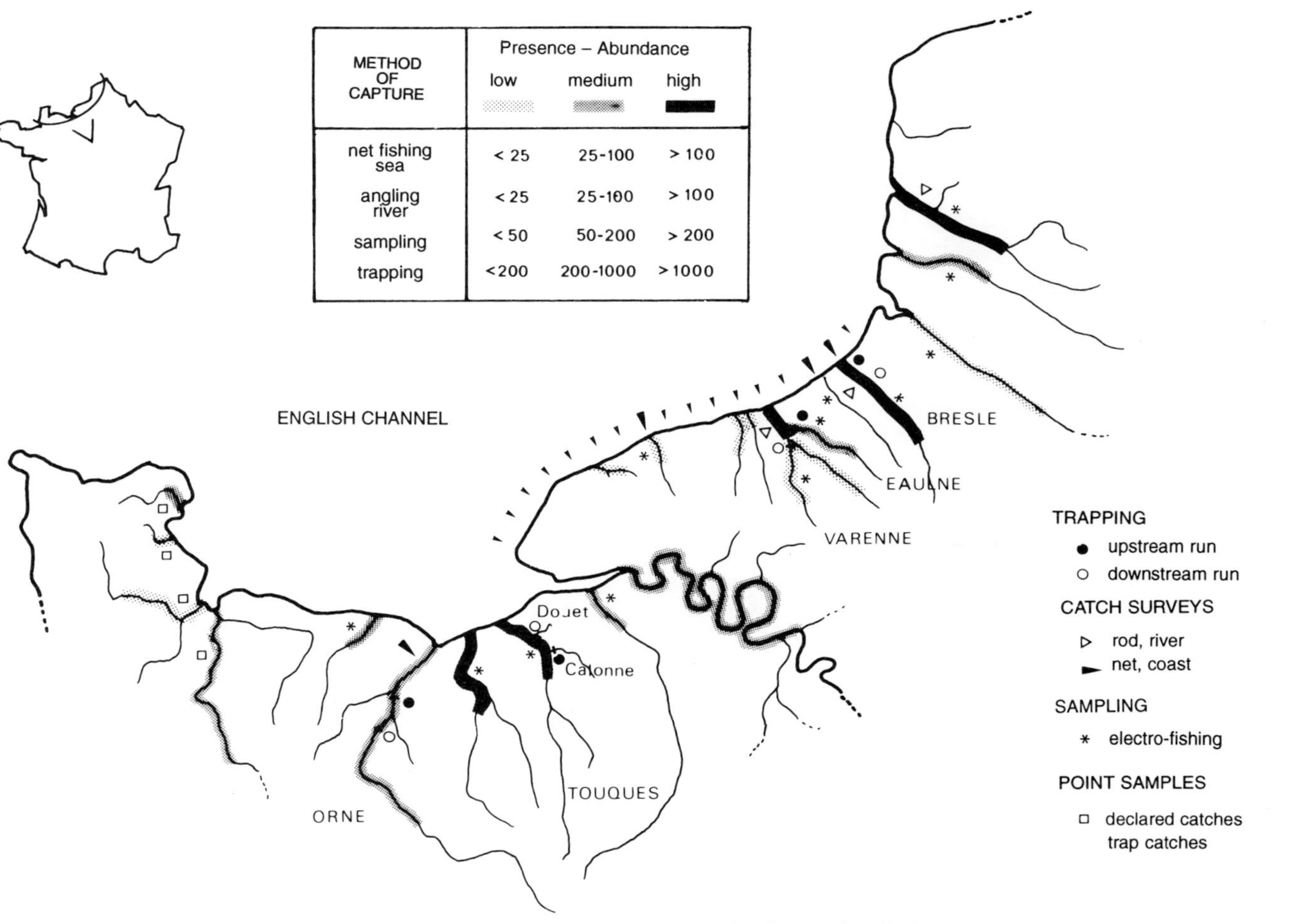

Fig. 1. Sea trout rivers in Normandy–Picardy: presence–abundance and study sites.

Table 1. Main physical characteristics of the rivers studied.

Characteristics	Upper-Normandy			Lower-Normandy	
	Bresle Bresle	Arques Eaulne	Arques Varenne	Orne Orne	Touques Calonne
Size of catchment area (km^2)	748	325	364	2 900	190
Geology	Calcareous	Calcareous	Calcareous	Schist/sandstone	Calcareous
Length (km)					
• main river	72[(1)]	43	43	175	46
• tributaries[(2)]	40	5	—	280	37
Slope (‰)	2.4	3.3	2.9	1.1	3.9
Flow (m^3/s)					
• mean	6.5	3	4	22	1.8
• min.–max. monthly means	5.8	1.7–4	2.5–6	3–55	1–2
Temperature (°C)	(3)	(4)	(5)		
• mean annual	9	11	10.8	12.7	—
• min.–max. monthly means	2.5–16.2	7.3–15.5	5–15	–20.5	–16.5

(1) Not taking into account that the main river splits in two along 50% of its length.
(2) Only those wider than 2 m.
(3) Average over 3 years 1985–86–87 (cold years).
(4) Average over the years 1980–81–82.
(5) Average over the years 1979–80.

- partial or full trapping, with or without efficiency evaluation;
- up and down stream trapping, carried out simultaneously or not;
- trapping facilities installed on the main stream or in tributaries, at a distance from the sea of between 3 and 22 km.

These operating methods are summarized in Table 2 and the location of traps is indicated in Fig. 1.

All the fish from samples were subjected to biometric measurements and scale sampling. Their numbers, all used in various calculations, are shown in Table 2.

Trapping is sometimes accompanied by migrant tagging and branding, either occasional or systematic.

Spot sampling is also carried out on fish upstream fish trapped:

- by electro-fishing in holding zones or below impassable dams;
- by anglers in the river and net fishermen in coastal waters.

Exploitation is studied in varying degrees, depending on the site (Fig. 1); in Upper-Normandy/Picardy,

- surveys concern exploitation in both the river (angling) and in the sea (net fishing, on foot and from boats, sporting and commercial);
- surveys are carried out over an extended geographical area, although clearly centred on 10 km around the Bresle.

The following methods of investigation are used;

- catch records
- field interviews during the fishing season
- interviews, at home or by telephone, at the end of the season.

—in Lower-Normandy, field interviews have been limited to the Orne estuary (rod and net fisheries).

These different study techniques are used alone or together, depending on the site. They have been combined on the Bresle over several consecutive years, in the population dynamics study carried out in this catchment:

- adult trapping and evaluation, over four years, of the facility efficiency by marking/recapture method, ending with estimation of the whole reproductive potential (years 1982–1987);
- smolt trapping, with evaluation of the efficiency of the trapping device, ending with the estimation of recruitment (years 1982–1986);
- marking of smolts and recapture of returning adults to evaluate the rate of returns to the river (Fournel *et al.*, 1990);
- catch monitoring, by field surveys, on the coast and in the river, to evaluate fishing mortality (years 1985–1988).

IV. RESULTS

1. Upstream run size

At present, the sea trout run up almost all the coastal rivers of the Eastern Channel; the run size varies, depending on the river, from a few dozen to several thousands of fish. In rivers without fish assessment facilities, the numbers of return fish are not known with accuracy; but the use of different indicators (catch by anglers, sampling by electro-fishing below dams, observation of spawning redds) allows their level of abundance to be estimated and the rivers can be divided into three abundance classes (Fig. 1).

The richest rivers are, from north to south, Canche, Bresle, Arques, Touques, Dives, Orne and Vire; the annual returner numbers there are of the order of thousands of fish.

The annual numbers of fish in traps are shown in Table 2.

2. Migration

(a) Migratory rhythms

When all stages are considered the periods of migratory activity in the sea trout cover the whole year (Fig. 2):

- the downstream migration of juveniles occurs in spring, starting at the end of February and ending in mid-May, with a peak usually in the first fortnight of April. Downstream migration in autumn has never been observed on the rivers studied;
- the upstream migration extends from May to January; it occurs in two very distinct runs, separated by a slack period in August–September. While the relative importance of the two runs remains quite constant in one river, it can vary quite significantly among rivers.

Thus on the Bresle, it is the first run which is always the most important, with an average of 73% of the numbers trapped (nearly 90% if the fresh return fish, carrying sea lice or sea-lice scars are included).

Roughly the same pattern appears in the Calonne (a tributary of the Touques), where the autumn peak corresponds mainly to the movement of spawners returning to the Touques.

By contrast in the Orne, the autumnal run, mostly made up of fish arriving straight from the sea, is preponderant (Fig. 3).

These modulations of migratory activity appear to be linked to the river flows: the Bresle, like the Calonne, with many fish running in the spring/summer are both rivers with stable flow (a maximum factor of 2 between monthly low and high waters) while the Orne, with a preponderance of autumn fish, is characterized by an irregular flow pattern, with marked periods of low water (a factor of 15–20 between low and high waters);

- the downstream migration of kelts starts at the end of November and finishes in April, with a very irregular migratory activity depending on the year, strongly dependent on temperature which, at this time of year, can be a limiting factor (inhibition during periods of frost and ice).

Table 2. Numbers of sea trout caught annually, from 1978 to 1987, in traps set in Normandy/Picardy. Location and working conditions of traps. (1): (M.R.) trap set in main river; (T.1) trap set in a tributary of order shown. (2): (U.Ad): upstream trap—adults; D.Ke: downstream trap—kelts; D.Sm: downstream trap—smolts. Trapped numbers shown in parentheses: incomplete trapping. Bold outline: years when trap efficiency was evaluated (estimated efficiency shown as a percentage)

Catchment area	Rivers Tributary + order (1)	Trapping type (2)	Distance to the sea (km)	Year of trapping									
				78	79	80	81	82	83	84	85	86	87
Bresle	Bresle (M.R)	U.Ad.	3				(127)	(389)	(430)	1 020 59%	770 70%	820 65%	819 60%
		D.Ke	15							(218)	(249)	(185)	(213)
		D.Sm	15					3 410	4 440	1 360	4 410	4 370 60%	1 850
Arques	Eaulne (T.1)	U.Ad	10		(11)	(36)	(143)	(31)	(95)	—	—	—	—
		D.Ke	10			(49)	(44)	(56)	(168)	—	—	—	—
	Varenne (T.1)	D.Sm	7	(594)	(729)	—	—	—	—	—	—	—	—

Table 2. *(continued)*

Catchment area	Rivers Tributary + order (1)	Trap type (2)	Distance to the sea (km)	Year of trapping									
				78	79	80	81	82	83	84	85	86	87
Touques	Calonne (T.1)	U.Ad.	22						(2 970)	4 046 100%			
	Douet (T.1)	D.Sm	15								437	328	252
Orne	Orne (M.R.)	U.Ad.	22				739	930 77%	535	614	—	—	488
		D.Sm	36					(360)	(748)	—	—	—	—

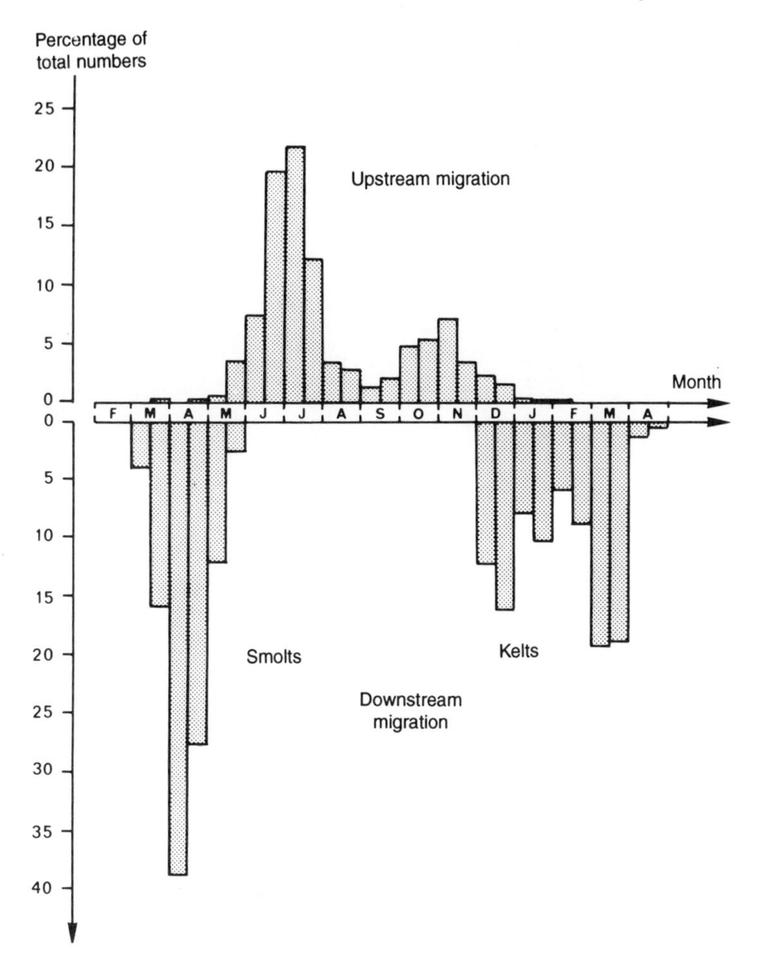

Fig. 2. Migratory rhythms of sea trout: upstream migration of adults and downstream migration of smolts and kelts (catches per fortnight expressed as a percentage of the total annual numbers). River Bresle: 1982–1987 year average.

(b) Migration in the sea

Information from capture of tagged fish, by all methods (Fig. 4) suggests that:

- during the first months after their downstream migration, post-smolts head north, making possible incursions into the lower parts of other rivers, at least until they reach the Meuse estuary;
- the feeding areas of Normandy/Picardy sea trout are situated to the north, in the English Channel and North Sea, the most distant recaptures coming from the west coast of Denmark.

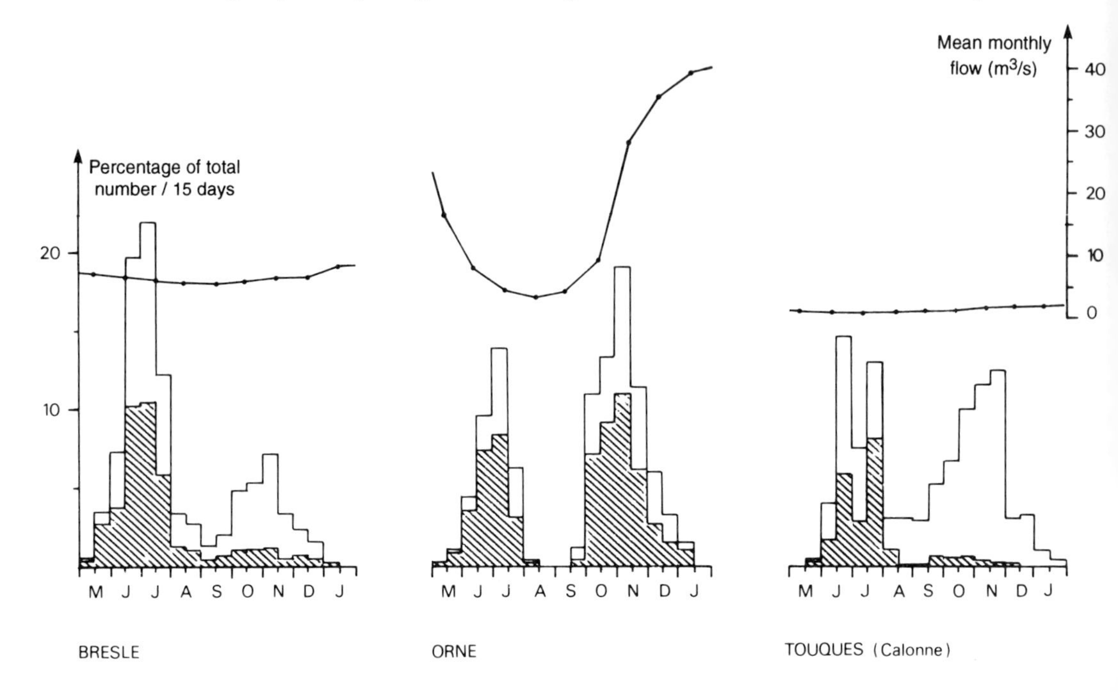

Fig. 3. Upstream migration pattern of adults and flow regime in the rivers Bresle, Orne and Touques. Shaded area indicates newly migrating fish, carrying sea lice or sea lice scars.

It should be noted that the migratory route outlined here coincides with the main direction of marine currents in the eastern English Channel.

3. Biometric and demographic characteristics of sea trout in the North-West

(a) Size and age

Length, weight and age data of sea trout, smolts and adults, in the four rivers studied, are presented in Tables 3 and 4.

Smolts

Smolts range in length from 11 to 33 cm, the mean being 20 cm for a weight of 90 g. Although there is wide variation in length within a river population, there is little difference among them (Fig. 5); the averages by river remain between 18 (on the Orne) and 21 cm (on the Arques).

The age at migration is 1 or 2 years, exceptionally 3, the relative length of the age classes varying according to the river:

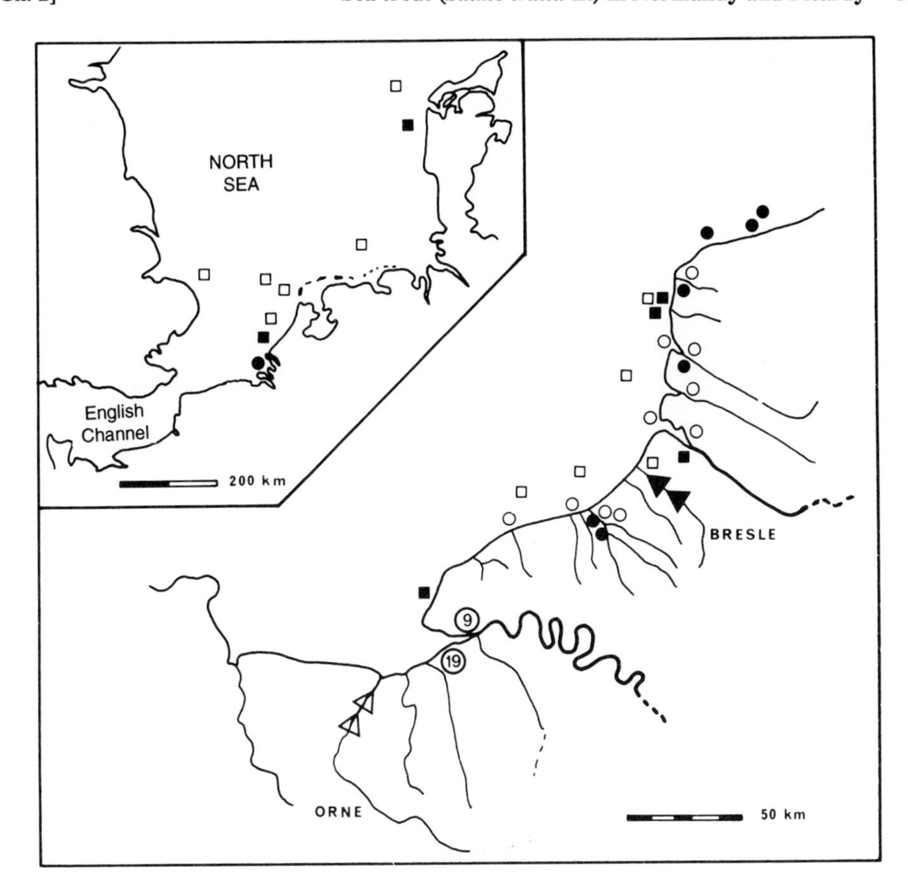

Fig. 4. Location of recaptures of marked sea trout at the finnock (●○) and adult (■□) stage on the rivers Bresle (in black) and Orne (in white). Each circle or square corresponds to a recapture. The circled numbers correspond to numbers of fish from the Orne recaptured at the finnock stage in the Seine (9) and Touques (19) estuaries. Only recaptures outside the original river are shown here.

- when spawning occurs mainly in the main rivers (Arques, Bresle, Orne), migration occurs mainly at the age of 1 year; the mean age at smoltification* is low: 1.15 to 1.22;
- in contrast, spawning in the tributaries (e.g. the Touques) produces mainly 2-year-old smolts, the mean age at smoltification thus reaching 1.6.

In addition, the characteristics of the smolts change markedly during migration: examination of length distribution per fortnight (Fig. 6) shows a shift towards the shorter lengths, and so a switch in age classes—clearly dominated by 2-year-olds at the start of the period, the migratory population is composed only of 1-year-old fish by the end of it.

* Mean age at smoltification or M.A.S. (Fahy, 1978): $(\%S_1 + (\%S_2 \times 2) + \ldots (\%S_n + n))/100$.

Table 3a. Length (fork length) and mean weight (with max. and min.) and mean condition factor (K) of sea trout smolts in Normandy/Picardy

Catchment	River	Length (mm)			Weight (g)			K
		min.	mean	max.	min.	mean	max.	
Bresle	Bresle	130	195	310	25	82	330	1.10
Arques	Varenne	140	209	325	30	100	370	1.08
Touques	Douet	135	191	255	24	77	180	1.11
Orne	Orne	115	180	285	16	63	250	1.08

Adults

Adult sea trout in rivers of north-western France vary widely in size, between 25 and 90 cm and 0.2 to 9 kg; mean sizes vary, according to the river, between 46 and 60 cm and 1.3 to 2.9 kg (Fig. 7).

Their condition factor* is high, around 1.20 to 1.30 on average. It changes significantly during the course of the year (Fig. 8): more than 1.30 at the start of up-migration, it drops to 1.10–1.15 in the weeks just before spawning and then to approximately 0.9.

Condition factor also depends to a large extent on the length and age of the fish: very close to 1 in young fish or finnock†, it can exceed 1.50 in larger and/or older fish.

The sea-age structure of sea trout is spread, from 0^+ to VI^+, with a general predominance, more or less marked according to the river, of the I^+ age group (45–75% of numbers). The fair variability of length among the rivers is linked both to different sea-age structures (although they are dominated everywhere by the I^+ age group) and to growth variations leading to significant length differences at an equivalent sea age.

In Upper-Normandy/Picardy, inter-river variability seems to be limited; the mean length is 56 cm (53–58 depending on the river) for a weight of 2.3 kg; the sea-age

* Condition factor (Allen, 1951): $K = 100 + W/L^3$, where W = weight in grams, L = length in centimetres.

† According to the English (Scottish) terminology.

Table 3b. Fork length, mean, minimum and maximum weights of adult sea trout in Normandy/Picardy

Catchment	River	Length (mm)			Weight (g)		
		min.	mean	max.	min.	mean	max.
Bresle	Bresle	255	551	860	200	2 200	9 000
Arques	Eaulne	285	574	815	270	2 460	8 000
Touques	Calonne	235	465	780	200	1 270	5 800
Orne	Orne	225	605	870	180	2 870	9 050

structure remains relatively stable from one year to the next; the I$^+$ age group represents 60–75% of the numbers; the II$^+$ age group is second in importance with 15–25%.

In Lower-Normandy, the situation appears to be more heterogeneous:

- on the Orne, the sea trout is large: on average 60 cm and weighing 2.9 kg; the I$^+$ and II$^+$ age groups share dominance in the sea-age structure, with 43% and 33% respectively of the numbers on average;
- on the Touques, however, the fish is of relatively small size: 46 cm and 1.4 kg; always clearly dominated by the I$^+$ age group, the age structure here contains an important part (20%) of early upstream migrants or finnock of sea age 0$^+$; in contrast, the II$^+$ age group is poorly represented.

Length also changes throughout the run, in relation to a modification of the relative importance of the age groups. While the I$^+$ age group fish spreads out over the whole season, large fish run up at the start of the period (May–June); II$^+$ and III$^+$ fish in the Bresle and Orne; repeat spawners in the Touques; the migration of finnocks (25–40 cm) only occurs from mid-July to mid-August.

Growth in the sea is rapid (Fig. 9) with, however, marked differences between the Upper-Normandy and Orne rivers on the one hand, where the mean sizes attained at different sea ages are very similar (on average, 54 cm/2 kg at I$^+$ and 65 cm/3.4 kg at II$^+$),

Table 4a. Mean age structures (expressed as a percentage of the total number, by age class and group) of sea trout smolts and adults in Normandy/Picardy. In parentheses, age structure of smolts established after reading of adult scales

Catchment	River	Smolt Age class		Adult Age group				
		1	2	0	I	II	III	IV
Bresle	Bresle	78 (60)	22 (40)	6.5	74.5	15	3	1
Arques	Varenne	85	15					
	Eaulne			7.5	61.5	23.5	6	1.5
Touques	Calonne	(56)	(44)	20	72	4	3	2
	Douet	30	70					
Orne	Orne	87	13	1	43	39	13	5

and on the other hand, the River Touques, which is conspicuous for its constantly smaller sizes, whatever sea-age is being considered (48 cm/1.3 kg at I^+; 56 cm/2.3 kg at II^+).

(b) Reproduction rate

First spawning can take place from the first (sea age 0^+) to the fourth (age III^+) winter after seaward migration, but it occurs mainly at sea-age I^+ on all fivers (65.83%) of numbers (Fig. 10).

The proportion of early (age 0^+) or late return fish (age II^+) varies quite a lot depending on rivers but remains quite stable within any given river. First spawning at age III^+ remains exceptional.

A mean sea age at first reproduction (M.R.A.)* can be used to characterize sea trout populations: this is 1.05 on the Bresle, 1.33 on the Orne and 0.89 on the Touques.

The return in the same year as smolt migration, either with or without spawning, is widespread in the sea trout.

* M.R.A. (Fahy, 1978): formula derived from that for M.S.A. $[\%R_1 + (\%R_2 \times 2) + \ldots (\%R_n \times n)]/100$, where R_1, R_2, R_n: are the first reproduction during the second, third, ... winter after the downstream migration of smolts.

Table 4b. Length (Lf in mm) and mean weight (g) by age class and group, of sea trout in Normandy/Picardy

Catchment	River	Smolt Age class		Adult Age group				
		1	2	0	I	II	III	IV
Bresle	Bresle	180/64	235/140	408/780	540/2 060	649/3 580	684/4 190	725/4 990
Arques	Varenne	200/86	255/180					
	Eaulne			370/580	550/2 160	632/3 280	680/4 090	730/5 050
Touques	Calonne			345/445	480/1 300	565/2 300	620/3 180	670/3 900
	Douet	163/48	199/88					
Orne	Orne	180/64	235/140	408/780	540/2 060	649/3 580	684/4 190	725/4 990

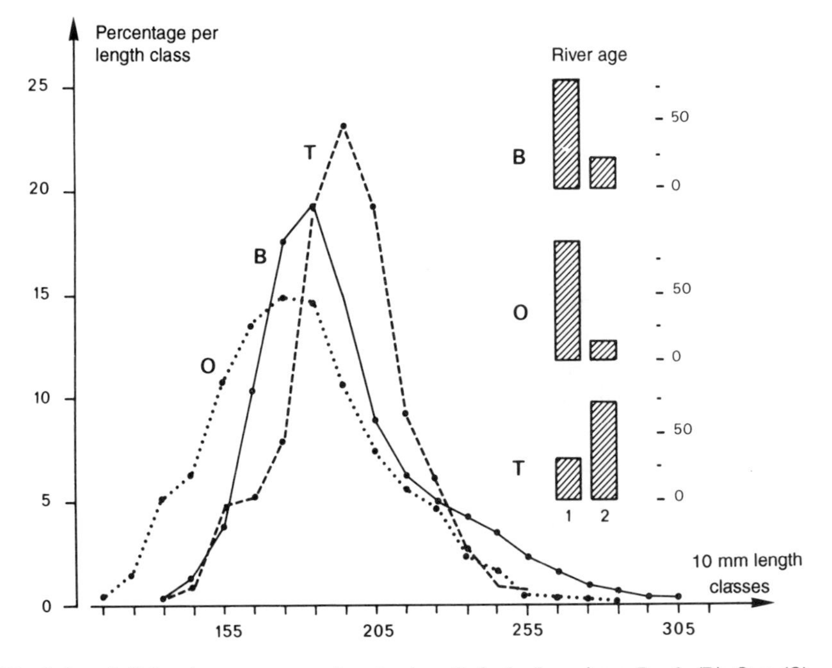

Fig. 5. Length (Lf) and age structures of sea trout smolts in the three rivers; Bresle (B), Orne (O) and Touques (Doet) (T). Expressed as a percentage of the total number, by 10-mm length class and by age class.

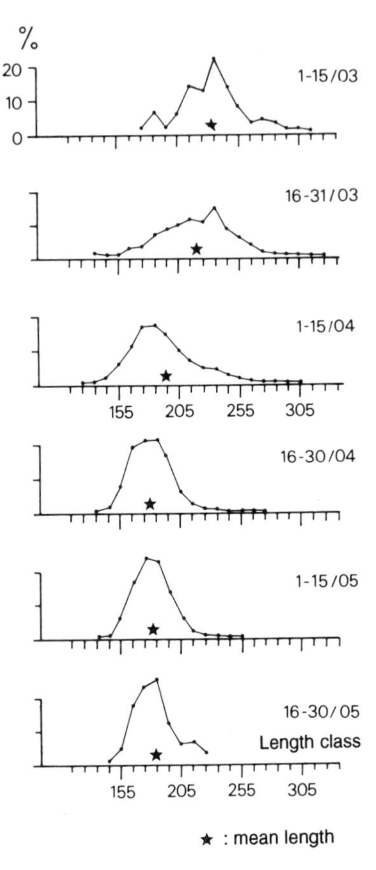

Fig. 6. Change in length distribution of sea trout smolts during downstream migration. River Bresle; 1985.

On the Touques, this behaviour is exhibited by more than 80% of individuals, in estuarine or lower parts of the river with upstream forays in summer or winter, of variable durations and sometimes several up-and-down movements. This reduces the effective time at sea and affects the growth, leading to the smaller sizes observed in this river.

In contrast, few finnock ascend the Bresle and the Orne rivers, although the traces of finnock-type behaviour can be observed on the scales of 20–40% of adults. These brief forays do not usually take place in the river of origin but mostly in the estuaries or lower parts of other rivers during the northward migration. Reproducing finnocks are recruited principally (80–85%) from the older juveniles which have smoltified as 2-year-olds.

As a result of the good post-spawning survival in the sea trout (30–50% as return rate of tagged kelts), a large fraction of the population can complete several successive spawnings. The proportion, in one run, of previously spawned fish, is thus 12% on the Touques, between 10 and 25% on the Bresle and Arques and 25–40% on the Orne.

The maximum number of successive spawnings that have been detected in the area is seven.

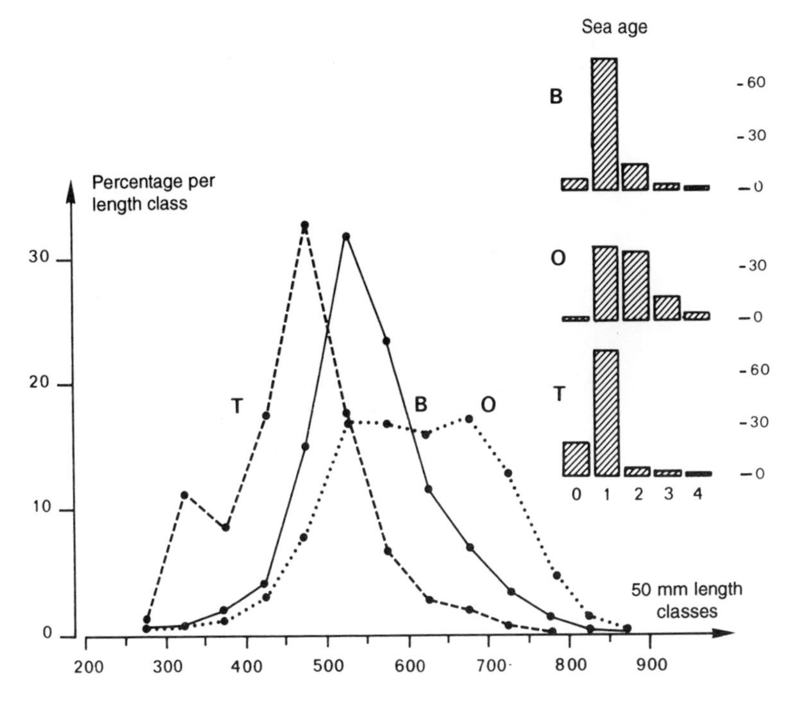

Fig. 7. Length (Lf) and age distributions of adult sea trout in three rivers; Bresle (B), Orne (O) and Touques (T). Data expressed as percentage of the total number of individuals by 50 mm length class and sea age group.

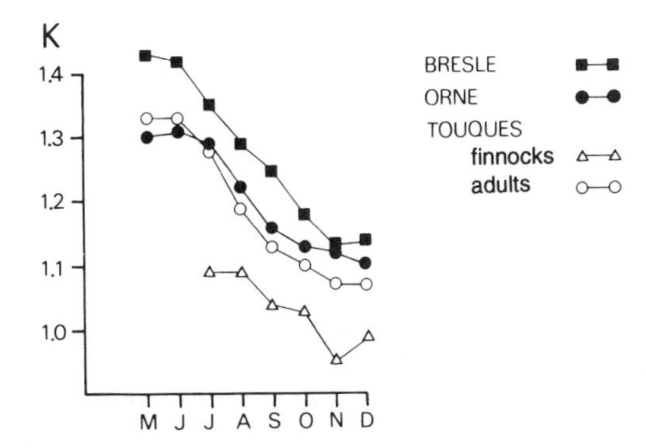

Fig. 8. Monthly change in mean condition factor of adult and finnock sea trout in the rivers Bresle, Orne and Touques.

The sex ratio is always unbalanced in favour of the females. Depending on the river, the females are thus on average 2 (Bresle) to 2.5 times (Orne) more numerous than the males. This said, the relationship fluctuates greatly during up-stream-migration: very unbalanced in favour of the females at the start and end of the ascent, it is close to equilibrium in September/October.

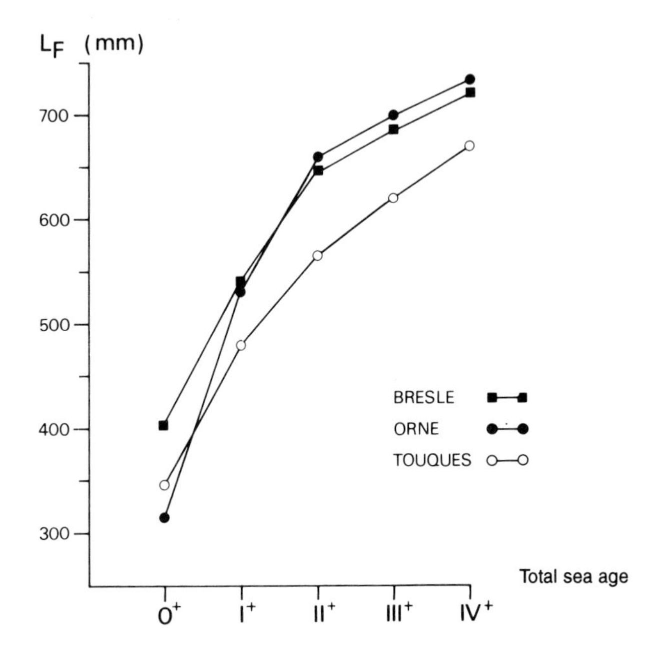

Fig. 9. Mean length by sea-age of sea trout in rivers Bresle, Orne and Touques. (Means calculated from lengths of all caught fish.)

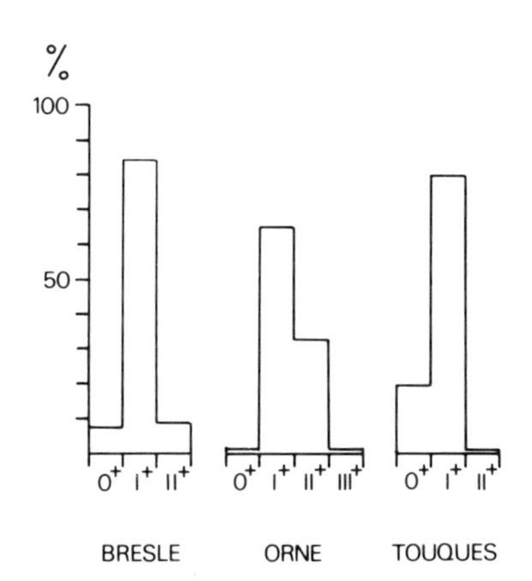

Fig. 10. Sea age distribution at first spawning of sea trout in rivers Bresle, Orne and Touques (percentage of population reaching sexual maturity at sea ages 0^+, I^+ or II^+).

This predominance of females exists as early as the smolt stage where the sex ratio is 1.7 to 2. It is subsequently accentuated by successive spawnings which preferentially eliminate males—and more so since the sea-age at first reproduction is high. By consequence, in the older fish also, the females are 4 to 9 times as numerous as the males (Fig. 11b).

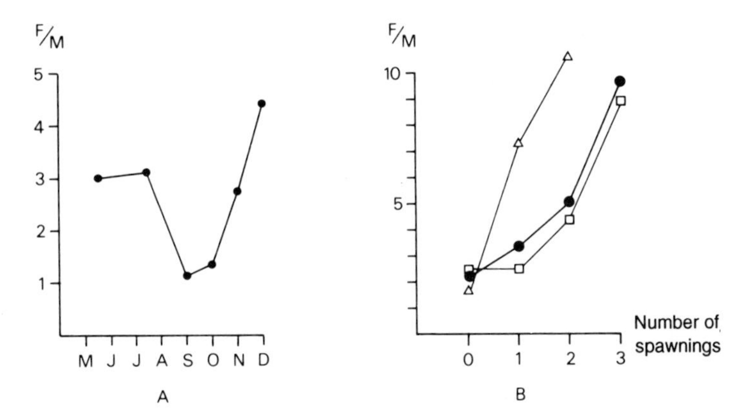

Fig. 11. Change in the sex ratio in adult sea trout, A, during upstream migration; B, in relation to number of spawnings and age at first spawning. □, first reproduction after one winter at sea; △, first reproduction after two winters at sea; ●, average. Data from River Orne; 1981 to 1984.

The mean individual fecundity is 4200 eggs on the Bresle, 6200 on the Orne and 3500 on the Touques, related to the different sizes of the females. The mean relative fecundity is 2240 eggs per kg of female (ranging from 2120 on the Bresle to 2450 on the Orne). Individual fecundity is strongly correlated with female length (Fig. 12); from a sample of 92 sea trout from these three rivers, the relationship is: $\log F = 2.92 \log L - 4.39$ (fork length in mm); ($r = 0.97$).

4. Functional approach

These different demographic aspects have been applied to the well-monitored Bresle system on which the in-numbers (adults and spawners) and out-numbers (smolts) have been evaluated over several consecutive years.

The potential number of eggs introduced into the system by the 830 to 1150 females which run in each year is thus estimated at 3 700 000 on average for the period 1984/1987 (range 3–5 million).

Linked with the available production area, including redd and juvenile development zones (riffles and rapid runs), that is a maximum of 30 ha over the 40 km of river actually accessible; this potential represents a minimal density egg of 125 000/ha.

The mean annual production of smolts, estimated from numbers in traps (see Table 2) is, for the period 1982–87, between 5500 and 8300 (taking into account trap efficiency of about 40–60%), giving a mean egg-to-smolt survival rate of 0.15–0.22%.

This survival rate remains below 0.5%, even if the most extreme hypotheses are considered, as below:

- the efficiency of the monitoring of up-run is total, leading to a reproductive potential of 2 300 000 (while double trapping carried out over four consecutive years shows the efficiency to be between 59% and 72%);
- the efficiency of the monitoring of the down-run does not exceed 30%, leading to an annual smolt production of around 11 000 fish on average even though double trapping has confirmed the fair control of migration.

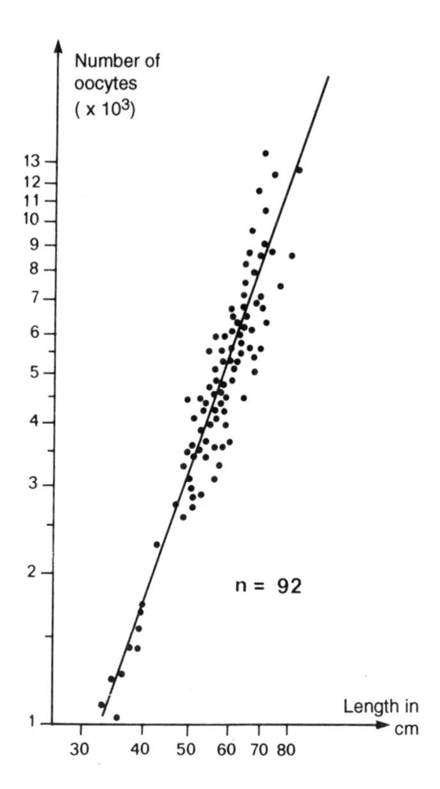

Fig. 12. Length (Lf)/fecundity relationship in the sea trout. Determined from a sample of 92 individuals collected from the Bresle, Orne and Touques rivers.

These results must be treated with caution, because the sea-trout is a species with more or less optional migratory behaviour; the interrelationships, still poorly understood, between the resident and migratory forms, can effectively introduce a bias in the estimations of runs: a part of the progeny of 'resident' fish may migrate and a part of the progeny of 'migratory' fish may become resident. An experiment carried out on juveniles produced by migratory parents, kept in a fish farm for one year and then released into a controlled natural environment, shows a propensity in the males to become resident (Fig. 13). More recent work allows us to characterize the smolt stage in the sea-trout. Smolting finds expression in physiological changes (mainly an increase in branchial Na^+/K^+ ATPase activity), morphological changes (silvering), and migratory behaviour, as for Atlantic salmon. However, smolting never seems to be necessary for seawater adaptability (high survival rate in seawater for individuals with a size >14 cm) even if it represents a vital advantage. Otherwise, the seawater physiological adaptation is not directly connected to the migratory character in as much as parr exist downstream. Furthermore, it is not synchronous with silvering. Lastly, smoltification seems to be dependent on parental inheritance (growth rate, migratory character) and individual size. So, the variability and reversibility of the smoltification phenomenon confirm well the great adaptability of the species and makes more complex the relationships between the two forms of the brown

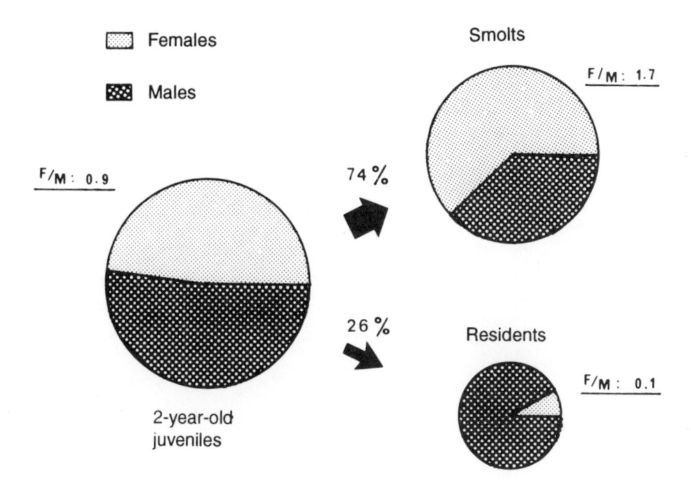

Fig. 13. Ratios of resident/migrant and female/male 2-year-old juvenile sea trout emigrating from the Touques, released at 1 year old (mean length: 85 mm) into a controlled natural environment (tributary of the Orne, 1982).

trout living together in the same stream (Ombredane *et al.*, 1996). It would therefore be more sensible to use the term 'apparent survival rate' from egg to smolt stages.

This being so, it must be noted that the trout densities found in different sites on the Bresle, chosen as 'production' sites are, on average, low-to-medium and this, from stage 0^+ (0.2–4 fish/100 m^2), leads to the concept that the sedentary way of life is quite limited and also confirms the efficiency of spawning is low.

This poor survival in fresh water is subsequently offset by the high marine survival: 14.5–20% of fish marked at the smolt stage return home to the Bresle (Fournel *et al.*, in press).

While assuming that these numbers could be multiplied by two to take into account mortality associated with marking and smolt handling (based on observations on salmon (Saunders and Allen, 1967; Hansen, 1981; Prouzet, 1983; Hay, 1985)), and the net catches downstream subtracted, the ratio between the spawning stock and the initial number of smolts reaches 14–20%, which assures the equilibrium of the system and the replacement of the parental generation (Fig. 14).

Exploitation by fishing has been surveyed along the coast north of the Seine, but the focus is more on the 'entire Bresle area'. This being so, the results obtained can be extrapolated at least to the whole Seine Maritime region. The investigations carried out show that:

- the greatest exploitation occured in the sea: thus in the Bresle area, 750–1000 fish (800 on average in the period 1985/1988) were caught annually by nets within a 2 km radius of the river mouth, as opposed to 100–200 by rod and line (140 on average), giving an exploitation ratio of sea/river of 6 to 1.

This differential is found again, to a greater degree, in all areas of the region: 3000 captures on average in the sea every year as opposed to 350 by rod and line:

- catches from the sea were mainly by amateurs, fishing with fixed nets, concentrated near the river mouth.

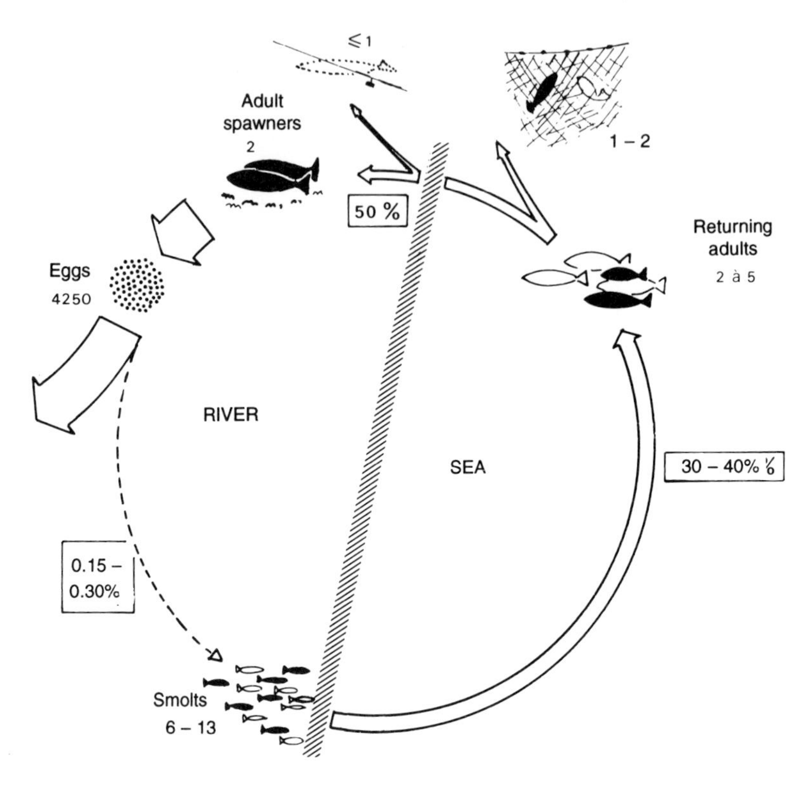

Fig. 14. Life cycle of the sea trout: functional view. Parent/offspring relationship based on initial spawning of 4250 eggs (mean fecundity of a female in the Bresle). Survival rates between main stages and surviving numbers are shown. Data from Bresle catchment area, 1982–1987.

In the Bresle area, fixed nets make on average two-thirds of the sea catch, the remaining one-third being attributable to fishing from boats, by professionals and pleasure sailors. However, the difference is probably less significant than is shown from the investigations, as there is probable under-declaration of catches from small fishing boats, which is much more difficult to control than fixed nets:

- the mean exploitation rate (Fig. 15) calculated for the Bresle area (1985/1988) is 46%, varying between years from 42% to 53%:
- 39% on average for net fishing, very probably underestimated as it does not take into account the catches made in neighbouring fisheries, of which part will have come from the 'Bresle' stock;
- and 7% for rod and line (11.5% if catches are related to available stock, i.e. to fish which penetrate the river after successfully passing the sea nets).

In other words, the ratio between exploited stock and spawning stock is at least 0.8 (0.75–1.15). Thus, when 2200 fish (mean size of return stock) are present around the mouth of the Bresle, 920–1160 are caught (830–960 by nets, 100–200 by rod and line on the river) and 1050–1300 remain available for reproduction.

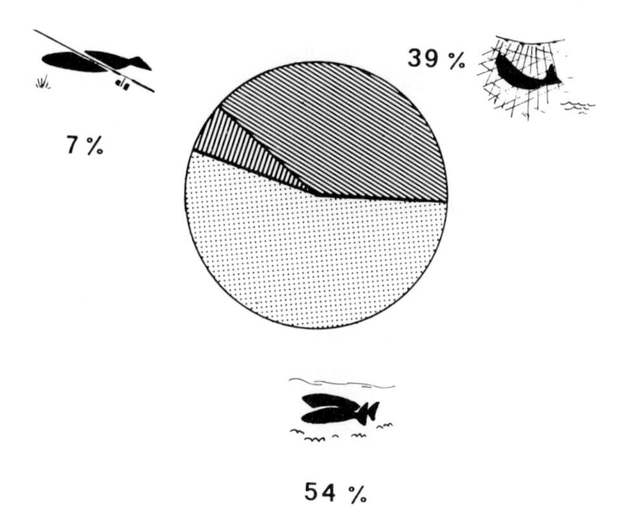

Fig. 15. Exploitation rate of sea trout, by rod and line in the river and by net in the sea, in Normandy/Picardy. Data from Bresle catchment area, average from 1985 to 1988.

More limited data collected in lower-Normandy show a reasonably similar situation in the river context with an exploitation rate by rod and line not exceeding 10%. Numbers of sea trout caught in the sea are not known; they seem to be taken mainly by amateurs, sometimes specialized, and as by-catches by fishermen.

V. CONCLUSION

The work carried out over the past decade on the coastal rivers of Normandy–Picardy has led to a fairly precise understanding of the sea trout, as a result of the intense study of both the migratory phases on a catchment scale, over a period of several years.

Still poorly understood at the end of the 1970s, the sea trout appears today to be the most important migratory salmonid of the north-west:

- present in all coastal rivers, it is often abundant there and always clearly preponderant over the Atlantic salmon;
- it is generally large in size and often similar to the salmon in both physical characteristics and biological cycle.

This conjunction between numerical preponderance and large size of the sea trout highlights the north-west area where knowledge of stocks and regulations mostly concern Atlantic salmon.

In the absence of local historical records and comparable quantitative data from other parts of France, it is rather difficult to come to a decision about the actual numbers in the region. They can be considered as being important compared to similar numbers for the salmon in France, such as the number caught by rod and line: 70 on average per river, less than 200 in 95% of cases (Kermarec, 1980; Anon., 1988) or the numbers trapped in a 'good' salmon river, the Elorn (Finistère)" 1200–1400 per year (Tellier, 1987; Nihouarn and Porcher, 1989).

They appear more modest, however, if compared to the numbers of sea trout observed in other European countries, the British Isles especially, often around several thousand individuals (Harris, 1970; Anon., 1984, 1986, 1987; Pratten and Shearer, 1985).

The British sea trout are generally small fish (below 45 cm and 1 kg in 75% of cases reported); this dominance of the 'small size' type is found elsewhere on the whole maritime front of the Western English Channel/Atlantic/Norwegian Sea (Nall, 1930; Jensen, 1968, Harris, 1970, 1972; Piggins, 1976; Fahy, 1978, 1979; Pratten and Shearer, 1983; Jonsson, 1985; Le Cren, 1985).

In contrast, the Norman sea trout, with its fast growth in river and sea, mostly belong to the 'large size' type, found in the North Sea and the Baltic (Zarnecki, 1960, 1973; Chelkowski, 1969; Sych, 1970; Palka and Bienarz, 1983). They are quite similar to the Atlantic salmon grilse: in France, 60–70 cm, 2–3 kg on average, depending on river (Prévost, 1987). Its abundance in some rivers of this region may appear surprising with regard to the real environmental constraints, notably decrease in surface and quality of spawning areas, degradation of water quality, high exploitation in the sea—especially since the origin of the populations has not truly been elucidated at present.

Is the sea trout an ancient form in our rivers, which has remained misunderstood for a long time because of its similarities to the brown trout and its confusion with the salmon? If so, it has been successful in its survival due to its intrinsic eco-ethological advantages:

- short life-cycle ensuring the rapid renewal of populations:
- high reproductive capacity linked to the preponderance of females, to their high fecundity, and to repeated spawnings;
- staggered sea-age structure allowing the 'buffering' of environmental variations;
- high marine survival, which can be explained at least in part, by the relative proximity of its feeding areas.

All these features are reported in foreign studies, both for sea trout (Alm, 1950; Went, 1962; Campbell, 1972; Pratten and Shearer, 1983; Le Cren, 1985) and steelheads (Ward and Slaney, 1988).

The other alternative is that sea trout emerged more recently, partly initiated by the stocking carried out by angling clubs in the last decades.

- Polish sea trout released in the 1960s, although the imports have been modest compared to the scale of each river stocked;
- stocking trout with a more-or-less migratory nature; the genetic identification studies of Guyomard (1989) suggest that their contribution could be a determining factor in some rivers, such as the Orne.

Lastly, the third possible explanation one must also consider is the possibility of the intervention of an 'anadromization' process in the brown trout in response to environmental modifications. The sea trout therefore would represent a resistant form of the species, allowing it to escape from environmental constraints and to maximize its reproductive potential by the preferential passage of females to the sea. That is still a hypothesis at our latitudes (Campbell, 1977), but this escape mechanism has been observed several times in populations of resident stocks introduced into oligotrophic rivers in southern countries (Arrowsmith and Pentelow, 1965; Davaine and Beall, 1982).

The current populations of normal sea trout very likely result from a mixing, in varying degrees according to the river, of these different processes, allowed by the great plasticity of the species.

Whatever the option, whatever the origin, it is now necessary to protect, develop and exploit the of sea trout in French rivers on the basis of available knowledge; management measures will consist, therefore:

- in offering the potential 'sea trout' an area for its full expression, and in restoring the migration routes and spawning zones, in both quality and quantity; regarding the successive spawnings and the important contribution made by kelts to the return migrations (and therefore to egg production) in the following year, particular attention should be paid to safeguarding them and their free passage downstream.

This being so, the great behavioural flexibility of the trout does not allow the prediction of the effects of improving the environment such as increasing the egg–smolt survival, or modification of the relationship between resident and migrant fish to the advantage of the former. The interrelationships between the two forms are effectively common: mixing during spawning (Campbell, 1972), optional migratory character of the progeny (Anon., 1984) and a well-known tendency towards sedentary lifestyle in reared fish (Piggins, 1983, 1988). All this leads us to question the effectiveness of artificial stocking of sea trout stocks.

The fact remains that restoration of the environment, will always be beneficial to the species;

- by adapting catches to the present and future holding capacity: increasing exploitation in areas where the spawning stock appears to be in excess in relation to available spawning grounds; moderating exploitation in under populated or recovering rivers.

Compared with foreign studies on the subject, the results obtained in Normandy could give the impression that egg deposition is excessive and exploitation insufficient in certain rivers; thus the Bresle, with its 125 000 eggs laid per hectare and an exploitation rate of 40–50%, is to be compared to the 15 000 to 25 000 eggs per hectare, classically given as optimal values in Canadian salmon rivers (Elson, 1975; Symons, 1979; Chadwick, 1982) and an exploitation rate of salmon reaching up to 75% (Kerswill, 1971, in Canada; Jensen, 1981, in Norway; Larsson, 1984, in Sweden; Piggins, 1986, in Ireland). But these references must always be used with great care, coming from a different species and taking into account the characteristics of Norman rivers, strongly influenced by man and with lower production capacities.

The definition of an optimal level of abundance and consequently the adjustment of the exploitation rate, must necessarily be made at the catchment level, on the basis of:

- a knowledge of stock dynamics and more particularly the factors influencing recruitment (environmental constraints and parental stock variations respectively);
- a monitoring of the exploitation level in rivers and at sea.

The rapid progression, during the last decade, of knowledge relating to the sea trout also allows the detection of gaps and, as a result, the proposal, in terms of research and development, the most appropriate directions to be taken in terms of real concerns and management requirements:

- to investigate the functional aspect, notably the factors affecting spawning efficiency and the interrelationships between the sedentary and migrant sub-populations;
- to study the relations between species and habitat, characteristics of the environment/ characteristics of the fish, and the responses of populations to modifications of the environment (whether or not there is human intervention);
- to determine the relationship between fishing and the species, considering both stocking and catches, both on the national coast and in the North Sea.

BIBLIOGRAPHY

Allen K. R., 1951. The Horokiwi stream: a study of a trout population. *Fish. Bull. N.Z.*, **10**, 238.

Alm G., 1950. The sea-trout population in the Ava stream. *Ann. Rep. Inst. Freshwater Res., Drottningholm*, 31, 26–51.

Anon., 1984. Triennial review of research, 1979–1981. Freshwater Fish. Lab. Pitlochry Rep., 45 pp.

Anon., 1986. Triennial review of research, 1982–1984. Freshwater Fish. Lab. Pitlochry Rep., 39 pp.

Anon., 1987. Annual review of research, 1985. Freshwater Fish. Lab. Pitlochry Rep., 31 pp.

Anon., 1988. Captures de saumons atlantiques en France en 1987 (domaine fluvial). Mise au point, gestion, traitement des fichiers. Rapport Cons. Sup. Pêche, Centre Rég. Et. Biol. Soc. Rennes, 43 pp.

Arrignon J., 1967. Comportement de l'espèce *Salmo trutta* dans le bassin de la Seine. *Bull. Fr. Pisc.*, **227**, 56–71.

Arrignon J., 1968a. Comportement de l'espèce *Salmo trutta* dans le bassin de la Seine. *Bull. Fr. Piscic.*, **228**, 77–101.

Arrignon J., 1968b. Comportement de l'espèce *Salmo trutta* dans le bassin de la Seine. *Bull. Fr. Piscic.*, **229**, 117–122.

Arrowsmith E., Pentelow F. T. K., 1965. The introduction of trout and salmon to the Falklands Islands. *Salm. Trout Mag.*, **174**, 119–129.

Backiel T., Sych R., 1958. Resorption and spawning marks in the scales of sea trout and lake trout from Polish waters. *Rocz. Nauk. Roln.*, B, **73**(2), 119–158.

Bartel R., 1977. Variability of sea trout returns as shown from many years tagging experiments with hatchery-reared parrs and smolts. I.C.E.S. CM/M, 9, 19 pp.

Campbell J. S., 1972. A comparative study of the anadromous and freshwater forms of brown trout (*Salmo trutta* L.) in the River Tweed. Ph.D. Thesis, Univ. Edinburgh, 247 pp.

Campbell J. S., 1977. Spawning characteristics of brown trout and sea trout *Salmo trutta* L. in Kirk Burn, River Tween, Scotland. *J. Fish Biol.*, **11**, 217–229.

Chadwick E. M. P., 1982. Stock recruitment relationship for Atlantic salmon (*Salmo salar*) in Newfoundland rivers. *Can. J. Fish. Aquat. Sci.*, **39**, 1496–1501.

Chelkowski Z., 1969. The sea trout (*Salmo trutta trutta* L.) of the Pomeranian coastal rivers and their characteristics. *Przeglad. Zool.*, **13**(1), 71–91.

Davaine P., Beall E., 1982. Acclimatation de la truite commune, *Salmo trutta* L., en milieu subantarctique (Îles Kerguelen). II. Stratégie adaptive. *Colloque sur les écosystèmes subantarctiques*, Paimpont, C.N.F.R.A., 51, 399–412.

Demars J. J., 1976. Contribution à la connaissance des Salmonidés migrateurs de la rivière Bresle, *Bull. Fr. Piscic.*, **261**, 187–197.

Elliott J. M., 1984. Numerical changes and population regulation in young migratory trout *Salmo trutta* in a Lake district stream, 1966–83. *J. Anim. Ecol.*, **53**, 327–350.

Elliott J. M., 1985. The choice of a stock-recruitment model for migratory trout, *Salmo trutta*, in an English Lake District stream. *Arch. Hydrobiol.*, **104**, 145–168.

Elson P. F., 1975. Atlantic salmon rivers, smolt production and optimal spawning: an overview of natural production. *Int. Atl. Salm. Found.*, **6**, 96–119.

Fahy E., 1978. Variation in some biological characteristics of British sea trout, *Salmo trutta* L. *J. Fish Biol.*, 13, 123–138.

Fahy E., 1979. Sea trout from the tidal waters of the river Moy. *Ir. Fish. Invest.*, Ser. A., **18**, 3–11.

Fournel, F., 1998. La truite de mer en France. Année 1997. Rap. Conseil Supérieure de la Pêche, Paris, 47 pp.

Fournel F., Euzenat G., 1979. Etude sur les Salmonidés migrateurs du bassin de l'Arques (Seine maritime) réalisée en 1978 (Parts 1 and 2). *Bull. Inf. C.S.P.*, **114**, 25–49, 67–90.

Fournel F., Euzenat G., 1982. La truite de mer en Haute-Normandie: caractéristiques, pêche fluviale et côtière, perspectives. *Colloque sur la production et la commercialisation du poisson d'eau douce*. Association Internationale des Entretiens Ecologiques, Dijon, 100–111.

Fournel F., Euzenat G., 1987. Truite de mer et saumon en Haute-Normandie/Picardie. Connaissance et gestion. *Truite, Ombre, Saumon*, **124**, 13–18.

Fournel F., Euzenat G., Gagard J. L., 1986. La pêche des Salmonidés migrateurs en Seine maritime; pêche fluviale et côtière. Rap. Cons. Sup. Pêche D.R. 1, 26 pp.

Fournel F., Euzenat G., Fagard J. L., 1987. Rivières à truites de mer et à saumons de Haute-Normandie. Réalités et perspectives. Le cas de la Bresle, 315–325. In M. Thibault and R. Billard (Eds), *Restauration des Rivières à Saumons*, INRA, Paris, 445 pp.

Fournel F., Euzenat G., Fagard J. L., 1990. Evaluation des taux de recapture et de retour de la truite de mer sur le truite de mer sur le bassin de la Bresle (Haute-Normandie/ Picardie). *Bull. Fr. Pêche Piscic.*, **318**, 102–114.

Guyomard R., 1989. Diversité génétique de la truite commune. *Bull. Fr. Pêche Piscic.*, **314**, 118–135.

Hansen L. P., 1981. Returns of Carlin-tagged, fin-clipped and unmarked wild smolts of Atlantic salmon (*Salmon salar* L.) from the river Imsa, SW Norway. I.C.E.S. CM/M, 13, 6 pp.

Harris G. S., 1970. Some aspects of the biology of Welsh sea trout (*Salmo trutta* L.), Ph.D. Thesis, Univ. Liverpool, 264 pp.

Harris G. S., 1972. Specimen sea trout from Welsh, English and Scottish waters. *Salm. Trout Mag*., **196**, 223–234.

Hay D. W., 1985. Overwinter post-tagging mortality and tag loss among emigrating juvenile Atlantic salmon (*Salmo salar* L.) held overwinter in tanks. I.C.E.S. CM/M, 13, 7 pp.

Jarvi T. H., Menzies W. J., The interpretation of the zones of scales of salmon, sea trout and brown trout. Cons. Perm. Internat. Expl. Mer, 47, 53 pp.

Jensen K. W., 1968. Sea trout (*Salmo trutta* L.) of the river Istra, Western Norway. *Rep. Inst. Freshwater Res., Drottningholm*, 48, 187–213.

Jensen K. W., 1981. Survival estimates of wild smolts of Atlantic salmon from river Ismsa, S.W. Norway. I.C.E.S. CM/M, 14, 5 pp.

Jonsson B., 1985. Life history patterns of freshwater resident and sea-run migrant brown trout in Norway. *Trans. Am. Fish. Soc*., **114**, 182–194.

Kermarec J. Y., 1980. Le retour du saumon (dossier): la saison 80. Bilan des captures en Bretagne et Basse-Normandie depuis 1955. *Eau Rivières*, **36**, 15–16.

Kerswill C. J., 1971. In Paloheimo and Elson (1974).

Larsson P. O., 1984. Effects of reducing fish for feeding salmon (*Salmo salar* L.) in the Baltic on home water fisheries according to simulations with the Carlin–Larsson population model. *Fish. Mngt*., **15**, 97–105.

Le Cren E. D., 1985. The biology of the sea trout. Symposium Plas Menai, Oct. 1984, Atlantic Salmon Trust Ltd, 42 pp.

Nall G. H., 1930. *The Life of the Sea-trout*. Seeley Service, London, 335 pp.

Nihouarn A., Porcher J. P., 1989. Station de contrôle des migrations de Kerhamon (Finistère): bilan des premières observations sur les populations de saumon atlantique de l'Elorn réalisées de 1986 à 1988. Rapport Cons. Sup. Pêche, 16 pp.

Ombredane D., Siegler L., Baglinière J. L., Prunet P., 1996. Migration et smoltification des juvéniles de truite (*Salmo trutta*) dans les cours d'eau de Basse Normandie. *Cybium*, **20** (suppl.), 27–42.

Palka W., Bienarz K., 1983. Migration, growth and exploitation of sea trout (*Salmo trutta* L.) from the Dunajec river. *Rocz. Nauk Roln*., Seria H.T. **100** (Z.2), 72–94.

Paloheimo J. E., Elson D. F., 1974. Effects of the Greenland Fishery for Atlantic salmon on Canadian stocks. I.A.S.F., Spec. Publ. Ser., 5(1), 34 pp.

Paterson D., 1973. Observations on the sea-trout (*Salmo trutta* L.) spawning populations from light Tweed tributaries. B.Sc. Thesis University of Edinburgh, 110 pp.

Piggins D. J., 1976. Stock production, survival rates and life-history of the sea-trout of the Burrishoole River system. *Ann. Rep., Salm. Res. Trust Ireland Inc*., **20**, 45–57.

Piggins D. J., 1983. *Salm. Res. Trust Ireland Inc., Ann. Rep*., **27**, 64 pp.

Piggins D. J., 1986. *Salm. Res. Trust Ireland Inc., Ann. Rep*., **30**, 15 pp.

Piggins D. J., 1988. *Salm. Res. Trust Ireland Inc., Ann. Rep*., **32**, 23 pp.

Pratten D. J., Shearer W. M., 1983. Sea trout of the North Esk. *Fish. Mngt*., **14**, 49–65.

Pratten D. J., Shearer W. M., 1985. The commercial exploitation of sea trout, *Salmo trutta* L. *Aquac. Fish. Manage*., **1**, 71–89.

Prevost E., 1987. Les populations de saumon atlantique (*Salmo salar* L.) en France:

description, relation avec les caractéristiques des rivières: essai de discrimination. Thèse Doc. Ing. Sci. Agron., ENSA Rennes, 103 pp.

Prouzet P., 1983. Le pacage en mer du saumon atlantique en Europe. *Pêche marit.*, **202**, 202–208.

Richard A., 1981. Observations préliminaires sur les populations de truite de mer (*Salmo trutta* L.) en Basse-Normandie. *Bull. Fr. Piscic.*, **283**, 114–124.

Richard A., 1982. La truite de mer (*Salmo trutta* L.) en Basse-Normandie; premières observations. *Colloque sur la production et la commercialisation du poisson d'eau douce*. Association Internationale des Entretiens Ecologiques Dijon, 89–99.

Richard A., 1986. Recherches sur la truite de mer, *Salmo trutta* L., en Basse-Normandie. Thèse Doct. 3e cycle, Univ. Rennes, 66 pp.

Saunders R. L., Allen K. R., 1967. Effects of tagging and fin-clipping on the survival and growth of Atlantic salmon between smolt and adult stages. *J. Fish. Res. Board Can.*, **24** (12), 2595–2611.

Skrochowska S., 1969a. Migrations of the sea trout (*Salmo trutta* L.), brown trout (*Salmo trutta* M. *fario* L.) and their crosses. Part III—Migrations to, in and from the sea. *Polsk. Arch. Hydrobiol.*, **16** (29), 2, 149–180.

Skrochawska S., 1969b. Migrations of the sea trout (*Salmo trutta* L.), brown trout (*Salmo trutta* M. *fario* L.) and their crosses. Part IV.—General discussion of results. *Polsk. Arch. Hydrobiol.*, **16** (29), 2, 181–192.

Sych R., 1970. Some comparisons on the background of an eleven-year study on the growth of sea trout (*Salmo trutta* L.). *Acta Hydrobiol.*, **12** (2–3), 225–249.

Symons P. E. K., 1979. Estimated escapement of Atlantic salmon (*Salmo salar*) for maximum smolt production in rivers of different productivity. *J. Fish. Res. Board Can.*, **36**, 132–140.

Tellier L., 1987. Mise en service d'une station d'étude des migrations de Salmonidés migrateurs sur la rivière Elorn (Finistère): premières observations sur les populations migrantes d'adults et de juvéniles de saumon atlantique (*Salmo salar* L., 1758). Mémoire E.N.I.T.E.F., 41 pp.

Ward B. R., Slaney P., 1988. Life history and smolt-to-adult survival of Keogh River steelhead trout (*Salmo gairdneri*) and the relationship to smolt size. *Can. J. Fish. Aquat. Sci.*, **45**, 1110–1122.

Went A. E. J., 1962. Irish sea-trout, a review of investigations to date. *Sci. Proc. R. Dublin Soc.*, **1** (10), 265–296.

Went A. E. J., 1967. Salmon and sea trout in the Foyle system. 16th Rep. Foyle Fish. Com., 4 pp.

Zarnecki S., 1960. General conclusions on the scale-reading of salmon, sea trout and brown trout originating in the Vistula. I.C.E.S. CM/M, 128, 1–3.

Zarnecki S., 1973. Differentiation of Atlantic salmon (*Salmo salar* L.) and of sea trout (*Salmo trutta* L.) from the Wisla (Vistula) river into seasonal populations. *Acta Hydrobiol.*, **5** (2/3), 255–294.

3

Genetic diversity and the management of natural populations of brown trout

R. Guyomard

I. INTRODUCTION

Data accumulated over the past twenty years have clearly shown the presence of significant genetic variability within most species. Part of this variability confers different adaptive values on individuals making up the species. The objective of genetic improvement is to select and propagate the genotypes showing the best adaptive value in a given production system. Selective breeding methods have been established for most domesticated species or populations with, in parallel, the setting up of gene banks. In contrast, this step has seldom been adopted for species which are directly exploited in the environment, such as fish. This tardiness results, on the one hand, from the very recent consideration of genetic concepts of populations in fishery management and, on the other hand, from the difficulty in defining the selection criteria and in carrying out controls for performance in the natural environment. Genetic improvement, as conceived in domesticated species, appears utopic when considering natural populations. It is sometimes possible to experiment with management plans for natural populations, with respect to the genetic diversity of species and the principles of population genetics, and to formulate schemes which, without being ideal or definitive, can constitute progress in relation to current practice.

The object of this chapter is to present the accumulated data concerning the genetic diversity of natural and domestic populations of brown trout and then to analyse, in the light of these results, the main genetic problems linked to the exploitation of this species in the natural environment and the solutions which can be put forward.

II. DESCRIPTION OF THE GENETIC DIVERSITY OF TROUT POPULATIONS

1. The population, unit of evolution

The genetic diversity of a species is classically described in terms of inter- and intra-population variability. This procedure depends on the relative importance accorded to the concept of genetic populations or the theory of evolution. Ideally, a population is a community in which individuals of both sexes and sexually mature, mix randomly to reproduce.* This mode of mating is called panmixia. In the population, subject to the side effects of various factors (selection, genetic drift, mutation, gene recombination and migration), multiple combinations of genes appear and are 'trialled', at each generation. Each population can therefore constitute an original gene pool co-adapted to the environmental conditions in which it evolved. Estimation of the inter-population genetic variability allows the degree of similarity of populations, compared to each other, to be determined.

2. Methods of description of genetic variability

The description of the diversity of a species usually calls on three categories of methods: biometry, which measures variations in quantitative characteristics, caryology which describes chromosomal variations, and molecular techniques which analyse the polymorphism of proteins or DNA.

Quantitative characters are those with continuous variation or providing relatively large numbers of values. A considerable number of close species or sub-species of the brown trout (Behnke, 1968, 1972; Dorofeyeva *et al.*, 1981) have been described from morphological and meristic characteristics. Their genetic determinism is not known and is assumed to be polygenic. In addition, these characteristics can be sensitive to environmental effects. The differences observed between populations sampled from different environments may thus reflect environmental, as much as genetic, variations. We should be particularly suspicious of the conclusions from these characteristics, especially in fish which appear to have considerable phenotypic plasticity. Under rigorous conditions, these characteristics would have to be measured from populations reared under identical environmental conditions and according to experimental protocols which would be relatively difficult to carry out since high numbers would be needed to represent populations satisfactorily, leading to a requirement to rear separately a significant number of batches up to a size at which marking is possible. In addition, the results obtained may only be valid under the environmental conditions tested, if genotype–environment interactions exist.

Interest in caryology for describing intra-specific genetic diversity in fish has been very low until recently. This is mainly due to the difficulties encountered in the employment of techniques which are known to work in the higher vertebrates (banding techniques). Certain caryological studies dedicated to the brown trout found both inter- and intra-population chromosomal variations (Dorofeyeva and Rukhyan, 1982).

Much of the relevant data on the genetic structures of fish species, including the salmonids, has been provided by electrophoretic analysis of enzyme systems. These systems have many advantages over quantitative characters. They are little affected by environmental variations and can be used to compare populations sampled *in situ*. In addition, each system is controlled by genes at a small number of loci (one to four, in general) and the observed electrophoretic variation can easily be interpreted in terms of allelic frequencies at each locus (Fig. 1). Electrophoretic procedures, principles of

* The exact definition also assumes the absence of gamete selection.

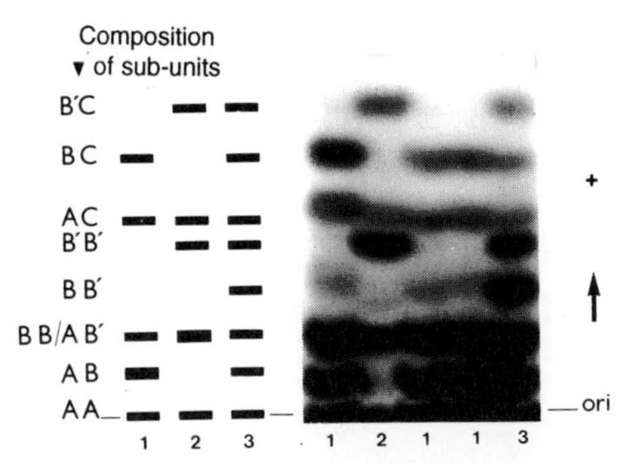

Fig. 1. Electrophoretic variations in a dimeric enzyme, MDH (malate dehydrogenase) in the liver of brown trout: electrophoregram is shown on the right and its interpretation on the left. In the liver, MDH is under the genetic control of two loci; Mdh-1 which is fixed and 'produces' sub-unit A, and Mdh-2 which is variable. In the collection of populations studied, three different alleles have been found at the Mdh-2 locus: Mdh-2 (100), Mdh-2 (200) and Mdh-2 (0). The variations observed in the figure are due to Mdh-2 (100) and Mdh-2 (200) which code for sub-units B and B′ respectively (the Mdh-2 (0) allele does not occur in Corsican populations (see Krieg & Guyomard, 1983)). The C sub-unit is 'produced' by the Mdh-3 and 4 loci which are specific to muscle and are expressed at low levels in the liver. Genetic interpretation: 1 = Mdh-2 (100/100); 2 = Mdh-2 (200/200); 3 = Mdh-2 (100/200). The validity of these genetic interpretations has been shown by inheritance studies (see Guyomard & Krieg, 1983).

interpretation of electrophoretic patterns and methods of validating proposed genetic determinisms have already been clearly and widely described (May, 1980; Krieg, 1984; Pasteur *et al.*, 1987).

The analysis of enzyme polymorphism in the brown trout has been undertaken in Scandinavia (Ryman, 1983), in the British Isles (Ferguson and Fleming, 1983) in the Soviet Union (Osinov, 1984) and in France (Krieg and Guyomard, 1985). In this chapter, only the results of studies carried out in France are given. These studies had as their objective (1) to analyse the genetic diversity of natural domesticated populations (2) to study the differentiation between migrant and resident phenotypes living in sympatry (3) to examine the effects of stocking on native populations (this point is shown in the second part).

3. Geographical differentiation

The geographical location of the 29 natural populations studied is shown in Fig. 2. Eleven of the domesticated stocks most used for stocking as well as three Polish stocks raised for aquaculture were also studied. Fifteen to thirty individuals per sample were examined. Both theoretical and experimental studies showed that the precision of estimation of genetic distances (inter-population variability) and rates of heterozygosity (intra-population variability) only depends, in practical terms, on the number of loci examined when the number of individuals studied per population is greater than 10 (Nei, 1978; Gorman and Renzi, 1979). Forty-seven loci were examined for each individual (Krieg and Guyomard, 1985). We have estimated the intrapopulation genetic diversity or calculated rate of heterozygosity (Hs), the genetic diversity between two populations (Dst) and the

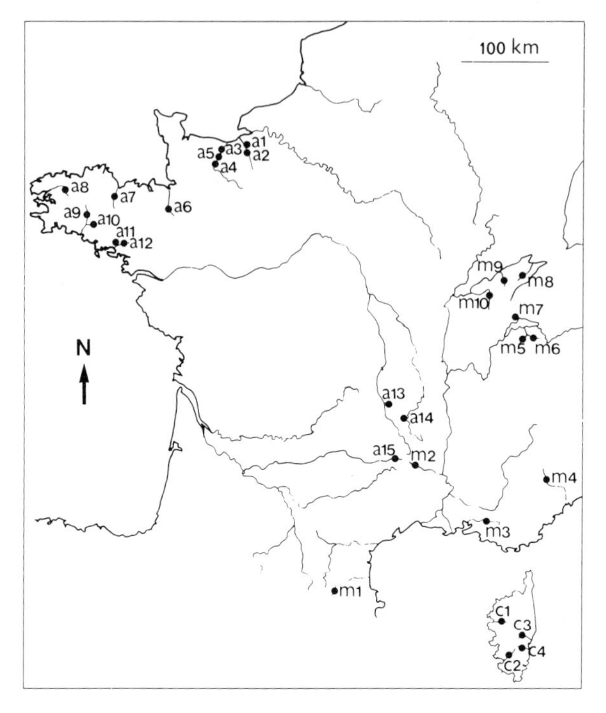

Fig. 2. Geographical location of natural populations analysed. a1: Calonne (sea trout); a2: Touques (smolts); a3: Orne (smolts); a4: Orne (resident trout); a5: Orne (sea trout); a6: Avion; a7: Leffe; a8: Elorn; a9, a10: Scorff; a11: Lay; a12: Montgolérian; a13; Allier (tributary); a14: Loire; a15: Lot; m1: Tech (tributary); m2: Cèze (tributary); m3: Touloubre; m4: Var (tributary); m5, m6: Drance (tributaries); m7: Aubonne; m8, m9, m10: Doubs (tributaries); c1: Aïtone; c2: Rizzaneze; c3: Travo (tributary); c4: Solenzara.

total genetic diversity (Ht), as defined by Nei (1975). Dendrograms of genetic distances were constructed using UPGMA numerical taxonomy (Sneath and Sokal, 1973).

Fig. 3 represents the geographical distribution of allelic frequencies observed at two Ldh-5 (lactate dehydrogenase) and Tfn (transferrin) loci. There appears to be a very clear dichotomy between the Atlantic and Mediterranean populations. This subdivision is confirmed by the dendrogram which summarizes the information provided by all 47 loci (Fig. 4). Natural populations divide up according to a very simple plan, partly into the populations of the Atlantic side and partly those of the Mediterranean. The appearance of a population (m2, Fig. 2), situated at the edge of the Mediterranean area, in the Atlantic group, probably results from the hydrographic catchment of a river. The degree of genetic differentiation between these two groups is equivalent to those observed between morphological sub-species in other salmonids (Loudenslager and Gall, 1980; Leary *et al.*, 1987). It is therefore possible that significant morphological differences exist between Mediterranean and Atlantic populations. It has been shown, for example, that Corsican populations can be distinguished by their low number of pyloric caeca (Olivari and Brun, 1988). Recent investigations on variation of DNA (Bernatchez *et al.*, 1992) and microsatellite loci (Presa *et al.*, 1995) have confirmed the dichotomy between Mediterranean and Atlantic populations. They have also identified additional divergent lineages at

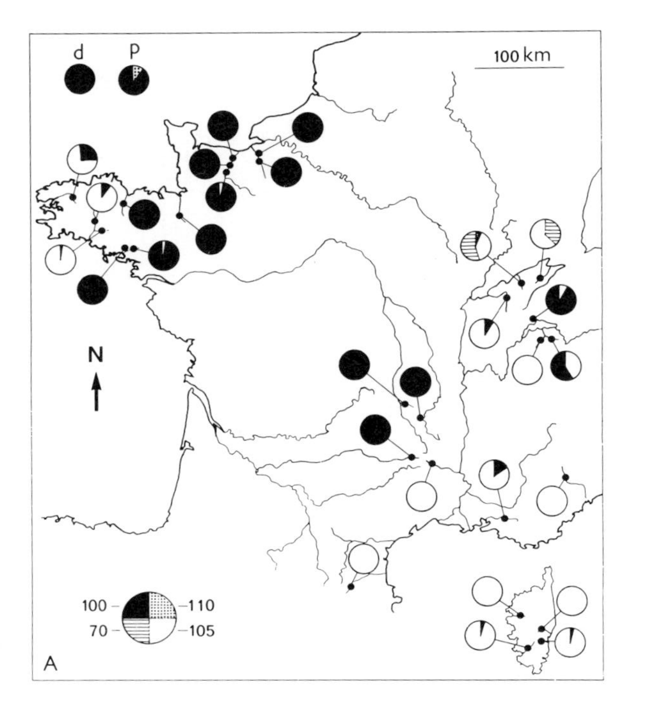

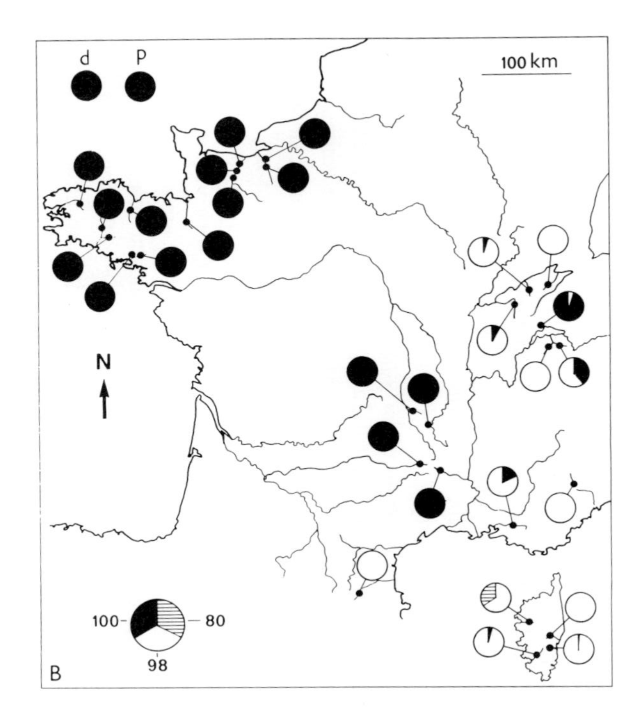

Fig. 3. Geographical distribution of allele frequencies at the Ldh-5 (Fig. 3A) and Tfn (Fig. 3B) loci in the natural populations studied. d: domestic stocks; p: Polish stocks. Four alleles (100, 70, 105 and 110) were observed at Ldh-5 and three at Tfn (100, 80 and 90). The most common allele is named 100; the others are identified by their electrophoretic mobility in relation to that of the most common allele. The sector sizes are proportional to the observed allelic frequencies.

the European scale (Bernatchez *et al.*, 1992; Giuffra *et al.*, 1994; Presa *et al.*, 1994; Presa, 1995).

All the domesticated stocks examined are related to the Atlantic group (Fig. 4). It is difficult to be more precise about the origin of these stocks which must certainly result from mixed populations and exchanges between fish farms as suggested by their high rates of heterozygosity and low degrees of differentiation (Fig. 5). The three samples from Poland were also of the Atlantic type. The total variability of the group is high (0.10; Fig. 5); within the group, the degree of differentiation between populations is sometimes significant (on average, 40% of the total diversity) and a tendency for geographical regrouping seems to be an underlying trend. This is the case, for example, for the samples from Brittany (a8, a9 and a10 in Fig. 2). The hypothesis of a genetic structuring by hydrographic basin remains to be proved using more complete sampling. The populations on the Mediterranean side present a radically different situation. They form a very homogeneous group within which the total variability does not exceed 0.04 (Fig. 5) and the dendrogram does not reveal any geographical subdivisions at all.

Paradoxically, the variability is greatest within the Corsican populations, in spite of the small size of the geographical zone (0.08; Fig. 5). This great variability, which has no connection with stocking, could be the result of an ancient colonization of the island by two genetically differentiated populations. The splitting of the four Corsican populations into two very distinct groups may possibly not be just an artefact due to insufficient sampling of the populations.

The comparison of data obtained by different laboratories proves to be difficult in the absence of exchange of samples. It appears nevertheless to be quite clear that the alleles observed in French populations from the Atlantic side are also found in populations from the British Isles. From the geographical distribution of two alleles Ldh-5 (105) and LDh-5 (100), it has been suggested (Ferguson and Fleming, 1983; Ferguson, 1985) that

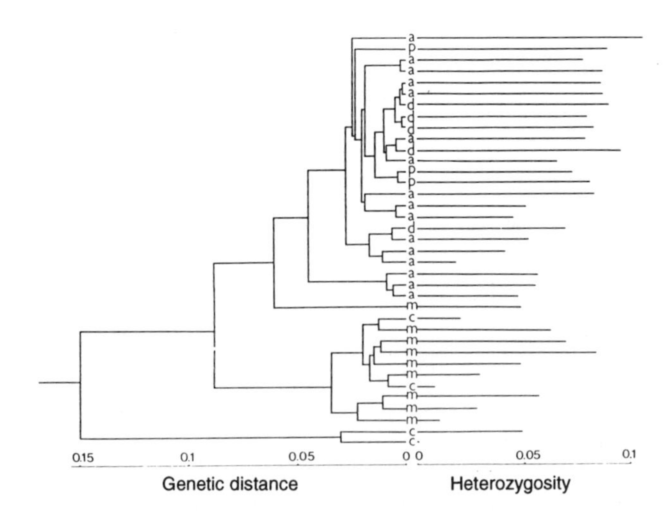

Fig. 4. Dendrogram of genetic distances and rates of heterozygosity calculated at 46 loci. Domestic (d), Polish (p), Mediterranean (m), Corsican (c) and Atlantic stocks.

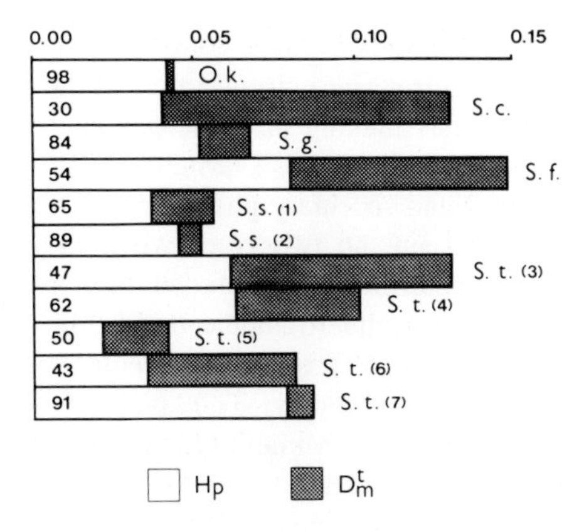

Fig. 5. Breakdown of genetic diversity into intrapopulation variability (Hp: mean calculated heterozygosity or mean intrapopulation genetic diversity (Nei, 1975)) and interpopulation variability (D^{tm}: mean interpopulation genetic diversity (Nei, 1975)) in the brown trout and several salmonid species. o.k.: *Oncorhynchus keta*, Okazaki, 1982; 54 populations, 22 loci. SSc.: *Salmo clarki*, Loudenslager and Gall, 1980, 1980; 24 populations, 35 loci. S.g.: *Salmo gairdneri*, in Ryman, 1983; 38 populations, 16 loci. S.f.: *Salvelinus fontinalis*, Stoneking *et al.*, 1981; 8 populations, 39 loci. S.s.: *Salmo salar*, Guyomard, 1987; (1): 10 populations, 32 loci; (2): 6 European Atlantic populations, 32 loci. S.t.: *Salmo trutta*, this chapter; (3): 37 populations, 46 loci; (4): Atlantic populations, 46 loci; (5): Mediterranean populations, 46 loci; (6) Corsican populations, 46 loci; (7) fish farmed populations, 46 loci. The figures given for each species represents the intrapopulation variability expressed as a percentage of the total variability.

the British Isles experienced two colonizations in post-glacial times, one by a form which was fixed for allele 105 (*Salmo ferox*) and the other, later on, by a form which was fixed for allele 100, and had a marked tendency towards anadromy. No electrophoretic data concerning the Mediterranean populations has been available. The Corsican populations appear to share particular meristic characteristics, e.g. a low number of pyloric caeca (Olivar and Brun, 1988), with the sub-species *Salmo trutta macrostigma*, but these similarities may have no phylogenetic significance.

4. Ecological differentiation

As with most of the salmonids, sympatric ecological forms are observed in the brown trout (Behnke, 1972; Dorofeyeva *et al.*, 1981). The main problem raised by the existence of these sympatric ecological forms is to know if they represent reproductively isolated and genetically distinct populations. In the brown trout this appears to sometimes be the case in certain lakes (Ryman *et al.*, 1979; Dorofeyeva and Rukhyan, 1982; Ferguson, 1985) but not always (Crozier, 1983). The sympatry most often described is that associating the sea trout (*Salmo trutta trutta*) with the river (brown) trout (*Salmo trutta fario*). This cohabitation is found in almost all the hydrographic basins which flow into the Atlantic Ocean. Electrophoretic studies have not revealed any significant genetic difference between samples of migrant and resident fish from the same river, despite the existence of significant enzymatic polymorphism within these samples (Fleming, 1983). Therefore the individual migrant and sedentary fish present in sympatry, probably belong to the same population.* This hypothesis is confirmed by studies which we have carried out on populations of brown trout into virgin waters in the Kerguelen Islands (Guyomard *et al.*, 1984). The individuals making up these populations, produced from one single introduction of fertilized eggs from a fish farm, differentiated themselves from the first generations into migratory and sedentary phenotypes. Electrophoretic analyses found no genetic differentiation between the two types when they came from the same water course (Fig. 6). It can therefore be concluded that the presence of sedentary and migratory fish within the same hydrographic basin probably does not result from the sympatry of two genetically differentiated species or sub-species, but from an intrapopulation phenotypic variability. The available data do not allow it to be determined whether this phenotypic variability is partly of a genetic origin or whether it is exclusively caused by environmental variations. It is possible that an analogous situation could occur for other characteristics such as the staying time in fresh water or in the sea.

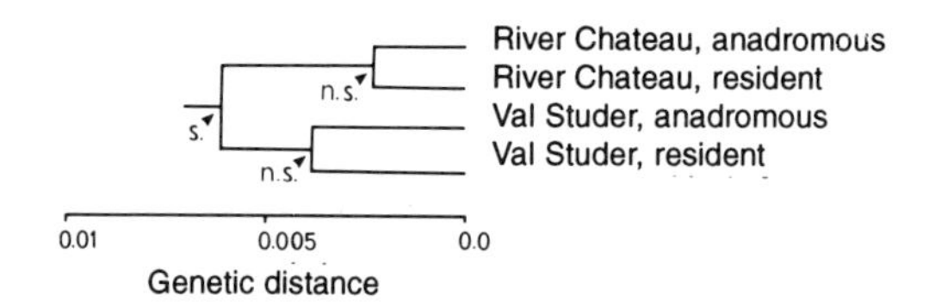

Fig. 6. Dendrogram of populations in the Kerguelen islands studied by electrophoresis (31 loci). s, allelic frequency statistically significant at 5% level; n.s. not significant.

5. The significance of differentiation

The many factors which are likely to cause the appearance of genetic differences have already been listed (Chapter II, §1). These factors are, on the one hand, those which generate new variability, mutation and gene recombination and, on the other hand, those which modify existing variability, genetic drift and selection. It is important to note that very low migration rates† (10^{-4}–10^{-5}) are sufficient to prevent all differentiation, even in the presence of strong selective forces (Li, 1976; Nei, 1987). Thus, even while 'homing' is very marked, as in the Atlantic salmon, the existence of a few erratic individuals can assure the maintenance of strong genetic similarity of the populations. In contrast, where there is complete geographical isolation, two populations can diverge relatively quickly (Nei, 1975). In the brown trout, geographical isolation is evident and probably very old (possibly intermittent) between Mediterranean and Atlantic populations. It is obviously impossible to say whether selection has played a part in the establishment of the observed geographical differentiation or whether this is the result of random evolution. The observed differentiation does not necessarily constitute an adaptive response. On the contrary, it is not certain that this would be the best response for the species to local

* A case has, however, been observed where the migratory and resident stocks in a river are genetically distinct (in the Orne, see Chapter II, §2); this is an artificial situation and is probably due to restocking.
† Percentage of individuals migrating at each generation.

selection. For example, perhaps an Atlantic population might have a better selective value than a Mediterranean one in a zone from which it originated (or vice versa).

6. Comparison with other salmonid species

The scope and structure of the intraspecific variability differs significantly from one species to another (see Fig. 5 for the salmonids). Some of these differences may be the consequence of varying strictness applied to the morphological description of the species (several freshwater fish species, in particular in the salmonid family, have been defined according to morphological criteria and not by the application of the biological concept of species) or from incomplete electrophoretic studies. It is nevertheless interesting to compare the brown trout and the Atlantic salmon, since their areas of distribution overlap so greatly. The results show (even if the comparison is limited only to Atlantic populations) that two phylogenetically close species, with numerous common features with regard to their biology, can present very different genetic structures over their identical areas of distribution.

III. EFFECTS OF RESTOCKING

1. Demonstration of gene flow between domestic and natural stocks

Fig. 3 shows the important allelic differences between domestic stocks and certain natural populations. Thus the fish-farmed fish stocks are fixed for alleles Ldh-5 (100) and Tfn (100) which are not present in any Mediterranean population which has not been subject to stocking. These alleles constitute true genetic markers, unalterable and heritable, of stocked trout. In contrast to traditional physical marks, they have the advantage of allowing the fate of domestic individuals to be followed, regardless of their stage of development,* and above all to discover if they participate in spawning and to determine the rate of gene introgression in the population.

Two examples of analysis of the effects of restocking are illustrated in Fig. 7. These analyses included the study of a sample of individuals collected in an 'unstocked' sector, to determine the genetic type of the original population, of a sample of individuals collected in a restocked site close to the virgin site and individuals from the utilized domestic stock. In the first case (Fig. 7A), all the 0^+ individuals correspond to recently released fish-farmed fry; in contrast, individuals aged 1^+ or more were practically all wild. Fry released in previous years were therefore eliminated, perhaps soon after release. It is therefore preferable only to estimate the level of introduced domestic genes in the natural stock from individuals 1^+ and older; this level does not exceed 5%. Fig. 7B shows that, in the second case, individuals in the restocked sector show a range of variations, from wild to domestic genotypes. Detailed analysis shows that domestic and wild genes are associated randomly (no deviation from panmixia and no linkage disequilibrium) and that a significant proportion of individuals are hybrids of the second generation, at least. In this case, the level of introgression can be estimated from the percentage of domestic genes found in the sample, i.e. 50%.

* At the stages most commonly used for stocking, the eyed stage and first feeding stage, the individuals are too small to be marked.

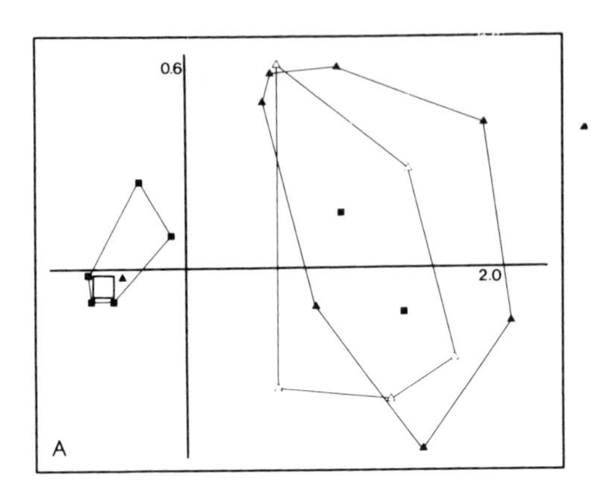

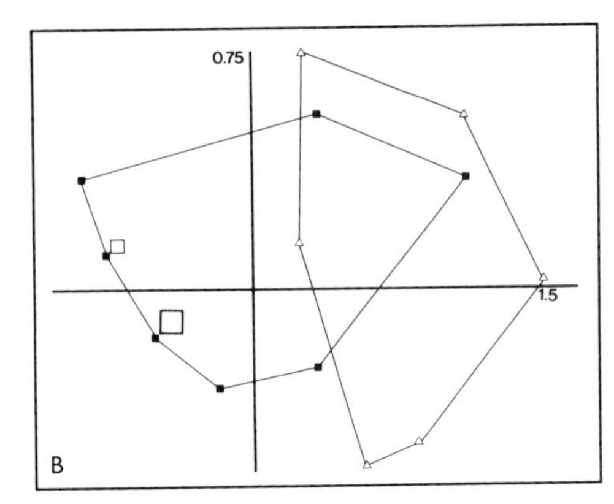

Fig. 7. Effects of restocking on the genetic structure of two natural populations of brown trout (results of an analysis of the main components are shown as the two principal axes in the diagram; after Barbat-Leterrier *et al.*). △: fish farmed stock (30 individuals analysed in each group); □: local population (unstocked sector, adults; 20/30 individuals per site); ■: local population (restocked sector, adults; 20 individuals/site; the symbols are approximately proportional to the number of individuals with the same co-ordinates); ▲: local population (restocked sector, fry; 40 individuals/site). A: river La Bernarde, tributary of the Var. With regard to restocking, almost all the fry analysed, with one exception (isolated black triangle), are included in a polygon which is superimposed on that of the domestic stock; these are the fry used for restocking. In contrast, individuals more than one year old (■) are almost the same as individuals from the pure local stock. B: river La Illas, tributary of the Tech. The polygon including the individuals from the restocked site extends from the symbols corresponding to individuals of the pure local stock to the polygon representing the domestic stock. This corresponds to a high level of introgression (around 50%).

The presence of the Ldh-5 (100) and Tfn (100) alleles in the Mediterranean samples (Fig. 3) is due to restocking (with the exception of the Cèze, m2 in Fig. 2).

Studies of restocked sites have demonstrated two interesting facts:

- firstly, the rate of introgression varied widely (from 0 to 50%, see Fig. 8), even when the extent of the restocking appeared to be identical. However, the mean annual rate introgression was low in all cases. Thus, a 50% rate of introgression indicated that the proportion of reared fish introduced annually, and surviving, did not exceed 15%, on average (Fig. 8) of the population *in situ*, after restocking over fifteen years (20% if carried out over 10 years). Restocking efficiency is therefore low.
- secondly, there appears to be no barrier to gene flow through reproduction between introduced, domestic stocks and wild stocks. This phenomenon is usually observed in the secondary contact (often induced by man), between species or sub-species of salmonids which attain, or even exceed, the degree of differentiation which distinguishes Mediterranean from domestic stocks (Busack and Gall, 1981; Halliburton *et al.*, 1981; Gyllensten *et al.*, 1985; Campton and Utter, 1985).

2. Effects of restocking on the biology of natural populations

While it is relatively easy to demonstrate the introgression of domestic stocks in natural populations, there are, on the contrary, no data available to clearly show that this

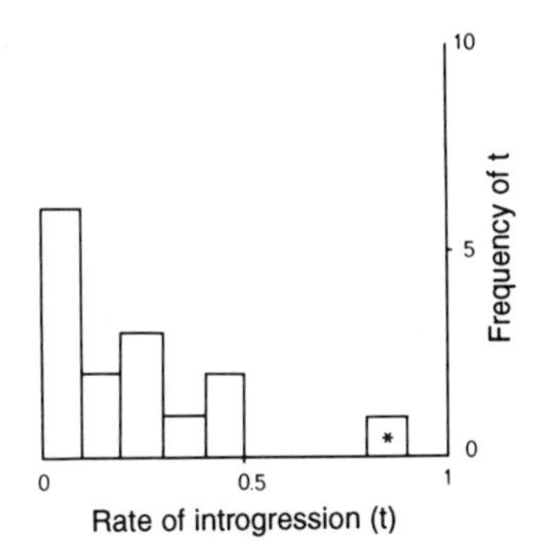

Fig. 8. Distribution of cumulative rate of introgression in different samples from sectors where restocking has been carried out (restocking effort was not identical at all sites). For a discriminant allele with initial frequency of 0 in the natural population and 1 in the domestic stock, the *mean annual* rate of introgression (m) was estimated using the formula $1 - (1 - p)^{1/n}$ where p was the observed frequency of the discriminant allele and n the number of generations elapsing from the start of restocking. This model also put forward the hypothesis that reproduction is assured by a single age class, which determines the duration of a generation. For a generation duration of 3 years and $p = 0.5$, after 15 years of restocking, m = around 15%. The sample marked with an asterisk must be regarded with caution, as the fish involved were reared in a fish farm, from several spawners (number not known) which had been taken from a tributary of Lake Léman.

introgression modifies the phenotypic expression of characteristics which affect the biology of populations (growth, fecundity, migratory behaviour etc.) and, most importantly, their mean selective value. The capacity of cultured fish stocks to 'produce' resident and migratory phenotypes, in the wild, has been clearly observed in certain cases such as in the Kerguelen Islands. The introduction of domestic stocks may therefore be at the origin of the emergence or reinforcement of migratory behaviour in natural populations. This may have happened for the population in the Orne. The first notable catches of sea trout in this river, whose population was previously described as resident, started in the 1960s, the years corresponding to the increase in the stocking rate in the Orne (Richard, 1981). The fish used for restocking had two very distinct origins, one a population of sea trout from the Dunajec (tributary of the upper Vistula basin) introduced at discrete times during the 1960s, the other cultured stock of fish (Etrun) used in a more regular and intensive way, during the same years. Moreover, smolts and adults returning to reproduce in the river after a period in the sea were identified as being the fish which had been restocked. This information suggests that the sea trout population of the Orne may be largely derived from restocking operations. This hypothesis is reinforced by electrophoretic studies which we have carried out. Fig. 9 shows the quasi-genetic identity of the domestic Etrun stock alongside the samples of smolts and sea trout from the Orne. In contrast, the Dunajec sample examined* remains very different from the migrants in the Orne. Lastly, migratory and resident stocks are genetically different in this river. The genetic originality of the 'Orne' population is confirmed by the presence of allele Mdh-3,4 (50) which is unique to it, and Ldh-5 (105) which is specific to the ancestral form of

* It is not possible to be absolutely certain that this sample is identical to the Dunajec sea trout, previously released in the Orne.

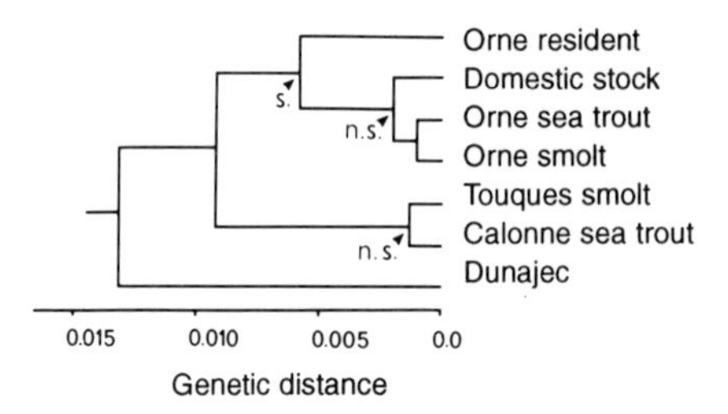

Fig. 9. Dendrogram of samples taken on the Orne and domestic stocks released into this river (46 loci). s, significant at 5% level; n.s., not significant.

the brown trout (Ferguson and Fleming, 1983). Sedentary individuals may therefore be representatives, at least in part, of the original sedentary population, genetically distinct from the introduced stock.

It cannot be excluded that the migratory stock is essentially maintained by annual restocking imports and not through natural reproduction. If this is the case, two remarks can be made:

- firstly, this migratory stock would not maintain itself effectively, if restocking was interrupted;
- secondly, the differences in migration aptitude observed between the resident and anadromous fractions are not necessarily genetic in nature, despite the existence of genetic differences for other characteristics such as enzyme systems; the migratory behaviour of the anadromous faction may be induced, for example, by the initial rearing phase prior to release or by the restocking method used (over-stocking, for example).

In all cases the introduction of restocked individuals appears to result in certain cases, in a modification of phenotypic characteristics of natural stocks.

The effect of transplantations on characteristics such as rate of return and spawning period has also been demonstrated for the chum salmon *Oncorhynchus keta* (Okazaki, 1982).

IV. GENETIC MANAGEMENT OF NATURAL POPULATIONS

1. Long-term management: conservation of genetic diversity

In domesticated species subjected to genetic improvement, the genetic diversity is preserved in order to be able to answer to changes in the selection criteria; populations, races or varieties to be conserved are defined from the variability of characteristics in relation to productivity (in general, quantitative characteristics) and Mendelian characters with visible effects, or from molecular polymorphism.

The protection of genetic diversity in the brown trout must be ensured, even if the effects of restocking are positive, in order to preserve the adaptive potential of the species to any change in the environment or to new habitats. In the brown trout, only electrophoretic data are available. Their use for defining populations to be conserved is justified if, for a wide electrophoretic diversity, there is more corresponding mean genetic variability. This appears to be the case in the salmonids in which the morphologically identifiable sub-species* are usually confirmed by electrophoretic studies (Loudenslager and Gall, 1980; Stoneking *et al.*, 1981).

Electrophoretic data allow the definition, on the one hand, of the genetic subunits from which the populations to be preserved will be chosen, and on the other hand, to only keep, within these subunits, populations which will have never been subjected to stocking. This sampling strategy is certainly much more efficient than random restocking. Conservation of a single population in each of the geographical subunits consisting of Atlantic, Mediterranean and Corsican populations, allows the essential elements of observed genetic diversity to be preserved. In practice, it is preferable to protect several populations from each group to take into account the part of the genetic diversity which escapes electrophoretic analysis. Moreover, the choice of 'protected' populations must be capable of being developed in relation to the acquisition of new electrophoretic or other data.

Lastly, these populations must be conserved in their original environment, a principle which should be easy to apply.

2. Short-term management

Short-term management of natural populations is considered in light of the two forms of intervention to which they are regularly and intentionally exposed; fishing and restocking.

(a) Effects of fishing

This can, in theory, reduce the diversity of a population, either by genetic drift if the catches are significant, or by selection if the fraction taken possesses particular genetic characteristics.

The loss of variability through genetic drift only becomes significant above a very small effective population size. Thus, for a locus with two alleles, it does not exceed 5% after 10 generations, when the reproductive population is composed of 25 males and 25 females (Fig. 10). It is almost certain that the number of breeding fish present in a water course is generally much higher than this. In fact, the rates of heterozygosity of natural populations of brown trout are relatively high and the cases of low intra-population variability appear to arise from natural genetic drift.

Numerous attempts have been made to try to demonstrate the selective effects of fishing by following changes in the characteristics of exploited stocks (Nelson and Soulé, 1987). This evolution is obviously difficult to interpret and the conclusions unconvincing, as the genetic effects (if they exist) cannot be dissociated from environmental effects. In theory, the mean effect of selection on the variability of a panmictic population should be low (see Fig. 11 for a specific selection model), when the characters subject to this selection are under the control of a significant number of loci, which is likely to be true in the majority of cases.

* Some morphologically identifiable sub-species show a degree of differentiation almost equivalent to that which characterizes the Atlantic and Mediterranean populations.

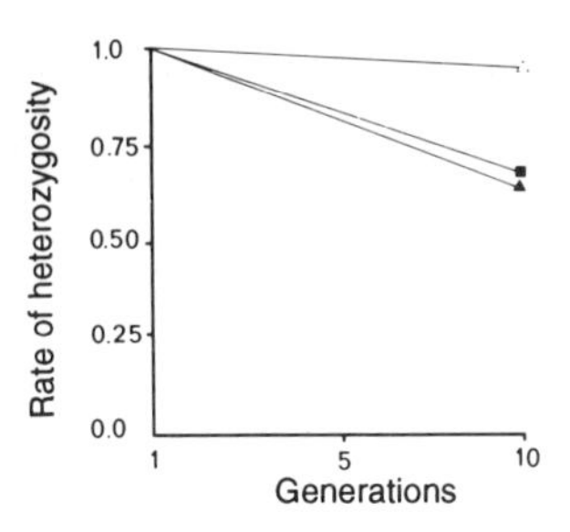

Fig. 10. Change in the genetic diversity in relation to the number of generations and number of spawning fish. △: 25 males × 25 females; ▲: 5 males × 5 females; ■: 3 males × 47 females. Hn = (1 – 1/2Ne)·Hn – 1; where Hn = rate of heterozygosity in generation n; Ne = number of individuals in population; here, we have calculated Ne from the formula Nf.Nm/(Nm + Nf) where Nf = number of spawning females and Nm = number of spawning males (see Falconer (1974) for more details.

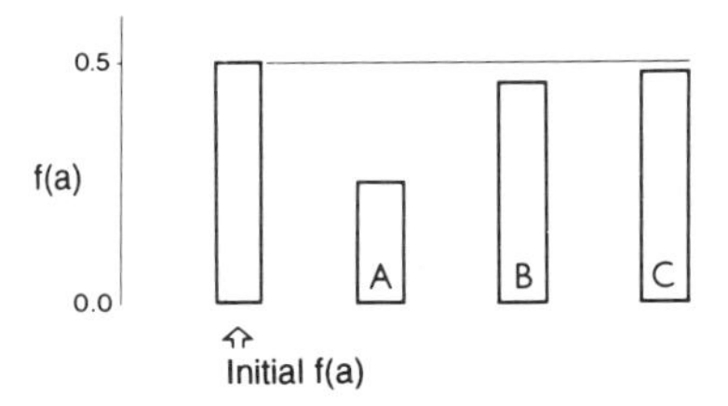

Fig. 11. Selective effect of fishing on a population when global fishing pressure (p) affects n loci identically, such that $p = \{(1 - s)\text{fAA} + (1 - s/2)\text{ fAB} + \text{fBB}\}^n$; where fAA, fAB and fBB are the initial frequencies of phenotypes AA, AB and BB (each locus having the same alleles, A and B) whose selective values are $1 - s$, $1 - s/2$ and 1 respectively. The figure shows the decrease in frequency of A for $p = 50\%$ in three cases: A = 1 single locus; B = 5 loci and C = 10 loci (the initial frequency of A is 0.50 and the population is assumed to be in Hardy–Weinberg equilibrium)

In addition, if one considers the probable existence of gene flow between populations, counteracting the effects of genetic drift and selection, the effects of fishing on the genetic diversity of a population can only be limited and reversible. It is very likely the same for characteristics of the stock.*

(b) Effects of restocking

The argument developed to condemn stocking operations using domesticated populations is often applied to the idea that observed geographical differentiation has an adaptive significance (in the sense that the local population is the best adapted to its environment). It must firstly be emphasized that neither theory, nor experimental observation, can sustain *a priori* such an affirmation. The low effectiveness of restocking, observed in some cases, is not necessarily due to a lower selective value of domesticated stocks in the

* The distinction must be made between variation in gene frequency at a locus, controlling a characteristic, and the resultant phenotypic variations.

natural environment, but may simply result from the restocking techniques themselves. At present, the only arguments which can correctly be invoked in favour of the cessation or reduction of restocking from domestic stocks are their ineffectiveness (although they are not always ineffective) and the precautionary principle.

Alternative solutions have been dreamt up, but rarely acted upon: these include the use of local stocks or the use of 'half-blood'. There is no indication that these two strategies would give better results, since it is not necessarily the genetic value of the domestic stocks which is the problem.

The use of wild stocks confronts two serious practical problems. The collection of male and female breeding fish from the wild does not ensure a significant restocking potential, as the observed fecundity of wild females is generally low and only a limited number of parents can be collected. Comparisons of performance (survival, Fig. 12; growth) in fish culture have clearly shown that wild fish are much more difficult to rear than domestic stocks and these difficulties are due, partly, to genetic factors (Maisse *et al.*, 1983; Guyomard and Chevassus, 1985). The make-up of farmed stocks, destined for restocking purposes, from natural populations can therefore be slow and ineffective; it can, in addition, turn out to be unsatisfactory regarding the genetic objectives, as the populations reared in such a way can rapidly evolve, by genetic drift or selection, towards an undesirable genetic type, if mortality is increased and fecundity varies widely between individuals. Taking into account the relevance of the constitution of the 'natural captive stocks' to restocking efforts, rearing techniques are being tried out to improve the performance of these stocks in fish farming. Thus, we have shown the positive effects of rough substrate on survival and early growth of a wild stock put into culture (Krieg *et al.*, 1989). The introduction of 'wild genes' via the male, at each generation is a possibility, in order to limit the undesirable genetic evolution of these new stocks transferred to aquaculture.

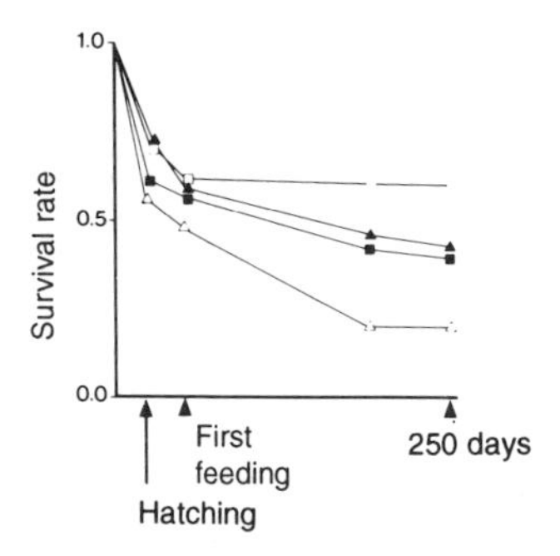

Fig. 12. Survival rate of groups of progeny from a diallelic crosss between a domestic and a wild stock (spawning fish taken from the Elorn, Finistère). Each group of progeny is the result of a cross between 10 females and 10 males. △: wild females × wild males; ▲: wild females × domestic males; ■: domestic females × wild males; □: domestic females × domestic males.

These difficulties in the production of wild individuals have led to the recommendation that offspring from fertilization of domestic females should be used, as they have higher fecundity, along with wild males bringing a certain 'rusticity'. This stratcgy of 'half-blood' does not resolve any of the problems posed by restocking, in either the short or the long term, as long as there are adaptive differences, in the natural environment, between domestic and indigenous stocks, in favour of the latter. As a matter of fact, in this case, inbreeding by the 'half-blood' would also reduce the adaptive value of the local population and this not necessarily slower than in the case of introduction of pure domestic stocks. The strategy of 'half-blood' is valid when there is no difference between domestic and natural stocks. However, in this case it only presents drawbacks with respect to the use of pure domestic stocks.

There are therefore only two consistent strategies regarding restocking: the use of domesticated or of local stocks. For the time being, only the first strategy is practicable, since wild stocks are no longer available for restocking. A variant strategy could be the use of domesticated stocks sterilized by triploidization. Neither of these strategies can be claimed to be preferable in relation to environment and mode of exploitation encountered. It is only through analysis of the quantitative and qualitative changes in natural stocks that the effectiveness and consequences of stocking can be assessed. It must, however, be noted that the use of wild stocks for restocking would provide an appropriate key to the problems of maintaining effective size in natural populations and the conservation of their genetic diversity.

V. CONCLUSION AND PROSPECTS

While the electrophoretic studies carried out are very likely to have demonstrated the most significant aspects of genetic differentiation of brown trout populations, two points merit further investigation; firstly, the diversity of the Atlantic populations and secondly, the morphological differentiation between the Mediterranean and Atlantic sub-species, differentiation probably indicative of the biological differences between these two sub-species.

Obviously, the most important questions and problems arise from the effects of stocking. The studies carried out do not yet allow a conclusion to be reached as to the genetic effects of restocking on the biological characteristics of natural stocks. This question could be partly answered by carrying out experiments simultaneously on fish farms and in the natural environment.

The first approach, carried out on fish farms, consists of finding out if there are genetic differences between natural and domestic stocks, with the use of crosses (see Fig. 12). This type of study has been applied to the Mediterranean population and was an opportunity to test performance in a culture situation.

The second possible route combines population dynamics with molecular genetics which allows the origin (wild, domestic, hybrid) of sampled individuals to be identified. It requires the analysis of a large number of individuals. This requirement can be achieved, now that molecular methods* are available which do not require the sacrifice of individuals examined.

* Methods based on the study of genomic DNA polymorphism.

This association between population dynamics and molecular genetics cannot pretend, *a priori*, to provide conclusions which can be extrapolated to all possible situations, taking into account their diversity and the requirements to restrict this kind of study to a few individuals within them. However, in certain cases it offers an interesting way of approaching, in certain special cases,* one of the questions which is central to the management of natural stocks: the genetic effect of restocking on the biology of natural populations.

BIBLIOGRAPHY

Barbat-Leterrier A., Guyomard R., Krieg F., 1989. Natural introgression between introduced domesticated strains and Mediterranean native populations of brown trout (*Salmo trutta* L.). *Aquat. Living Resour.*, **2**, 215–223.

Behnke R. J., 1968. A new subgenus and species of trout, *Salmo (Platysalmo) platycephalus*, from southcentral Turkey, with comments on the classification of the sub-family salmoninae. *Mitt. Hamburg Zool. Mus. Inst.*, **66**, 1–15.

Behnke R. J., 1972. The systematics of salmonid fishes of recently glaciated lakes. *J. Fish. Res. Board Can.*, **29**, 639–671.

Bernatchez L., Guyomard R., Bonhomme F., 1992. DNA sequence variation of the mitochondrial control region among geographically and morphologically remote European brown trout (*Salmo trutta*, L.) populations. *Mol. Ecol.*, **1**, 161–173.

Busack C. A., Gall G. A. E., 1981. Introgressive hybridization in populations of Paiute cutthroat trout (*Salmo clarki seleneris*). *Can. J. Fish. Aquat. Sci.*, **94**, 939–951.

Campton D. E., Utter F. M., 1985. Natural hybridization between steelhead trout (*Salmo gairdneri*) and coastal cutthroat trout (*Salmo clarki clarki*) in two Puget Sound Streams. *Can. J. Fish. Aquat. Sci.*, **42**, 110–119.

Crozier W. W., 1983. Population biology of Lough Neagh brown trout (*Salmo trutta* L.) Ph.D. Thesis, The Queen's University, Belfast, 478 pp.

Dorofeyeva Ye.A., Zinov'yev Ye.A., Klyukanov V. A., Reshetnikov Yu.S., Savvai-tova K. A., Shaposhnikova G. Kh., 1981. The present state of research into the phylogeny and classification of salmonoidei. *J. Ichtyol.*, **21**, 1–20.

Dorofeyeva Ye.A., Rukhkyan R. G., 1982. Divergence of the sevan trout, *Salmo ischchan*, in the light of karyological and morphological data. *J. Ichtyol.*, **22**, 23–36.

Falconer D. S., 1974. *Introduction à la Génétique Quantitative*. Masson, Paris, 284 pp.

Ferguson A., 1985. Lough Melvin, a unique fish community. *Occasional papers in Irish science and technology*, 1, Went memorial lecture, Royal Dublin Society, 17 pp.

Ferguson A., Fleming C. C., 1983. Evolutionary and taxonomic significance of protein variation in brown trout (*Salmo trutta* L.) and other salmonids. In G. S. Oxford and D. Rollison (Eds), *Protein Polymorphism: Adaptive and Taxonomic Significance*, 86–99, Academic Press, London.

Fleming C. C., 1983. Population biology of anadromous brown trout (*Salmo trutta* L.) in Ireland and Britain. Ph.D. Thesis. The Queen's University of Belfast, 475 pp.

* Previously restocked sectors or populations showing no evidence of introgression may be used for restocking trials.

Giuffra E., Bernatchez L., Guyomard R., 1994. Mitochondrial control region and protein coding genes sequence variation among phenotypic forms of brown trout, *Salmo trutta* L., from Northern Italy. *Molecular Ecology*, **3**, 161–171.

Gorman G. C., Renzi J., 1979. Genetic distance and heterozygosity estimates in electrophoretic studies effects of sample size. *Copeia*, 242–249.

Guyomard R., 1987. Differenciation génétique des populations de saumon atlantique: revue et interprétation des données électrophorétiques et quantitives. In M. Thibault and R. Billard (Eds), *Restauration des Rivières à Saumons*, INRA, Paris, 297–308.

Guyomard R., Krieg F., 1983. Electrophoretic variations in six populations of brown trout (*Salmo trutta* L.). *Can. J. Genet. Cytol.*, **25**, 403–413.

Guyomard R., Grevisse G., Oury F. X., Dvaine P., 1984. Evolution de la variabilité génétique inter et intrapopulations de Salmonidés issues de mêmes pools géniques. *Can. J. Fish. Aquat. Sci.*, **41**, 1024–1029.

Guyomard R., Chevassus B., 1985. Recherches sur la génétique des populations de truite commune et de saumon atlantique. Compte-rendu de contrat INRA-CSP, 15 pp.

Gyllensten U., Leary R. F., Allendorf F. W., Wilson A. C., 1985. Introgression between two cutthroat trout subspecies with substantial karyotypic, nuclear and mitochondrial genomic divergence. *Genetic*, **111**, 905–915.

Halliburton R., Pikpin R. E., Gall G. A. E., 1983. Reproduction success of artificially hybridized golden trout (*Salmo aguabonita*) and rainbow trout (*Salmo gairdneri*). *Can. J. Fish. Aquat. Sci.*, **40**, 1264–1269.

Krieg F., 1984. Recherche d'une différenciation génétique entre populations de *Salmo trutta*. Thèse de 3e cycle, Université de Paris-sud, Orsay, 92 pp.

Krieg F., Guyomard R., 1983. Mise en évidence électrophorétique d'une forte différenciation génétique entre population de truite fario en corse. *C.R. Acad. Sci.*, Paris, **296**, 1084–1089.

Krieg F., Guyomard R., 1985. Population genetics of French brown trout (*Salmo trutta* L.).: large geographical differentiation of wild populations and high similarity of domesticated stocks. *Genet. Sel. Evol.*, **17**, 225–242.

Krieg F., Guyomard R., Maisse G., Chevassus B., 1989. Influence du génotype et du substrat sur la croissance et la survie au cours de la résorption vitelline chez l truite commune (*Salmo trutta* L.). *Bull. Fr. Pêche Piscic.*, **311**, 126–133.

Leary R. F., Allendorf F. W., Phelps S. R., Knudsen K., 1987. Genetic divergence and identification of seven cutthroat trout subspecies and rainbow trout. *Trans. Am. Fish. Soc.*, **116**, 580–587.

Li W.-H., 1976. Effect of migration on genetic distance. *Am. Nat.*, **110**, 841–847.

Loudenslager E. J., Gall G. A. E., 1980. Geographic patterns of protein variations and subspeciation in cutthroat (*Salmo clarki*). *Syst. Zool.*, **28**, 27–42.

Maisse G., Porcher J. P., Nihouarn A., Chevassus B., 1983. Comparaison des performances en pisciculture d'un hybride intraspécifique (mâle sauvage × femelle domestique) et de la souche domestique chez la truite commune (*Salmo trutta* L.). Essais préliminaires d'implantation en ruisseau. *Bull. Fr. Piscic.*, **2911**, 167–181.

May B., 1980. The salmonid genome: evolutionary restructurating following a tetraploïd event. Ph.D. thesis, Pennsylvania State University, 199 pp.

Nei M., 1975. *Molecular Population Genetics and Evolution.* North Holland, Amsterdam, 228 pp.

Nei M., 1978. Estimation of average heterozygosity and genetic distance from a small number of individuals. *Genetics*, **89**, 583–590.

Nei M., 1987. Genetic distance and molecular phylogeny. In N. Ryman and F. Utter (Eds), *Population Genetics and Fishery Management*, 193–223, University of Washington Press, Seattle and London.

Nelson R., Soule M., 1987. Genetic conservation of exploited fishes. In N. Ryman and F. Utter (Eds), *Population Genetics and Fishery Management*, 345–368, University of Washington Press, Seattle and London.

Okazaki T., 1982. Genetic study on population structure in chum salmon (*Oncorhynchus keta*). *Bull. Far Seas Fish. Res. Lab.*, **19**, 25–116.

Olivari G., Brun G., 1988. Le nombre de cœca pyloriques dans les populations naturelles de trite commune *Salmo trutta* Linné en Corse. *Bull. Ecol.*, **19**, 197–200.

Osinov A. G., 1984. Zoographical origins of brown trout, *Salmo trutta (Salmonidae)*: data from biochemical genetic markers. *J. Ichthyol.*, **24**, 10–23.

Pasteur N., Pasteur G., Bonhomme F., Catalan J., Britton-Davidian J., 1987. Manuel technique de génétique par électrophorèse des protéines. Technique et Documentation, Lavoisier, Paris, 217 pp.

Presa P., 1995. Déterminisme et polymorphisme génétiques des séquences microsatellites de *Salmo trutta* et d'autres salmonidés. Comparaison avec le polymorphisme des locus enzymatiques. Thèse d'université. Orsay.

Presa-Martinez P., Krieg F., Estoup A., Guyomard R., 1994. Diversité et gestion génétique de la truite commune: apport de l'étude du polymorphisme des locus protéiques et microsatellites. *Génét. Sel. Evol.*, **26**, suppl. 1, 183s–202s (Elsevier/INRA).

Richard A., 1981. Observations préliminaires sur les populations de truite de mer (*Salmo trutta*) en Basse-Normandie. *Bull. Fr. Piscic.*, **283**, 114–124.

Ryman N., Allendorf F. W., Stahl G., 1979. Reproductive isolation with little genetic divergence in sympatric population of brown trout (*Salmo trutta*). *Genetics*, **92**, 247–262.

Ryman N., 1983. Patterns of distribution of biochemical genetic variation in salmonids: differences between species. *Aquaculture*, **33**, 1–21.

Sneath P. H. A., Sokal R. R., 1973. Numerical taxonomy, W. H. Freeman, San Francisco, 573 pp.

Stoneking M., Wagner D. J., Hildebrand A. C., 1981. Genetic evidence suggesting subspecific differences between northern and southern populations of brook trout (*Salvelinus fontinalis*). *Copeia*, 810–819.

Part III

The management of natural populations of brown trout

The management of natural populations of brown trout in France, analysed from an historical perspective (1669–1986)

M. Thibault

I. INTRODUCTION

Management of brown and sea trout can be defined from the perspective of one (or several) objective(s) and the means and methods used to attain this (or these) objective(s). It is therefore important, according to this definition, to verify, at regular intervals, if the objective has been reached, to examine the consequences by comparing the results obtained and those predicted.

It is not possible, however, to plan for the future without knowledge from the past. Retrospective studies, by critical analysis of means and methods used in the past, can provide the tools for such verification. Such a study must comprise both as extensive as possible a presentation of the various documents used and a review of information integrating past mechanisms of development with cultural, social and economic conditions. This allows the main trends to be identified and the effects of different management practices determined.

For river, maritime and coastal fisheries for freshwater and migratory fish, the term 'management' is recent. It was introduced by the law of 29 June 1984. However, access to the natural resources of brown and sea trout has been regulated since ancient times. The history of this regulation is dominated by two key events; the mid-seventeenth and the mid-nineteenth century:

- The ordinance of August 1669 on Waters and Forests was the first law on a national scale (title document XXXI, in fisheries). It has therefore been chosen as the starting point of this chapter. 'This law constitutes a preferred guide, a reference law for the decades to come' (Cocula-Vaillieres, 1979).

> Even though it reflects the bias of the era, towards forestry, it brings, as well as potential policy accompanied by a sense of justice, a definition of the aquatic domain which increases the standing of the jurisdiction of Waters and Forests, and a set of laws for fishery practice. This harmonization of jurisdiction, bending the idea of privilege, shows clearly the royal will to bring into line those who, until then, played tacitly or legally with prejudiced impunity upon the richness of nature. (Cocula-Vaillieres, 1979).

Three main elements were put in place concerning access to the resource. These have been constantly altered over a long period (Fig. 1):

- open season for fishing: no fishing allowed during spawning or, outside this period, on certain days (Sundays and public holidays) and at certain times of day (at night);
- types of equipment which are authorized and prohibited;
- minimum size of capture, below which fish must be returned to the water.

These elements reflect the predominant attitude in relation to the fish, constituting the fear 'of sacrificing, through short-lived gain, the resources and expectations of the future' (De Bouthillier, 1828). In the Dordogne, 'the managers of the waters and forests, at the start of the eighteenth century, constantly predicted the exhaustion or depopulation of the waters' (Cocula-Vaillieres, 1979). According to this author, 'if one judges them by the confiscation of fish and by slaughters, causing the poisoning of the waters, the predominant impression is, in contrast, that of a continuously renewing abundance'.

The mid-nineteenth century brought two significant modifications:

- on the one hand, after the law of 15 April 1829 (largely inspired by the forestry law of 1827, since three-quarters of its 84 sections were copied from there (Buttoud, 1983)):
  - — the state presided from then on over judgements on fixed fisheries and over mobile fisheries split into stations along the length of the river,
  - — the main objective displayed was that of conservation of the resource (Dralet, 1821; De Bouthillier, 1828);
- on the other hand, the law of 31 May 1865, the consequence of the announcement of the discovery of artificial fertilization of trout in 1849 (Thibault, 1989). In addition to the restocking of eggs, practised since 1852, three new measures appeared (Fig. 1):
  - — creation of fishery reserves,
  - — prohibition of the sale, hawking or transport of fish during the spawning period,
  - — installation of fish ladders.

Two other modifications were added during the course of the century: the idea of harmful species and the classification of water courses into two categories (decrees of 21 March 1913 and 29 August 1939, respectively). They were implicit in the terminology used previously.

This chapter proposes to describe and quantify, in as thorough a way as possible, the measures which codify the access to brown and sea trout in the natural environment, both in the fluvial zone (lakes and water courses) and in coastal and maritime zones (mainly estuaries). It also attempts, in relation to these measures:

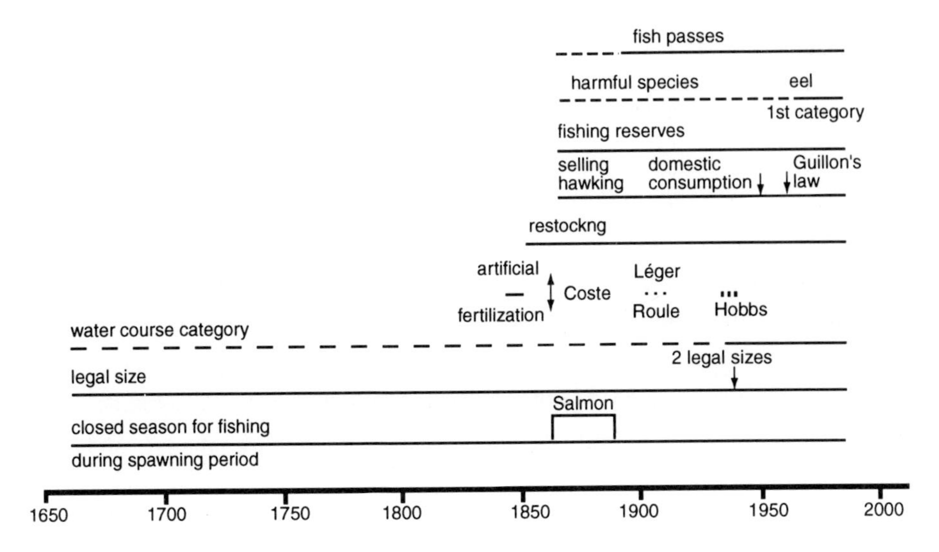

Fig. 1. Historical development of the main measures used in the management of natural populations of brown trout in France since the ordinance of 1669.

- firstly, to place them in relation to human activities carried out on watercourse, in relation to the reproduction of the three forms of brown trout (sea, lake and river trout);
- secondly, to determine their impact on the natural populations of this species.

II. MATERIALS AND METHODS

1. Documents consulted

The documentary basis of this chapter is made up of laws, ordinances, decrees and orders which have been collected throughout the study period. To these have been added accounts of background and discussions of certain laws (1829, 1865, 1961), reports carried out for ministeries, some circulars, articles and other works.

These documents were obtained mainly from Rennes, departmental archives (law bulletins, official maritime bulletin, universal journals and official journals) and law libraries (law bulletin and collections of annotated laws). The remainder came from Ministry of Agriculture libraries in Paris, from the Conseil Supérieur de la Pêche at Paraclet, from the ENGREF library at Nancy, the National Library and departmental archives in Tarn and Garonne (papers concerning five fisheries). Around 230 legal texts were used.

There are many studies of the regulation of fisheries. However, they are all exclusively descriptive and limited and do not question either the development of regulations or the impact of regulations in relation to fisheries protection.

Two of the regulatory measures have required some calculations to be done before collation of the data to allow comparisons to be made over the period studied; the legal size and the duration of the closed season during the spawning period.

2. Legal size

Over the entire study period, the minimum fish size, below which capture is prohibited, has been measured in three ways:

- from the eye to the tail or the base of the tail, from the ordinance of 1669 to the decree of 29 August 1939 in the river zone, and from the law decree of 9 January 1852 to the decree of 15 December 1952 in the maritime zone;
- from the tip of the snout to the middle of the tail fork after the decree of 29 August 1939 to the decree of 17 March 1952 in the river zone;
- from the tip of the snout to the end of the tail fin from 1952 in river and maritime zones.

Two particularities are observed for water courses on borders:

- legal size was measured from the eye to the base of the tail after the law of 11 June 1859 for the Bidassoa. It has been measured from the tip of the snout to the end of the tail fin since the decree of 4 March 1965;
- legal size has been measured from the tip of the head to the tip of the tail in Lake Léman and in the Doubs since the decree of 4 February 1905.

To facilitate comparisons over the study period, all sizes have been converted to the last measurement, from the tip of the snout to the tip of the tail fin. To do this, the three types of measurement were collected from a fisheries inventory carried out on the Elorn river on 31 August 1988, on 96 trout between 6.1 and 31.8 cm in total length.

From this sample, two relationships were calculated:

- between fish length measured from eye to tail base and length measured from snout tip to end of tail fin, or total length, the formula is:

$$\text{length from eye to tail base} = 0.80578 \times \text{total length} - 2.14883$$

- between total fish length and length measured from snout tip to tail fork, the formula is:

$$\text{total length} = 1.05655 \times \text{fork length} + 1.0892.$$

These two relationships and the graphs describing them allow the total length of brown and sea trout to be easily determined.

3. Duration of closed season for fishing during spawning time

This time period has varied since the decree of 1669. It has been established either on a national level or by region.

Since 1958, seven decrees[(1)] have divided the French regions into four to six sub-regions in relation to the duration of the closed season (Table 1 and Fig. 4). Since the same date, the opening of the season has always been on a Saturday. The closing date

[(1)] Decree No. 58814 of 16 September 1958 set up a group of five departments (Eure, Seine Maritime, Somme, Pas-de-Calais and Nord) in which fishing was prohibited from the 1st Tuesday in October to the last Friday in March. The decree of 9 January 1960 added two departments, the Oise and the Seine-et-Oise. This change is included in the calculations from 1958 (Table 1).

of the season has varied three times; Monday evening from 1958 to 1968, a fixed date in 1979 and 1982 and Sunday evening from 1985. From 1958, the mean annual durations of the closed fishing seasons on a national scale were calculated as follows:

- For each sub-group of regions this duration was calculated in relation to the open season dates, for example, the first Saturday in March or the Saturday following the 15th March, with closure, for example on the second Tuesday in October (Table 1A). For these calculations, it was assumed that 1st January could fall on any day of the week, from Monday to Sunday. However, leap years were not taken into account, so February was assumed to have 28 days throughout the estimation. While certain closed seasons are not affected by the day of the week 1st January falls on, others vary by up to 7 days (Table 1B);
- The total duration of the closed season was obtained by adding the sum of the different sub-regions. By dividing this total by 90 departments, the mean annual regional duration, on a national scale, of the closed season, was obtained (Table 1B).

In 1964, the two departments of Seine and Seine-et-Oise were divided into seven new departments: Paris, Seine-St. Denis, Hauts-de-Seine, Val de Marne, Val d'Oise, Yvelines and Essonne. To facilitate comparisons of closed season durations over the whole period, the calculations only take account, since 1964, of the two original departments.

III. MANAGEMENT METHODS USED SINCE THE SEVENTEENTH CENTURY

1. Closed seasons

(a) Prohibition during the spawning period

In the river zone

Since the decree of Louis XIV on Waters and Forests presented at St Germain-en-Laye in August 1669, three main stages can be identified:

- 1669–1831. The prohibition of fishing determined in a uniform manner for the whole kingdom, from 1st February to mid-March[(2)], i.e. 43 days on average, remained in operation for more than 150 years (Fig. 2). Two exceptions occurred in the eighteenth century:
  - the decree of Lorraine in 1707, in which fishing for trout in streams feeding into the Voges region, was prohibited from 1st November to 1st February (Dralet, 1821);
  - the King's declaration of 24 August 1773 which fixed, from 15th December to 1st February, the time during which fishing was allowed in certain rivers flowing into the Channel and in which trout were abundant. This concerned the fisheries established on the small rivers of the Eaune, Béthune or Neufchatel, Arque, Scie and Saâne (Baudrillart, 1821).

[(2)] Section XXXI on fishing, article VI. Fishermen may not fish during spawning times, in rivers where the trout is more abundant than all other fish, from 1st February to mid-March; in other rivers, from 1st April to 1st June.

Table 1A. The development of closed seasons and open seasons for fishing in waters of the first category in France since 1958

Duration of closed season during spawning time of the brown trout				Open season	
Decree no. 58-874 of 16 September 1958	Decree no. 63-98 of 11 February 1963	Decree no. 68-33 of 10 January 1968	Decree no. 79-993 of 23 November 1979	Decree no. 82-911 of 15 October 1982	Decree no. 85-1385 of 23 December 1985
(a) from the 2nd Tuesday October to 3rd Friday of February	(a) from the 2nd Tuesday October to 3rd Friday of February				
(b) from the last Tuesday of September to 3rd Friday of February	(b) from the last Tuesday of September to 3rd Friday of February	(a) from 2nd Tuesday of October to 1st Friday of March	(a) from 1st Saturday of March to 9th October	(a) from 1st Saturday of March to 4 October	(a) from 1st Saturday of March to 1st Sunday October
(c) from the 2nd Tuesday of September to 3rd Friday of February	(c) from the 2nd Tuesday of September to 3rd Friday of February	(b) from 2nd Tuesday of Septem-ber to 3rd Friday of February	(b) from 1st Saturday of March to 29 September		
	(d) from the last Tuesday of September to 1st Friday of March	(c) from the last Tuesday of Septem-ber to 1st Friday of March	(c) from Saturday following the 15 March to 29 September	(b) from Saturday following 15 March to 4 October	(b) from 1st Saturday of March to 3rd Sunday of September
	(e) from the 1st Tuesday of October to 3rd Friday of March		(d) from 1st Saturday of March to 14 September	(c) from 1st Saturday of March to 14 September	(c) from 3rd Saturday of March to 1st Sunday October
(d) from 1st Tuesday of October to the last Friday of March	(f) from 1st Tuesday of October to the last Friday of March	(d) from 1st Tuesday of October to the last Friday of March	(e) from Saturday following 15 March to 14 September	(d) from Saturday following 15 March to 14 September	(d) from 3rd Saturday of March to 3rd Sunday of September

Table 1B. Development of the durations of closed periods for fishing in waters of the first category, during the spawning period of the brown trout. Sub-groups of between 4 and 7 departments are arranged according to the mean annual duration of the closed season

1958	1963	1968	1979	1982	1985
(a) 131 days 4 departments 524	(a) 131 days 4 departments 524				
(b) 145 days 74 departments 10730	(b) 145 days 32 departments 4 640	(a) 145 days 4 departments 580	(a) 142–148 days 4 departments 568–592	(a) 147–153 days 9 departments 1323–1377	(a) 146–153 days 9 departments 1314–1377
(c) 159 days 5 departments 795	(c) 159 days 954 6 departments	(b) 159 days 6 departments 954	(b) 152–158 days 5 departments 760–790		
	(d) 159 days 30 departments 4 770	(c) 159 days 59 departments 9 381	(c) 166–172 days 22 departments 3652–3784	(b) 161–167 days 26 departments 4186–4342	(b) 160–167 days 51 departments 8160–8517
	(e) 166 days 6 departments 996		(d) 167–173 days 50 departments 8350–8650	(c) 167–173 days 50 departments 8350–8650	(c) 160–167 days 25 departments 4000–4175
(d) 173–180 days 5+2 departments 1211–1260	(f) 173–180 days 12 departments 2076–2160	(d) 173–180 days 21 departments 3633–3780	(e) 181–187 days 9 departments 1629–1683	(d) 181–187 days 5 departments 905–935	(d) 174–181 days 5 departments 870–905
13260–13309	13960–14044	14548–15499	14959–15499	14764–15304	14344–14974
147–148 days 147.5	155–156 days 155.5	162–163 day 162.5	166–172 days 169	164–170 days 167	159–166 days 162.5

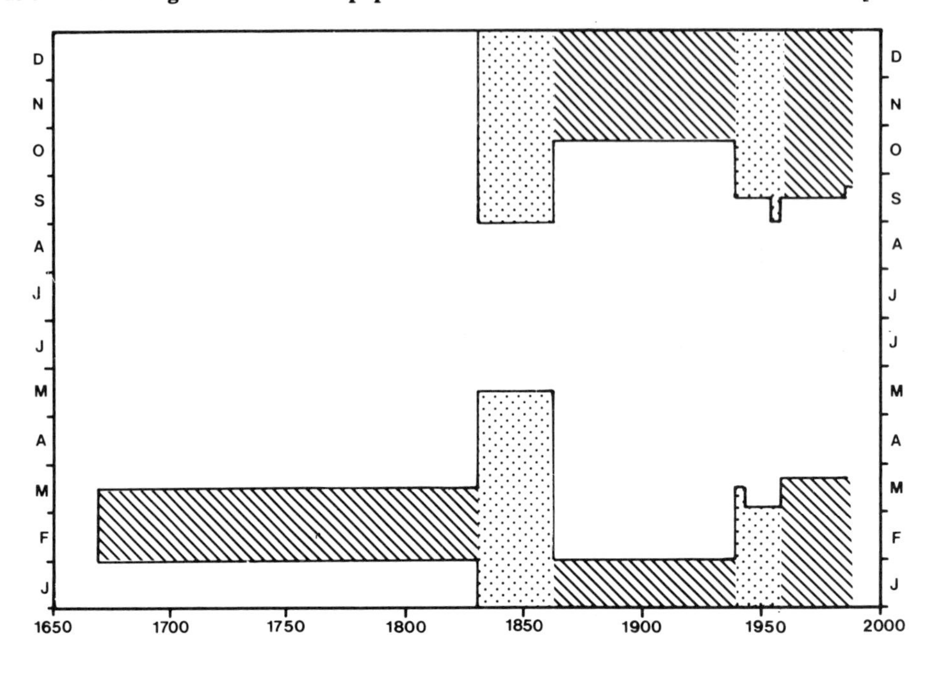

Fig. 2. The development of closed seasons for fishing in the riverine zone during the spawning season of brown trout in France since the ordinance of 1669. Hatched area, true prohibition: limits of prohibition since 1958 (cf. Table 1A). Dotted area, potential limits of prohibition: from 1831 to 1863, true prohibition of 22 to 153 days, depending on department (cf. Fig. 3); from 1939 to 1958, true prohibition of 100 days within this period (cf. 3111).

- 1831–1863. After the law of 15 April 1829, prefectoral decrees[3] put in place by the decree of 15 November 1830, ratified by royal decrees between 1831 and 1834 resulted in considerable diversity between the regions (Herbin de Halle, 1835).
  - 58 departments, in each of which the closed season lasts from 3 weeks to 3 months (22 to 153 days, i.e. 67.4 days on average), spread from 1st September to 15 May[4] for all regions put together (Fig. 3). Only two of these, Allier and Marne, retain the closed season of the decree of 1669. There are:
    - * 46 departments where the trout is designated, one of which has a date error (Ariège) and one where the law is neither fixed nor ratified (Morbihan). While the great majority of regions only have one closed season per year, four have two: Hautes Alpes, Pyrenees orientales, Haute Garonne and Tarn;
    - * in contrast, 12 departments where the trout is not designated, but in which the closed season covers at least part of the spawning season of this species. There are three closed dates, depending on the river, in the Jura;

[3] Except in Eure-et-Loire where the law was neither fixed nor ratified.

[4] The latest date is mid-April except for the Deux Sèvres (Fig. 3). The intention in this department appears to have been to include all species in one decision; 'closed season at spawning time extends from 15 March to 15 May but starts from 15 February in water courses where trout is the dominant fish species'.

  * 25 departments[5] in which trout is not considered and the closed season is for cyprinids;

- 1863–1986. There is simultaneously greater uniformity and lengthening of the durations of closed seasons in two distinct periods:
  - the closed season became uniform over all territories from 20 October exclusive to 31 January inclusive, i.e. 103 days, after the decree of 19 October 1863. This measure affected the brown and sea trout and the Atlantic salmon.[6] It was to remain in use for 75 years for the trout[7] until the decree of 29 August 1939;
  - during the following half-century:
    * from 1939 to 1958, the rulings are comparable to the preceding period, with some additional modifications. The closed season lasts 100 days, within a period of 5.5 months between 1st October and 15th March (decrees of 29 August 1939 and 17 March 1952). The latter was extended to 6 months from 15 September to 15 March, by the decree of 23 January 1954. However, this flexibility was withdrawn over the whole country and changed to the first Saturday in March (rectification of JO, 3 October 1943, to the decree of 14 August 1943).
- The decree of 16 September 1958 brought three modifications:
  - a significant lengthening of the duration of the closed season to 147–148 days on average (131–180 days depending on region and year: Table 1, Fig. 4), i.e. an increase of more than 1.5 months;
  - a desire for flexibility indicated by the introduction of four different periods of closed season: one for the great majority of departments and three others for three small regional sub-assemblies (west, north-north-west and south-east). The four closure dates (second Tuesday in September to second Tuesday in October) and the two dates of open season (third Friday in February to last Friday in March) extend over a month;
  - the days of opening and closing are fixed to Saturday morning and Monday evening respectively, taking into account the leisure aspect of the activity.

This same organization is still in place:

- the policy of increasing the duration of the closed season for fishing at spawning time Fig. 5) was pursued until 1979 until it reached 169 days, i.e. 3 weeks longer than in 1958. This duration decreased slightly thereafter; it has been 162.5 days since 1985, as in 1968;
- the regrouping into sub-groups. Since 1968, more than half of the French regions have been regrouped into a vast west-south-west subset, centred on various contours to the north and west, depending on the decree (Fig. 4);

[5] Aisne, Ardennes, Bouches-du-Rhône, Cher, Dordogne, Finistère, Gard, Gironde, Indre, Indre-et-Loire, Landes, Atlantic Loire, Loiret, Lot, Lot-et-Garonne, Maine-et-Loire, Meurthe, Nièvre, Nord, Oise, Seine, Seine-et-Marne, Seine-et-Oise, Vendée, Vienne.

[6] The two species remained linked until the decree of 27 December 1889 (Fig. 1).

[7] Decrees of 25 January 1869, 10 August 1875, 18 May 1878, 27 December 1889, 5 September 1897. However, this closed season for trout fishing was extended to 31 March for the maritime section of the river Liane (Pas-de-Calais) by the degree of 2 June 1886. This decree became in accord with the prefectoral order of 2 September 1885 for the inland part of this river.

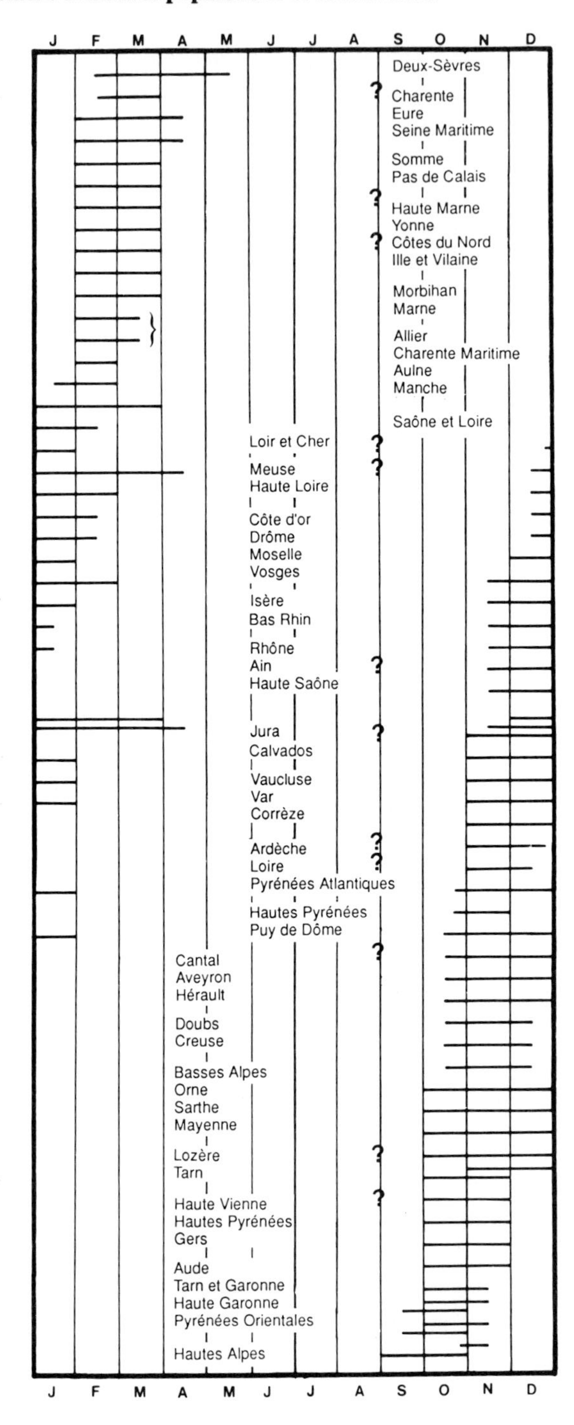

Fig. 3. Closed seasons for fishing during the spawning season of the brown trout in France. ? = the trout is not a designated species in these departments.

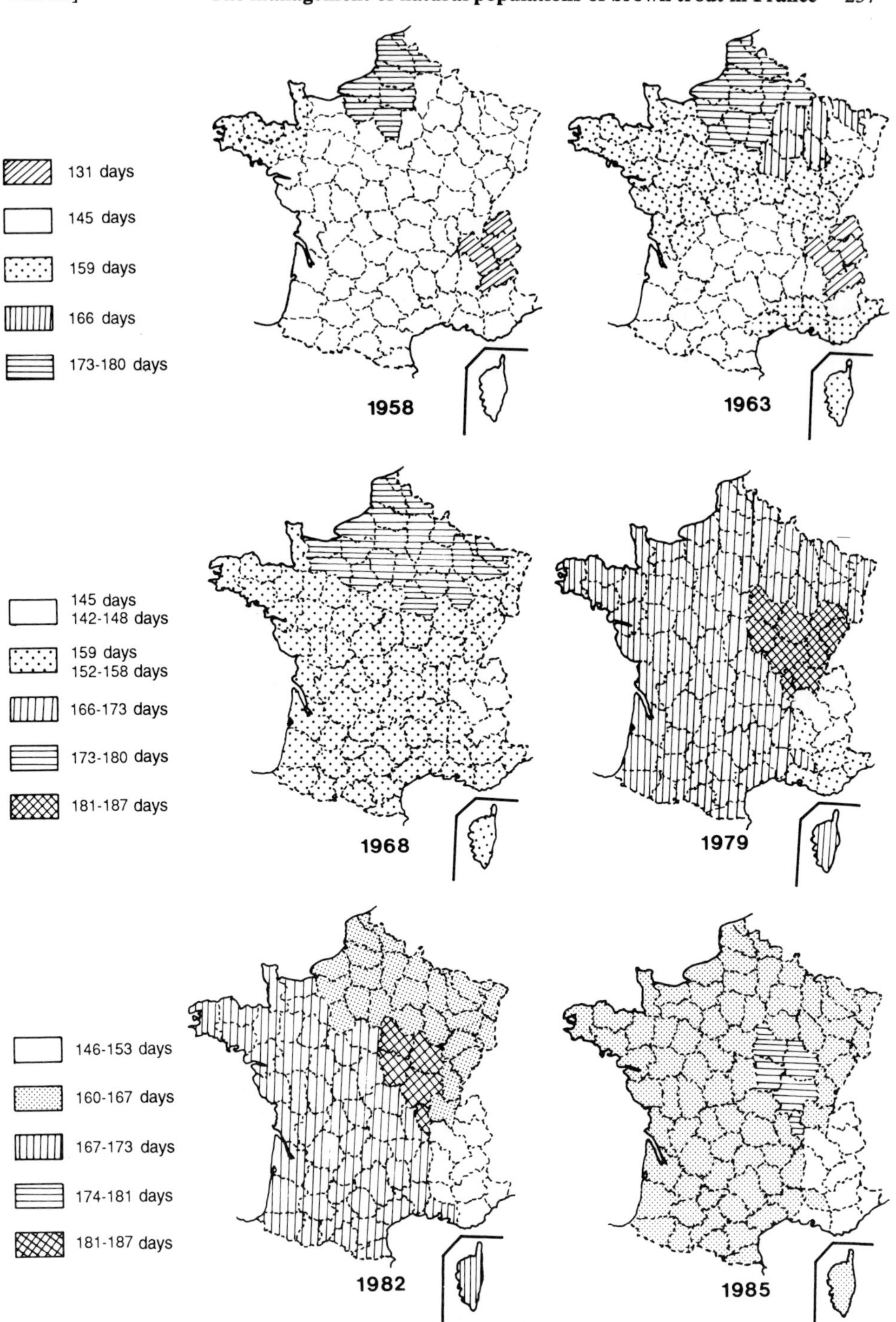

Fig. III.4. Development, by department, of different closed seasons for brown trout fisheries in the riverine zone in France, defined by decrees between 1958 and 1985.

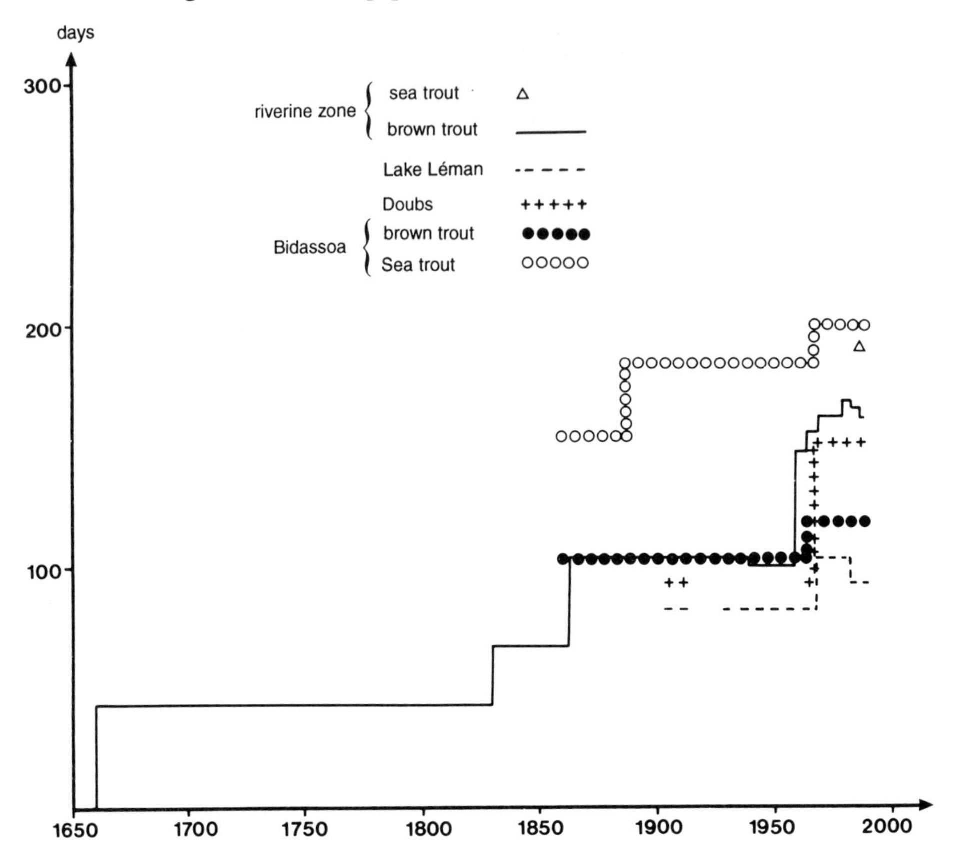

Fig. 5. Development of the mean duration of the closed season for a fishery during the spawning period of brown trout and sea trout in France from 1669 to 1986.

- opening and closing dates were established in relation to the ends of the week except in the decrees of 1979 and 1982, in which the closure date is a fixed day.

The sea trout appears in the regulations from the decree of 23 December 1985 onwards. The order of 21 February 1986 includes the closed season for sea trout in waters of the first category. The order of 4 July 1986 extended the fishing period for that year by about 3 months, on sections of watercourses in the Channel and Calvados regions. The order of 12 December 1986 introduced four closed periods for 1987; 139 days for Lot-et-Garonne, 181 days for Calvados, 193 days for Channel and 218 days for Landes and Atlantic Pyrenees.

The maritime region

The prohibition of fishing for trout during the spawning period, both in the sea along the coasts and in sections of estuaries, rivers, ponds and canals where the water is salty, was introduced by the decree of 24 October 1863. All subsequent decrees[8] confirm this closed season of 103 days from 20 October exclusive to 31 January inclusive. However, the decree of 26 July 1927 indicates that the maritime administrators can decide, in exceptional circumstances, after consultation with departmental officials, to increase the closed season, while retaining the same rules for all waters.

Border zone, Bidassoa, Lake Léman and Doubs

Bidassoa. Sea trout and brown trout have been considered separately since the law of 11 June 859. The duration of the closed season has increased regularly throughout the period. However, the closed season for brown trout is always shorter than that for sea trout:

- for brown trout, the closed season was 104 days from 1859 to 1865 (from 20 October to 31 January) and 119 days since the decree of 14 March 1965 (from 20 October to 15 February);
- for sea trout, the closed season increased in three stages: 153 days from 1859 to 1886 (end of August to 1 February); 184 days after the decree of 31 October 1886 to 1965 (end of July to 1 February) and 199 days since the decree of 14 March 1965 (from 31 July to 15 February).

Lakes Léman and Doubs, borders between France and Switzerland. The law of 31 January 1905 brought approval of the convention signed in Paris on 9 March 1904, between France and Switzerland, to regulate the fishing in the two border waters of the two countries. The decree of 4 February 1905 rendered this agreement binding. The Swiss federal council denounced this agreement on 31 December 1910. The agreement then ceased to be in force from 1 January 1912 until the decree of 15 March 1929, the dispositions of which were applicable in French waters of Lake Léman from the date of promulgation of this decree.

Lake Léman. The length of the closed season is 92 days (1 October to 31 December) after the decree of 4 February 1905. To this was added a closed season for fishing of all fish species, from 15 February to 5 March inclusive (paragraph a of section 8 of the decree). It was changed to 82 days (1 October to 20 December inclusive) in the decree of 15 March 1929. This duration was confirmed in the decrees of 25 January 1939 and 3 April 1958. It reached 103 days (21 October to 31 January) in the decree of 22 November 1968. It was finally reduced to 93 days (15 October to 15 January inclusive) by the decree of 1 September 1982, a measure retained in the decree of 17 November 1982.

Doubs. The length of the closed season is 93 days (20 October to 20 January), after the decree of 4 February 1905. It reached 151 days after the decree of 27 July 1966 (from 1 October to the last day in February).

[8] 20 November 1875, 1 February 1890, 26 July 1927, 23 November 1935, 15 December 1952.

(b) Closed seasons outside the spawning period

The measure of closing fisheries is taken at three different levels; specified days and specified times of day, and fixing of reserves annually or for 5 consecutive years (renewable).

Specific days and times

The regulations are general and do not specifically concern the trout, except in the decree of 23 December 1985. It is inscribed in the ordinance of 1669 that fishing is prohibited on Sundays and public holidays[9] and at night.[10]

The prohibition of fishing on Sundays and public holidays has been subject to various modifications since the start of the nineteenth century, to which have been added written dispensations in successive decrees, concerning certain fishing methods and certain species:

- it appears that there was a lapse in the prefectoral orders of 1831 to 1834, in the application of the law of 15 April 1829 and the ordinance of 15 November 1830. Herbin de Halle (1835) observed 'note that some regulations legally prohibit fishing on Sundays and public holidays, but this prohibition is not general and it would hamper fishing exploitation, to instigate such an inflexible rule'.
- this restriction reappeared in the decree of 25 January 1868 with a dispensation[11] which was removed by the decree of 5 September 1897.[12] Another dispensation reappeared in the decree of 29 August 1939,[13] then disappeared in that of 17 September 1945, which retained article 12 of the decree of 5 September 1897. Dispensations related to certain equipment were introduced in the decrees of 5 August 1946 and 18 July 1947.

The prohibition of fishing at night has been more stable although the general rules have been affected by dispensations relevant to certain species. The decree of 25 January 1869 only ever allows fishing between sunrise and sunset (article 6). That of 29 August 1939 increased this duration by an hour, as according to article 7, fishing cannot be practised more than half an hour before sunrise, nor more than half an hour after sunset. This regulation is still in use: enshrined in article 13 of the decree of 23 December 1985.

[9] Article 4: Fishing is prohibited on Sundays and public holidays ... to this effect, they are requested to bring, every Saturday and the evening preceding a holiday, straight after sunset, all their tackle and fishing equipment to the dwelling of the head of the community; this will only be returned the day after Sunday, or the holiday, after sunrise.

[10] Article 5: Fishing is equally forbidden on certain days and during certain seasons which are: at times other than during daylight, except at bridge arches, mills and duck ponds, where fishing can be carried out both at night and during the day, as long as the day is not Sunday or a public holiday or other prohibited day.

[11] Article 11: Fixed nets used for fishing will be lifted at the middle for 36 hours each week, from Saturday at 6 p.m. to Monday morning at 6 a.m., over a length equivalent to a tenth of their total length. This must be done in such a way as to leave a free space of at least 50 cm between the bottom and the lower boltrope. Article 12 of the decree of August 10 1875 is almost identical.

[12] Article 12: Fixed nets used for fishing must be taken out of the water and placed on land for 36 hours each week, from 6 p.m. on Saturday to 6 a.m. on Monday morning.

[13] Article 15: Nets and equipment of all kinds, either fixed or moveable ... must be taken out of the water and placed on land for 36 hours each week, 6 p.m. on Saturday to a.m. on Monday, if they take up more than a third of the width of the water course.

However, the latter decree introduced a dispensation in relation to sea trout, the fishing for which is allowed from half an hour before sunrise until 2 hours after sunset (article 14) in water courses classed as sea trout water courses.

Fixing of fishery reserves for five consecutive years

This measure is mentioned in purely historical terms for riverine zones. While the law of 31 May 1865 only concerned trout and salmon at first; it was widened, after discussion within the law commission, to include all fish species. Prohibition of fishing was announced for a period of five years in the navigable sections of estuaries, rivers and canals, by the decrees of 25 January and 20 September 1868, 30 January and 17 July 1869. Taking into account the localization of the reserves, the first decrees mainly concerned species other than salmonids. Such decrees were made regularly thereafter, according to circumstances in both navigable and non-navigable waters.

In the maritime zone, orders have created reserves for fish since the middle of the 20th century, for two Breton estuaries in the Morlaix, Penzé and Dossen areas[14] and for rivers in Upper Normandy.[15]

Fishing prohibitions are made without reference to fish species in all of the orders consulted, except that of 31 October 1961: 'considering the interest attached to the prevention of poaching of migratory fish and more especially the salmonids, during their movements in the estuaries of the rivers Yères, Arques, Scie, Saane (Dieppe district) and Durdent (Fécamp district)'.

Setting of annual closures

Annual prohibitions of fishing for trout or salmonids which may or may not include salmon, were made by orders during the mid-1960s for the river sections and from the mid-1970s for the section under maritime and coastal regulations.

In the river zone, these orders are applied to certain water courses, streams and lakes and to certain sections of the water course. The justification shows some variation in the way it is formulated:

- either the fishing of all species of trout is forbidden without any clarification (orders of 7 January 1965 and 7 January 1966);
- or the requirement for protecting the trout is at the forefront (order of 13 May 1966) with two main variations:
  - 'In view of the requirement for continued, assured protection of certain other salmonids: all species of trout'.[16]

[14] Orders of 29 December 1949, 6 February 1950, 15 December 1955, 11 February 1961, 27 July 1967, 18 January 1972. The decree of 11 January 1865 prohibited fishing the maritime part of the Dourduff, until a new order arose.

[15] Orders of 31 October 1961, 13 January 1967 and 18 May 1984.

[16] Orders of 17 January 1967, 27 December 1967, 30 December 1968, 8 January 1970, 5 January 1971, 4 February 1972, 14 February 1973, 23 January 1974, 5 February 1974, 24 February 1975, 29 January 1976, 13 January 1977, 22 February 1978, 8 January 1979, 16 January 1980, 15 January 1981, 11 February 1982, 31 January 1983, 14 February 1984.

— 'In view of the requirement for assured protection of trout species' in locations where restocking of juveniles is carried out, in spawning ground streams,[17] in lakes[18] and in sections of water courses.[19]

From 1979, reference to trout disappears from the orders. The formulation becomes more general: 'In view of the requirement to assure the directed protection of fish populations in certain water courses, lakes or ponds in certain regions'.

It is impossible to make a distinction between water courses and water bodies frequented by trout and those frequented by other species, even when certain water courses were cited in previous orders.

In the maritime zone, these prohibitions were localized exclusively in Brittany and Normandy. They included an increasing number of estuaries in Brittany, from 3 in 1976 to 12 in 1985.[20] The orders refer to salmonids in the Breton estuaries and Fécamp harbour and to trout in Tréport harbour (mouth of the Bresle).[21] For the estuaries and Fécamp harbour, fishing is prohibited without any supplementary notes, while for Tréport harbour it is noted: 'considering that it complements the regulation of the fishing in the maritime zone, protecting the salmonids of the Bresle'.

2. Legal size

The sizes cited in this chapter refer to the total length of the trout, measured from the tip of the snout to the end of the tail (caudal) fin (cf. Part II, Chapter 2). Three cases are examined:

- the river zone
- the estuary zone, including the Bidassoa
- border waters between France and Switzerland

[17] In view of the requirement to ensure the particular protection of streams developed by local angling clubs in nursery streams (Yonne), the order of 4 February 1972; ...special protection...orders of 14 February 1973, 23 January 1974, 29 January 1976, 23 June 1977, 22 February 1978.

[18] In view of the requirement to ensure the protection of trout in the Tech reservoir (Hautes Pyrenees) in which the federation...envisages progressing towards intensive rearing, order of 13 June 1972...proceeding towards intensive rearing, order of 14 February 1973. In the St Peyre (Tarn) reservoir where the federation proceeded to intensive rearing, order of 13 June 1975.

[19] In view of the requirement to ensure the protection of trout:

— in the Salles stream (Vienne) in which the federation proceeded with intensive rearing in 1973, intended to rebuild the fishery stock which had been destroyed by accidental pollution, order of 23 January 1974.

— in a section of the Maronne (Cantal) where the federation...proceeded with intensive rearing, order of 7 October 1974.

— in a section of the Senouire and some of its tributaries (Haute Loire), where the federation proceeded with intensive rearing, order of 9 June 1976, 23 June 1977, 22 February 1978.

[20] Orders of 13 April 1976, 27 April, 1977 and 12 January 1978 (Trieux, Jaudy, Scorff), of 21 January and 26 December 1980 (ibidem and Leff), 14 January 1981 (Gouët, Gouessant, Trieux, Jaudy, Leff, Aber Wrac'h, Scorff), 30 April 1982 (Elorn), 7 January 1983 (Gouët, Gouessant, Trieux, Jaudy, Leff, Aber Wrac'h, Elorn, Scorff), 19 January 1984 (ibidem and Léguer) 18 February 1985 (ibidem and Goyen, Laïta, Blavet), 31 October 1985 (ibidem).

[21] Tréport harbour at the mouth of the Bresle: orders of 15 May 1962, 25 April, 1979, 15 April 1980, 16 March 1981, 30 March 1982 and 15 July 1983. Fécamp harbour: orders of 3 August 1981, 27 April 1982, 17 January 1984.

(a) River zone

Four periods can be distinguished (Fig. 6):

- a period of around 160 years during which the legal size was fixed at 20.4 cm, after the ordinance of 1669;

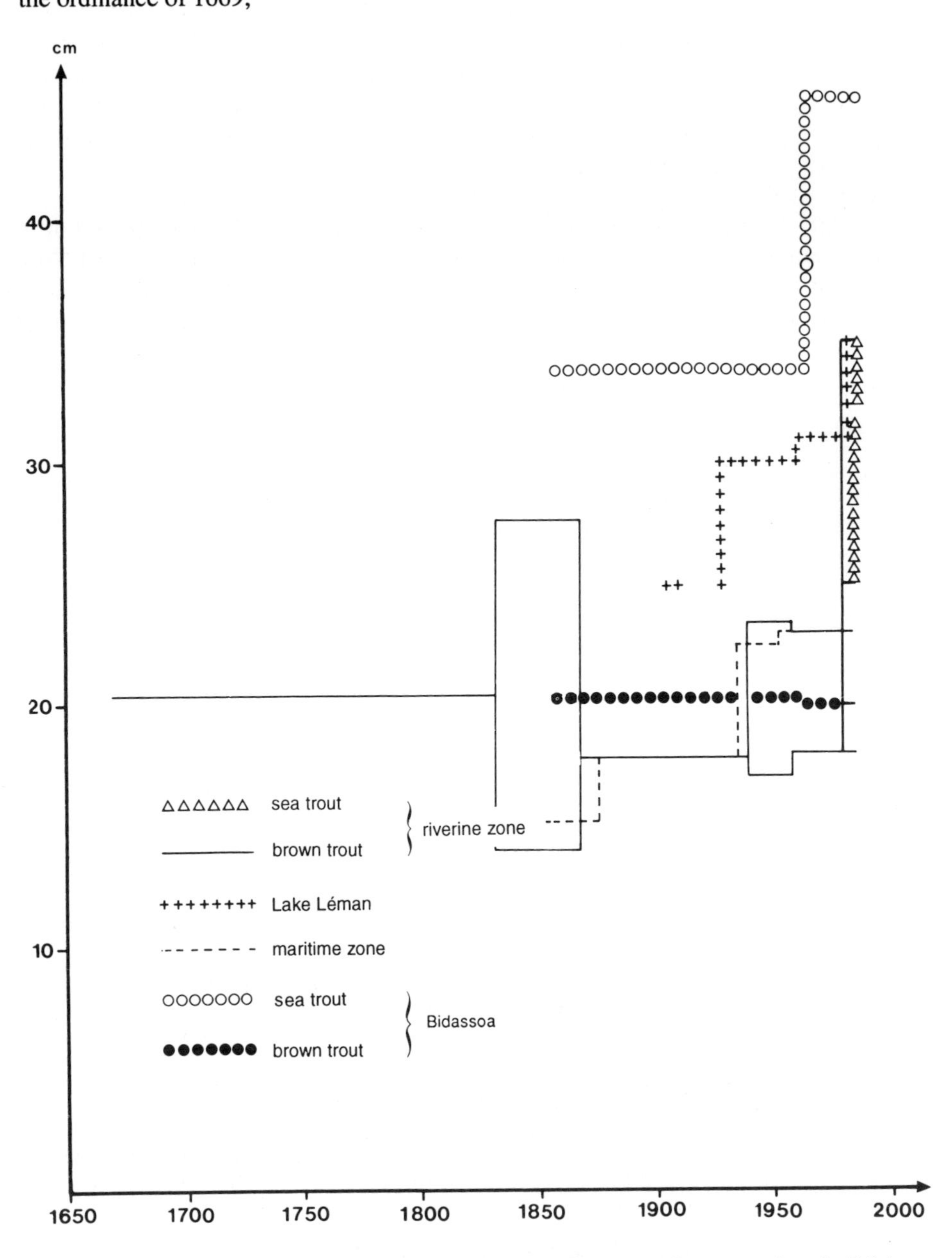

Fig. 6. Development of the legal size (length measured from snout tip to extremity of tail fin) of the brown trout and sea trout in France since 1669.

- a period of around 35 years (1831–1867) of great diversity during which the legal size varied, depending on department, from 14 cm (Vosges) to 27.6 cm (Meuse). Three patterns appear in the 37 departments studied:
  - — the legal size remains the same over the entire period, in almost all cases;
  - — the size of 20.6 cm was lowered to 15.2 cm in the Channel district from 1855. This measure was taken in order to have an identical size in the river zone to that proposed by the decree of 4 July 1853, for the maritime zone (cf. subsection (b));
  - — two sizes are indicated in the Vosges district: 20.2 cm and a lower one in certain mountain streams (14 cm in 1837 and 16.5 cm in 1858).
- a period of nearly 75 years during which the legal size was lowered uniformly on a national scale to 17.8 cm, from the decree of 25 January 1869 to the decree of 29 August 1939;
- a period of around half a century since the decree of 29 August 1939 which introduced two legal sizes: 23.4 and 17.1 cm in relation to water course types.(22)

Since then, some modifications have been made:

- the smallest legal size is limited to trout destined for domestic consumption by the decree of 5 August 1946.(23) The list of departments where trout can be caught from 18 cm is given in the ministerial order of 16 September 1958 (43 departments). The list of water courses, by department, is noted in subsequent orders;(24)
- the aim of increasing the legal size has been constant for more than thirty years. Initially, the ability to increase this size was left to the prefects by the decree of 23 January 1954.(25) Then, decree No. 58.874 of 16 September 1958 introduced two new measures. The minimum size of capture increased from 17.1 cm to 18 cm and when a water course was shared between two districts, the legal size was the higher one.(26)

(22) Article 9. Trout must not be fished for and must be returned to the water if their length is less than 23.4 cm...the prefects can lower this to 17.1 cm, the length below which trout fishing is forbidden in water courses in the mountains (JO of 31 August 1939) and those with granite bedrock (dispensation, JO of 30 May 1941).

(23) Article 2. In mountainous water courses and those running over granite terrain, the prefects can lower the minimum size for brown trout to 17.1 cm, where they are being caught for domestic consumption; the hawking and putting up for sale of trout less than 24 cm in length remains punishable by penalties dictated in article 30 of the law of 15 April 1829.

(24) 25 February 1963, 24 February 1964, 17 February 1970, 14 April 1971, 14 June 1973, 26 January 1974, 21 June 1974, 6 November 1978, 21 February 1986, 10 July 1986.

(25) Article 3. The prefects can, while ensuring the special protection of one of the species listed in the current article, increase the minimum size below which these fish cannot be caught.

(26) Article 12. In mountainous districts or those with soil poor in lime, a list of which is held by the Minister of Agriculture, trout...can be caught above a length of 18 cm. When a water course is shared between two departments and, as a result of the dispositions setting legal sizes, the legal sizes in the two districts are different, the lower size should be applied on both banks. The prefects can, for areas of water with borders on other countries and for these only, increase the minimum length below which fish...cannot be caught.

From 1979 to 1985, the two legal sizes could be increased by 2 cm, 18 to 20 and 23 to 25 cm.[27] Since 1985, the legal size for sea trout has been 35 cm and that for brown trout, 23 cm, with the possibility of being decreased to 20 or 18 cm.[28] The order of 21 February 1986 (disposition order of 10 July 1986) gives the list of water courses where the minimum capture size was restored to 18 cm.

(b) The maritime zone

The legal size for trout designated as sea trout (or salmon trout) was fixed at 15.2 cm in the first three maritime districts (Cherbourg, Brest and Lorient) by the decrees of 4 July 1853. The trout is not mentioned in the regulations for the two other maritime districts, Rochefort (4 July 1853) and Toulon (19 November, 1859).

The legal size increases progressively to:

- 17.8 cm in the decree of 20 November 1875. This length is used again in the decrees of 1 February 1890, 3 December 1923 and 26 July 1927;
- 22.6 cm in the decree of 23 November 1935;
- 23 cm in the decree of 15 December 1852 and in the order of 19 October 1964.

It should be noted (Fig. 6) that the tendency to increase the length has been constant since the mid-nineteenth century. It was the same as that for the river zone during two periods; for 60 years from 1875 to 1935 and during the past 30 years.

In the Bidassoa estuary, the minimum legal lengths of brown trout and sea trout have remained distinct since the law of 11 June 1859 and their development has been very different (Fig. 6):

- the size of brown trout is almost constant over the entire period, at 20.2 cm since 1859 and 20 cm since the decree of 4 March 1965;
- the size of sea trout was fixed at 33.8 cm by the law of 11 June 1859, the same sort of size allowed in France since 1985 (cf. subsection III2a). It changed to 45 cm after the decree of 4 March 1965.

(c) Border waters between France and Switzerland

Lake Léman (Fig. 6)

The legal size for trout was increased by 10 cm in stages over a 75-year period; 25 cm in the decree of 4 February 1905, 30 cm since the decree of 15 March 1929 (size confirmed

[27] Decree of 23 November 1979, article 3. When it is necessary to ensure the particular protection of the trout...in certain areas of water, water courses or sections of water courses, the Minister in charge of river fishing can, in exceptional circumstances, after consulting with the Conseil Supérieur de la Pêche, increase the minimum length below which fish of this species cannot be caught by changing it from 23 to 25 cm or from 18 to 20 cm.

The decree of 3 March 1981, article 1, restated the above article and added: 'the prefects may, in areas of water bordering foreign countries, as well as in large interior lakes, notably those of Bourget and Annecy, increase the minimum length below which the fish...cannot be caught'.

[28] Decree of 23 December 1985, article 19. By dispensation of article 18 of this decree, the minimum size of trout other than sea trout can be, in relation to size at first spawning: (1) changed to 25 cm or to 20 cm by order of the Republic commissioner in certain water courses, canals and areas of water in the district; (2) changed to 18 cm by the Minister in charge of freshwater fisheries in water courses or parts of water courses and areas of water in some mountainous areas with acidic soil.

by the decree of 25 January 1939), 31 cm after the decree of 2 April 1958 (size confirmed in the decree of 2 August 1960 and its dispensation No. 60.827) and 35 cm since the decree of 1 September 1982.

Doubs
The increase in legal size is only 3 cm over the same period. It is 20 cm in the decree of 4 February 1905 and 23 cm since the decree of 27 July 1966.

3. Water course categories

The ordinance of 1669 implicitly separated water courses into two categories for the determination of two closed fishing season during spawning (cf. section 1(a)). The separation is only explicit in the decree of 29 August 1939 which states; article 1 'in order to assure the regulation and conservation of fishing, the ministries for Agriculture and Public Works classified water courses into two categories, in each district. The first consists of those which are mainly populated by trout. The second includes other water courses'.

However, circular no. 824 of 15 October 1913 from the Director-General of Waters and Forests, included an annex of 3 July 1913 which proposed, in the 1st article of the prefectoral order regulating fishing, to distinguish two categories of water course: those in which salmonids predominate, which spawn predominantly during the closed season in winter, and those where the most numerous species spawn during spring.

After the orders of 17 July 1941 (correction JO of 28 September 1941), a series of orders and decrees (29) have followed, bringing modifications to the list of water courses in the first category (salmonids dominant) and those of the second category (cyprinids dominant).

4. Characteristics of equipment used for fishing

(a) In the riverine zone

For over 250 years (ordinance of 1669 to the decree of 29 August 1939), the characteristics of authorized (fixed or moveable nets, hoop nets, floating lines...) and prohibited equipment have been described in various decrees. Taken into consideration have been their number and location, immersion period in relation to the wetted width of water course, the hours of use during the fishing period (days and times) and their mesh sizes, as well as the methods of verification used for the latter.

- from 1669 to the prefectoral orders of 1831 to 1835, in application of the law of 15 April 1829, three mesh sizes are indicated in relation to tackle type: 12 lines (28 mm), 14 lines (30 mm) and 18 lines (40 mm);
- mesh size was defined in relation to species in the decree of 25 January 1868. While the trout was not named, it can nevertheless be taken that the minimum mesh size of 27 mm applied to it: 'for large species other than salmon'. The checking method was defined in the decree of 26 August 1865 and a tolerance of one-tenth allowed. This dimension and tolerance were used again in the decrees of 10 August 1875, 18 May 1878 and 5 September 1897.

The decree of 29 August 1939 brought two fundamental changes:

- it prohibited the use of line and tackle in water courses of the first category;[30]
- it defined the mesh sizes for sea trout and brown trout (not specifically named).[31] The order of 14 August 1943 gave the verification conditions, as did the orders of 25 June 1852 and 26 May 1986.

Some changes were made later. They included:

- Hexagonal meshes. The side of a hexagonal mesh was used in the decrees of 29 April 1943 and 14 September 1950,[32] then one quarter of the perimeter of these meshes was used from the decree of 8 April 1952.[33] The latter measurement had the advantage of being identical to that of the edge measurement of nets with square or diamond-shaped mesh;
- A dispensation concerning the use of equipment in water courses of the first category, after the order of 13 June 1944.[34] This disposition was used again in the decrees of 5 August 1946,[35] 17 March 1952, 23 January 1954 and 16 September 1958. The locations of water courses and type of equipment used are defined in the orders of 16 September 1958, 26 July 1960, 6 April 1961, 26 January 1962, 12 November 1963, 22 March 1967, 2 April 1969, 7 July 1960, 20 November 1970, 12 October 1973, 3 July 1974, 23 June 1977. The use of nets and lines in first category water courses has been prohibited since the decree of 23 December 1985.

(b) The border zone

Lake Léman

There has been an increase in the mesh sizes allowed since the start of the century: 30 mm in the decree of 4 February 1905, 33 mm in the decree of 15 March 1929 (a size used again in the decrees of 25 January 1939 and 2 April 1958), 50 mm in the decree of 5 June 1970 and 48 mm since the decree of 17 November 1982, a measure which was reused in the decree of 13 February 1986.

Since the decree of 13 February 1986, fishing for lake trout has been prohibited, except between 16 January and 31 March; the number and types of nets are defined.

[30] Article 17. The use of nets and all types of tackle, including set lines, but excluding the worm and the crayfish drop-net, is prohibited in water courses or sections of water courses classed in the first category which are designated as such by ministerial order, on the advice of the river fisheries commission.

[31] Article 12. The side of square or diamond-shaped mesh, the small side of rectangular mesh: 40 mm at least for sea trout, 27 mm at least for brown trout; large diagonal for hexagonal mesh, 57 mm at least for sea trout, 38 mm at least for brown trout.

[32] 29 mm at least for sea trout and 19 mm at least for brown trout.

[33] 40 mm at least for sea trout and 27 mm at least for brown trout.

[34] In water courses or sections of water courses in the first category where only the use of floating lines, worm and crayfish drop-nets are allowed, by ministerial order, nets and equipment listed below may be used from the date of opening of fishing season for trout until 14 July inclusive.

[35] The following paragraph was added to article 17 of the decree of 29 August 1939 (cf. footnote 30). In other water courses or sections of water courses of the first category, ministerial orders laid down under the same conditions will determine the nature and conditions of use of authorized fishing equipment.

Bidassoa

Net length is 160 m maximum. Mesh size varies according to the species being caught; sea trout or brown trout: 57 mm and 20 mm respectively, according to the law of 11 June 1859. The former was decreased to 52 mm after the decree of 31 October 1886. The two sizes of 52 mm and 20 mm have been retained in subsequent decrees.

The decree of 4 March 1965 prohibited net fishing for sea trout to 'allow the repopulation of this species'. However, a dispensation is anticipated. This decree does not envisage net fishing for brown trout.

5. Prohibition of selling or hawking fishing during the closed season; establishment of fish ladders

These two measures were applied after the law of 31 May 1865, article 5[36] and article 1.[37]

(a) Prohibition of selling and hawking

The trout appears named for the first time in the order of 14 August 1943 (single article): 'it is forbidden to put up for sale, to sell, to transport, to hawk or to buy brown trout during the closed season for fishing this fish (3rd paragraph)'. A supplementary restriction was added; it concerned the size of fish at the time of fishing: 'it is forbidden at all times to put up for sale, to sell, to hawk or to buy brown trout smaller than 23.4 cm (2nd paragraph)'. This restriction was retained in the decrees of 5 August 1946, 17 March 1952 (23 cm from this date), 23 January 1954 and 9 January 1960. When trout were allowed to be caught at less than 23 cm (cf. section 2(a)), they were destined for domestic consumption.

The law of 21 November 1961 prohibited the hawking, offering for sale, selling or buying of wild trout from the free waters described in article 401 of the countryside code. This measure applied, with the exception of certain professional fishermen fishing with nets and other equipment, only in public waters or dammed reservoirs, whose fishing rights belong to the State, and to fish which were caught in private lakes.

(b) Fish ladders

Decrees appeared several decades after the law, 39 to 59 years later.[38] It is only in the order of 2 January 1986 that the trout is named and designated in various French water courses. Article 2 specified that 'all existing structures installed on the water courses included in the decrees (see footnote 38) should be fitted with mechanisms to allow fish

[36] In each department, it is prohibited to put up for sale, to sell, to buy, to transport, to hawk, to export or import various species of fish, during the closed season for fishing, according to article 26 of the law of 15 April 1829. This disposition is not applicable to fish from ponds or reservoirs defined in article 30 of the aforementioned law.

[37] Decrees produced by the Council of State with the advice of general district councils will determine: secondly, in which sections of rivers, canals and water courses containing dams fish ladders could be established, after investigation, to ensure the free movement of fish.

[38] 3 August 1904 (Seine basin), 1 April 1905 (Loire basin), 3 February 1921 (Canche basin), 15 April 1921 (Adour basin), 31 January 1922 (Breton water courses, Channel, l'Ille and Vilaine, north Coasts, Finistère and Morbihan department), 2 February 1922 (Authie basin) and 23 February 1924 (Calvados, Channel and Orne departments).

migration, within 5 years of the publication of the order. All new structures must be equipped with these mechanisms at their installation'.

6. Destruction of harmful species

The destruction of certain species is envisaged by the decree of 25 January 1868.[39] The term 'harmful' appears for the first time in the decree of 21 March 1913. Since then there have been some variations in the fish or species of fish which have been categorized as being harmful (decree of 29 August 1939), harmful or invasive (decree of 17 March 1952), particularly harmful (decree of 21 March 1913, order of 11 June 1954 and decrees of 23 November 1954 and 19 December 1964), or mainly harmful (order of 16 July 1953).

Species are named for the first time in the decree of 17 March 1952; they include the hotu, sun perch and catfish. Added to these are the Chinese mitten crab in the decree of 16 September 1958 and the eel in water courses of the first category, in the decree of 19 December 1964.

The term 'harmful' disappeared after the law of 29 June 1984.

The decree of 8 November 1985 gave a list of species likely to cause biological disequilibrium; the eel is not on this list.

7. Restocking

The use of artificial fertilization in order to obtain fertilized eggs for subsequent release into water courses commenced properly in 1852. Mainly salmonids, trout and salmon, were used. Until 1870, great enthusiasm prevailed in France for this practice (Deroye, 1903).

Doubts were expressed very early as to the efficiency of this method of repopulation (Haime, 1854; Blanchard, 1866) and tentative explanations were put forward. The mediocrity of the results and reasons for failures were blamed on overfishing, poaching and downgrading of the environment, industrial pollution and habitat modification (Paratre, 1894; Roule et Del Pere de Cardaillac de Saint-Paul, 1902; Deroye, 1903; de Drouin de Bouville, 1906).

At the start of the twentieth century, some authors demonstrated the lack of method. Amongst these, Léger (1910) was the first to propose the methodical restocking of a water course, in proportion to its nutritive value. He introduced the idea of biogenic capacity, the expression of the nutritive value of a water course from the point of view of the fish. He transformed this into a numerical scale from 1 to 10, then determined, using a formula, the number of fry to release:

- for one kilometre of water course, 5 m wide, this number is X (biogenic capacity between 1 and 10) $\times$ 100;
- for a water course wider than 5 m, this number become: $X \times 10 \times (1 + 5)$.

From this, Léger deduced the annual yield of 1 km of salmonid water course, rationally populated, to be: $R = X\,(1 + 5)/2$.

[39] Article 14: The prefects can authorize, at certain times and places, movements of water and special fisheries to destroy certain species in order to allow other more highly valued species to propagate.

Note that Léger's estimates are relatively cautious; he qualifies his formula as an initial one and indicates that the assessment of biogenic capacity is more a matter of judgement than measurement. But, in fact, he never states the determining criteria for measuring biogenic capacity.

Nevertheless, this formula has enjoyed a certain fame until the present day. It has been used to determine the number of fry to release and is therefore considered as a vital management tool (Ducret, 1982); it is also used to evaluate damage to fisheries (Vibert, 1948). It was subsequently modified by Huet (1949) and Arrignon (1970). The latter noted, however, its empirical nature and the subjectivity linked to the determination of biogenic capacity, leading to variations of 30–50% in stock assessment (Arrignon, 1970). However, it has never been tested again.

IV. DISCUSSION

1. General remarks

This retrospective study of the management of natural populations of brown trout in France has noted the five main trends which vary in both duration and significance.

- Two main trends which appear over the entire period, studied with different intensities:
  - —firstly, the constant primacy of economic activities concerning the river in relation to fishing. Even when the harmful effects on the fish of certain activities are known (e.g. the retting of flax and hemp, industrial and other wastes, dam construction etc.) and warnings often given, one is forced to conclude that the regulations, while taking into account these problems, are only palliative in nature. The environmental damage only stops when the activity ceases;
  - —secondly, the predominance of the role of forests in relation to waters. The preference for forestry, noted in the ordinance of 1669, is found again in the law of 1829. This was linked to the economic importance of the forest at that time (shipbuilding and ironworks). Even after this influence diminished, it reappeared at two particular points:
    - (i) the concerns of foresters, characterized by a catastrophe from which nuances and half-tone were excluded (Larrere, 1981), which heavily contaminated the fishing world at the end of the nineteenth and beginning of the twentieth century. This is confirmed by the assessment of the managers of Waters and Forests in the fishing domain, noted in the eighteenth century by Cocula-Vaillieres (1979);
    - (ii) the use of nursery streams and growing-on streams where trout were 'planted out' in a similar way to young trees.[40]
- Three trends occur over half the period, each one of around 150 years:

[40] Nursery streams, following the example of tree seedlings in forestry, are streams in which hatching of trout eggs, for example, is carried out, followed by the rearing of fry over a period of a year or less (Arrignon, 1970). Growing-on streams, in the manner of tree seedlings for pricking out, are streams into which trout fry are introduced, depending on their growth in the same period, either with 3/4 of their yolk sac or at a later stage (Arrignon, 1970).

— a trend from 1669 to 1829, dominated by the progressive control by central powers, over navigable water courses. There was a direct overlap between this aim and the general trend described above. Thus, Louis XIV, in the ordinance of 1669, 'took up the cudgels again for the dynamic use of water courses, against their static use, for seagoing vessels against the building of obstacles, whether they were paying owners or overlords, or mill owners' (Cocula-Vaillieres, 1979). In the fishing domain, the dominant characteristic is stability, the modifications made being minor;

— two trends have appeared since the middle of the 19th century:

(i) the first dominated by the objective of protection, inscribed in the general movement for protecting nature which flourished thereafter (Raumolin, 1984). It comprised two periods:
   — a short period of caesura, from 1849 to 1865;
   — a period which has lasted since 1865, characterized by three events which go together: increasing complexity of the regulations, progressive restriction of access to the resource and progressive increase in the role of sport fishing in relation to professional fishing.

These three events evolved slowly in parallel, but in a non-uniform manner, in stages of variable duration, sometimes with small backwards movements;

(ii) the second is characterized by the pre-eminence accorded to the trout and more generally the salmonids, in relation to other fish. This is proved, throughout, by the vocabulary used (these species are often described as precious) and by the regulations which favour these species. The trout is the object of particular consideration in this area, which goes together with the increasing favour for sport fishing.

2. The caesura in the mid-nineteenth century

This caesura was the consequence of the rediscovery in France of artificial fertilization of trout in the mid-nineteenth century. An extraordinary obsession arose, firstly in France in 1849 and then extended on a global scale in little more than a decade. Three types of people were jointly responsible for this in France, the most noteworthy and influential being scientists and engineers. Among the latter, Coste played an important role (Thibault, 1989):

- he attributed the unrealistic hopes placed on artificial fertilization to those responsible. Thus, the senior member of the Council of State, Director of Agriculture and Commerce (Heurtier, 1852), entitled his report to the Minister of the Interior, Agriculture and Commerce 'on the means of repopulating all waters in France by the artificial hatching of fish eggs'. Similarly, the Minister of the Interior De Morny intervened and demanded from the prefects, a greater crackdown on fishing offences in lakes and rivers, in order to assure the success of a new procedure of artificial fertilization in fish (circular of 19 January 1852);
- he became inspector-general of river fishing and maritime coastal fishing by the imperial decrees of 26 April and 24 May 1862 respectively;

- this followed his proposition made to the Emperor in 1859 (Coste, 1861) that responsibility for river fishing be taken away from Waters and Forests and assigned to Bridges and Roads, by imperial decree of 29 April 1862. Waters and Forests retrieved this responsibility in 1896;
- he proposed the prohibition of salmon and trout fishing from 1 October to 15 January on all water courses in the Empire, to prevent the destruction of these two species and favour their repopulation (circular of 2 September 1862 by the Minister for Agriculture, Commerce and Public Works);
- lastly, he was very probably, at least in part, the instigator of the law of 31 May 1865.

There followed a relentless pursuit, for more than a decade (1852–1865), after which, to compensate for the lack of success of restocking, it was necessary to resort to regulations. This was then a crash in the stocks, whose influence is still felt. Regulatory powers were put in place, which became more and more important, while all critical scientific views or doubts faded away.

In view of this situation, it must be concluded that the arguments were usually based on the use of arbitrary formulae[41] which were supported neither by continuous monitoring of field data nor by experiments or figures giving quantitative information. They were a reflection of the strongly predominant opinion at that time, characteristic of an unscientific standpoint (Bachelard, 1986). The world of scientists and engineers contented themselves with this attitude and shut themselves away with a blinkered vision which was expressed by:

- on the one hand, the use of Léger's formulae since the start of this century and their attempts to improve it, without questioning its principles or carrying out observations or experiments in the wild, until recently;
- on the other hand, by failing to take into account overseas studies which were literally overshadowed. For example, Hobbs showed in 1937 that natural spawning in trout was an efficient process in exploited water courses.

3. Complication of regulations

In addition to the three laws relevant to river fishing, since the first half of the nineteenth century (15 April 1829; 31 May 1865 and 29 June 1984), five decrees have been

[41] Salmon and trout are becoming increasingly rare in our water courses (circular of 2 September 1862).
— the disappearance of trout from our water courses has been noticed for a long time (Lestidoubois, 1865).
— depopulation has increased and a time can be envisaged, in the near future, when our water courses will be almost desert (Roule & Del Péré De Cardaillac de Saint-Paul, 1902).
— water courses which are not navigable can almost all be considered to be nearly completely ruined (Deroye, 1903).
— I was pleased to note their unanimity (professional fishermen) in recognizing the considerable increase in Salmonids in our region (Léger, 1907).
— most of the Dauphiné streams accommodate trout which has today certainly become more frequent, only a few years after the methodical restockings carried out by the aquaculture laboratory in Grenoble, angling clubs and the administration of Waters and Forests (Léger, 1909).
— the Lozère, and excellent district for trout...can only offer waters which are quasi-deserts to fishermen (annex 902)...resulting in an accelerated depopulation of rivers of the first category (annex 1188), law of 21 November 1961.

produced which follow the development of the codification of access to fisheries resources: 25 January 1868; 5 September 1897; 29 August 1939; 16 September 1958; 23 December 1985.

The increasing complexity of the regulations resulted in an increase in the number of articles in the decrees and the number of regulatory texts (Fig. III.7):

- the development of the number of articles occurred mainly in two stages; 1939 (34 articles) and 1985 (58 articles). The decree of 1955 contained more than three times the number of articles than that of 1868 (17 articles). The increase in number of articles is less in 1897 (24 articles) and 1958 (36 articles) in relation to the preceding decrees. In addition to increasing in number, the articles also increased in length, in order to take into account all factors;
- the development of the number of decrees and ministerial orders about brown trout is very revealing. By intervals of 12 years, note that the increase is most noticeable as the second half of this century goes by, the largest number of regulations being produced in the last interval between 1975 and 1986.

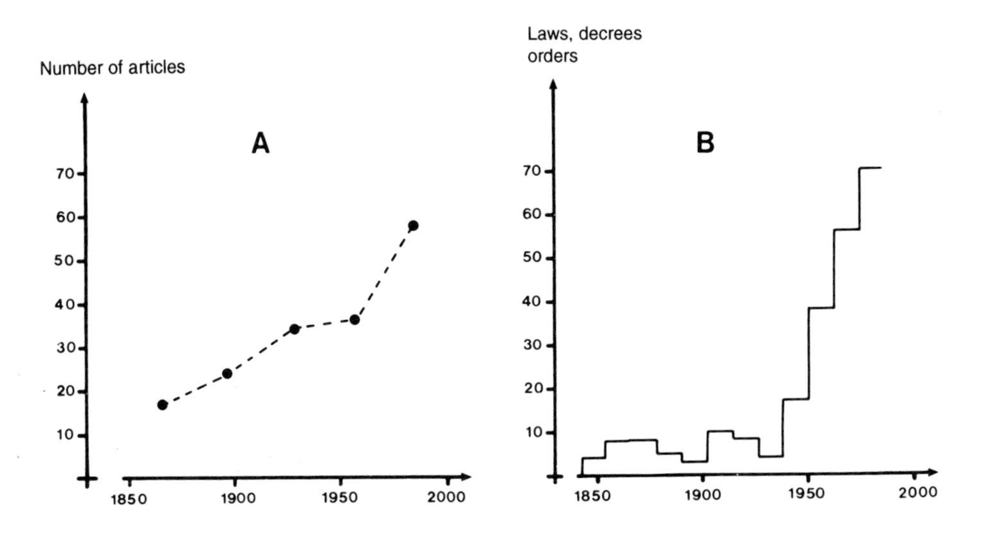

Fig. 7. Development of the number of articles (A) in the five main decrees relating to river fishing (25 January 1868, 5 September 1897, 29 August 1939, 16 September 1958) and in the number of regulatory texts (laws, decrees and orders) since 1850 (B) by 12-year period.

4. Progressive restriction of access to the resource

(a) Increase in fishing prohibitions

During the spawning season

The duration of the closed season for fishing at spawning time for trout has been increased by a factor of about 3.8 in the river zone over the entire period studied, changing from 43 days in 1669 to 162.5 days in 1985. It changed from 12% to about 45%, based on 365 days in the year, by stages, over three centuries.

The inclusion of sea trout in the regulations since 1985 has resulted in an increase in the duration of fishing prohibition which changed to 190 days in 1986.

Year-long in certain sections of water course

This restriction has been used only since the mid-twentieth century and includes:

- sections of water course or areas of water where juveniles are released;
- an increasing number of estuaries, particularly in Brittany.

(b) Increase in the legal size

The increase in legal size of trout in the river zone measured from snout tip to the extremity of the tail fin was 5 cm between 1669 and 1985. This increase occurred in stages:

- 20 cm, between 1669 and 1868;
- 18 cm, between 1868 and 1939. Note that, over a period of 7 years between the decrees of 1868 and 1875, the legal size for brown trout was removed for fishing with floating lines;
- since 1939, two separate patterns have arisen; water courses where the size has increased to 23 cm and mountain water courses or those in soil poor in lime, where the size was lowered to 17 cm, then 18 cm (since 1954). These two sizes can be changed to 25 and 20 cm by special dispensation since the decree of 3 March 1981.

In addition to this development in fresh water, there was the increase in estuaries, where the size for sea trout increased from 16 cm in 1823 to 23 cm in 1952 and in Lake Léman, from 25 to 35 cm from 1905 to 1982.

5. Increased role played by sport fishing

The importance of the role played by sport fishing become apparent for the first time in the law of 20 January 1902 (decree applied 17 February 1903), which included the following single article: 'there can be a departure from, in favour of angling clubs, the principle of adjudication determined by public administration rules'.

Sport fishing for trout became progressively more favoured in the regulations, mainly during the past half century:

- by restricting access to the resource for methods other than rod and line;
- by putting closed and open days in place to coincide with the weekends;
- by the restriction (since 1946), then the prohibition (since 1961) of selling fish from waters of the first category.

To these three criteria can be added a fourth, technique, privileging fly fishing within the rod and line fishing community. From the second half of the nineteenth century, fly fishing was considered as the only true sport, by its followers. Albert-Petit (1982),[42]

[42] Re-edition of the first publication in 1897.

amongst others, eulogized about this technique, copying the English. For him, 'fly fishing is incomparably superior to all other forms of fishing...what a difference between the vulgar methods (real insects, small live or dead fish bait, or artificial fish or even the humble earthworm) and the feathered fly! What heavy prose beside our poetry...only the fly can give the fisherman palpitations of excitement for this superior art'.

6. Coordination of regulations in maritime and river zones; potential for dispensations

National texts determine the general characteristics of the regulations over the entire period studied. Since the mid-nineteenth century, one can observe:

- a desire for homogenization of access to the resource between the maritime and river zones. Elsewhere, measures were taken first in fresh water;
- two contradictory trends on a national scale were concerned:
  - —on the one hand, uniform regulation, which was particularly clear in the mid-nineteenth century, after a period of wide diversity in fresh water, in relation to the closed season during spawning. The start extended over five and a half months (1 September to mid-February) and the end, over six and a half months (end of October to mid-May). Such diversity (cf. section 1(a)) did not reappear later on;
  - —on the other hand, local adaptation of the regulations to take into account ecological realities, both for the closed season during spawning and for the legal size.

Setting aside the increase in access to the resource constituted by the lowering of the legal size to 16 cm in certain water courses (decree of 29 August 1939), the prefects still had the potential to increase the restrictions. These included:

- the prohibition of fishing[43] and destruction of certain species,[44] from the decree of 25 January 1868;
- the prohibition of certain fishing methods,[45] or all fishing methods other than floating lines[46] since the decree of 29 August 1939;
- the increase in legal size in mountain water courses and those running off granite, since the decree of 23 January 1954.[47]

National regulation relied on the freedom of the assessment of the prefects in increasing access to the resource:

- the decree of 5 August 1946 restricted the consumption of trout smaller than the legal size to domestic use;

[43] Article 2. Each year, the bailiffs can, after taking the advice of the general councils, prohibit fishing of all fish species by special order, during one or other of the stated periods (20 October to 31 January; 15 April to 15 June) when this intervention is required to protect the predominant species.

[44] Article 14. ...the bailiffs can authorize, at certain times and places, movements of water and extraordinary fishing to destroy certain species in order to allow the proliferation of other, more precise species.

[45–46] Article 20. The bailiffs can, by special orders, prohibit the use of all lines, equipment, methods or means of fishing other than floating line and, for the latter, all baits which are harmful to the repopulation of the water course. These dispensations can be general or specific for certain species of fish, certain water courses or for certain periods.

[47] Article 3. The bailiffs can, in order to ensure the special protection of one of the species listed in the current article, increase the minimum size below which these fish cannot be caught.

- ministerial orders gave a list of districts (order of 16 September 1958) then defined water courses by district (orders of 28 February 1963, 24 February 1964, 17 February 1970) affected by the decrease in legal size;
- the decree of 23 December 1985 gave the responsibility for changing trout size to 20 cm to the prefect, but retained the potential with the Minister to change it to 18 cm.

V. CONCLUSION

This study is based mainly on the national documentation of regulations. It is probable that some texts have been overlooked, within the study period. It is nevertheless concluded that this should not alter the main thrust of the results presented here. Elsewhere, regional texts were not consulted apart from those taken from 1831 to 1834. Just as the freedom given to districts was in a sense only serving to increase restriction, the conclusions relevant to the different aspects studied are minimal.

It therefore appears reasonable to conclude that the objective of this study has been attained in so far as:

- management of natural populations of brown trout fall well within the trends described on a national scale, including river fishing in its widest sense. The regulations protect reproduction (prohibition of fishing during spawning time, reserves, legal size of fish and mesh size of nets);
- explanations can be determined for the different methods of intervention. Numerous and varied modifications, introduced since the mid-nineteenth century regarding access to the resource, express the main ways of thinking of the era, which is always an underlying factor: ecological Malthusianism and the Promethean vision of man dominating nature are expressed both through the regulations put in place and in the release of juvenile stages. However, the regulations respond more to social and cultural requirements[(48)] where moral considerations[(49)] are significant, than towards an improved scientific knowledge or understanding.

Have the means and methods employed for over 150 years allowed the objective of protecting the renewable resource, consisting of brown trout, to be reached? With current knowledge, this question cannot be answered for two reasons:

- it is not known to what extent the regulations have been applied and obeyed;
- one of the main elements of management is consistently absent. Measures have been applied in some sectors, without setting up means of monitoring their impact. Such monitoring would also allow baseline data to be collected, in relation to human activity on water courses.

[(48)] Plasticity of fishing periods according to district in 1831–1834. National unity in regulations in 1863, advocated thereafter with ecological considerations in circular No. 23 of the Minister of Public Works of 14 June 1878: 'The Council of State has noted... the importance of maintaining the view that, to avoid the depopulation of water courses, it is necessary to conserve as far as possible, if not for the whole of France, at least for each main basin, the unit of regulation which was established in 1868 and 1875'.

The progressive restriction of access to the resource for professional fishing in water courses of the first category ended with its total prohibition in 1985. The eel was declared a harmful species in water courses of the first category in the decree of 19 December 1964.

[(49)] Prohibition of the sale of wild trout caught by angling/sport fishing since 1961.

Three examples are given below where, in the absence of numerical data, scientific doubt has crept in:

- where new regulations are more restrictive, does this imply that preceding ones have not been sufficiently effective, or judged to be so, to protect the species?
- did the error in determining the spawning period from 1669 to 1831 have consequences 'most regrettable for the repopulation of rivers', just as this appeared in the explanatory notes of the prefectoral order of Tarn-et-Garonne of 26 May 1831,[(50)] which prohibited trout fishing during spawning time from 1 October until 15 November?
- fisheries to destroy eels were carried out after this species was declared harmful in water courses of the first category. The significance and extent of destruction of this species and its effect on trout populations could be analysed, since the prefectoral orders demanded that an account of the fisheries should be sent to the departmental Director of Agriculture and to the fisheries region at the end of the campaign. Even without such results, the results of fisheries inventories carried out over more than a decade in Brittany cast doubt on their real effect on trout populations.

Finally, the term 'protection', which is frequently used, is itself ambiguous. It is either qualified or not (in view of the need to protect...in view of the need to assure better protection...special protection...particular protection) without the significance of what it covers being clearly expressed.

A scientific problem arises in the area of management of natural populations of brown trout concerning the role played by scientific knowledge. A real trend and fundamental deterministic concept stands out in the phrase 'know in order to manage' and its variants 'know more to manage' and 'to know for better management'. Such phrases are ambiguous as they assume the primacy of the scientist over the manager and, to this end, they are not neutral. It is all the more erroneous to wait for knowledge in order to manage as this implies that without scientific knowledge, management will come to a standstill.

In fact, knowledge and management constitute two complementary, or even parallel processes, since they do not have the same objective; they are however indispensable to each other: from one perspective, the improvement in knowledge is justified in so far as it is necessary as reference for management. The knowledge gained during the past two decades, where certain qualitative aspects have allowed a glimpse of the complexity and diversity of the ecology of this species to be shown:

- distribution of age classes and differences between growth rates within hydrographic networks, both in time and space (Baglinière, 1979; Maisse *et al.*, 1987);
- relationship between growth and sexual maturity (Maisse *et al.*, 1987);

[(50)] 'All authors who have dealt with the natural history of fish have been mistaken about the time of year at which the male releases its fertilizing liquid on the eggs deposited by the female; and this error, on the part of the naturalists, having been included in the rules for fishing, has resulted in the most terrible consequences for repopulation of rivers over several centuries. This has been demonstrated by the author of considerations on the natural history of fishes, on fishing and the laws which regulate it, a work published in Toulouse in 1821' (A. D. Tarn and Garonne).

- adaptive strategy as a function of the location of the watercourse in the hydrographic network and characteristics of sedentary or migratory behaviour in one part of the population (Baglinière *et al.*, Elliott, 1987, 1988).

The desire to take into account certain aspects of the ecological characteristics of the species cannot be denied:

- the determination of the spawning periods since the first half of the nineteenth century;
- the instigation of different legal sizes in relation to length at three years of age, according to whether the water course is situated in the mountains or on the plains;
- recent extension of open season for sea trout fishing based on observations of upstream migration periods.

Should all new information be integrated into the regulatory process? It can be recalled here that the legal size for trout at 23 cm can be changed to 20 cm or even 18 cm. Results have shown that size at first reproduction can be a lot smaller in certain sub-tributaries (11 cm); a good part of the trout population can die of old age before reaching the legal size. Maximum sizes of trout measured in this sub-tributary over four consecutive years varied depending on the year from 19 to 24.5 cm for males and 19.5 to 21.5 cm for females (Baglinière, personal communication). In addition, only tens of metres away, in the same hydrographic network, there existed cohabitation and even mixing of trout of different sizes at first reproduction. In this respect there seems to be a contradiction between ecological diversity of the species and the regulations, which cannot take these elements into account. Is it also possible to increase the complexity of the regulations if the fishermen are not informed of the main aspects of the ecology of the species?

The quantitative and dynamic aspects, linked to the rate of renewal of generations, the capacity for resilience of the species and individual migrations have been more or less neglected. However, if one takes into account the results from observations in the natural environment and others, even partial, from the reconstruction of a population after its destruction, one can conclude, at least from a hypothetical point of view, that regulations and restocking only have a marginal effect on the protection of the species in relation to its ecological potential and effects of man on the fishery environment.

Knowledge about the size of annual fluctuations of natural populations, which may or may not be included in longer-term trends and regulatory capacities within populations seem promising avenues of future research. From another perspective, two views can be envisaged within management:

- On the one hand, management can contribute to improved knowledge and provide information in the long-term through data relevant to fishing ('manage to know'). This is an experimental idea in management.[51] In this regard, it remains to be done, e.g.

[51] Pérau (1931) indicated the usefulness of fishery statistics. Vibert (1961) regretted that so little experimental regulation of fisheries had been carried out and proposed that controls should be carried out on sampled sectors on a national scale, in order to be able to discuss production in French rivers, based on sufficiently precise data, obtained over several years.

testing the influence of legal sizes on the population dynamics of the species in a hydrographic network. One can also suppose that through the results obtained it would be possible to repeatedly test the hypothesis outlined above.

- On the other hand, the interpretation of the collected data will be facilitated by referring to the recent information.

VI. ACKNOWLEDGEMENTS

It is my pleasure to thank all those who helped me to collect all the documents necessary for this chapter, especially the personnel in the district Archives of Ille-et-Vilaine for research into regulations, including Mr Lattwein of the Ministry of Agriculture, Mrs Jantzen et Vigneux of the Conseil Supérieur de la Pêche, Ms Lionnet, librarian at ENGREF in Nancy, and the staff at the National Library, the Law Faculty library in Rennes and district Archives at Tarn-et-Garonne, Loir-et-Cher, Vienne and Maritime Affairs of Bayonne, who allowed me to complete the documentation.

VII. BIBLIOGRAPHY

1. BOOK AND JOURNAL ARTICLES

Albert-Petit G., 1982. *La truite de rivière. Pêche à la mouche artificielle.* 1 vol., Ed. de l'Orée, Bordeaux, 439 pp.

Arrignon J., 1970. *Aménagement piscicole des eaux intérieures.* 1 vol., SEDETC, SA, Ed., Paris, 643 pp.

Bachelard G., 1986. *La formation de l'esprit scientifique. Contribution à une psychanalyse de la connaissance objective.* Treizième ed., 1 vol., Libr. philosophique J. Vrin, Paris, 257 pp.

Baglinière J-L, 1979. Les principales populations de poissons sur une rivière à Salmonidés de Bretagne-sud, le Scorff. *Cybium*, **7**, 53–74.

Baglinière J-L, Maisse G., Lebail P-Y, Nihouarn A., 1989. Population dynamics of brown trout, *Salmo trutta* L., in a tributary in Brittany (France): spawning and juveniles. *J. Fish Biol.*, **34**, 97–110.

Baudrillart J-J., 1821. *Traite général des Eaux et forêts, chasses et pêches, première partie, Recueil chronologique des réglements forestiers.* Tome 1. Mme Huzard, Impr. Libr., Paris, 723 pp.

Blandard E., 1866. *Les poissons des eaux douces de la France.* 1 vol., J-B. Baillière et fils, Libr., Paris, 656 pp.

Bouthillier (de), 1828. Expose des motifs de la loi sur la pêche fluviale. *Bull. des lois, loi relative à la pêche fluviale du 15 avril 1829*, 87–90.

Buttoud G., 1983. *L'Etat forestier. Politique et administration des forêts dans l'histoire française contemporaine.* Thèse, Doct., Univ. Nancy II. Fac. Droit Sciences Econ., 1 vol., 691 pp.

Cocula-Vaillieres A.M., 1979. *Les gens de la rivière de Dordogne, 1750 à 1850.* 2 vol., Thès. Doct., Univ. Bordeaux III, 740 pp.

Coste V., 1861. *Voyage d'exploration sur le littoral de la France et de l'Italie. Deuxième édition suivie de noueaux documents sur les pêches fluviales et marines.* 1 vol., Impr. impériale, Paris, 291 pp.

Deroye F., 1903. *La pêche fluviale et l'administration des eaux et forêts.* 1 vol., Impr. Jacquot et Floret, Dijon, 313 pp.

Dralet E-F., 1821. *Considérations sur l'histoire naturelle des poissons, sur la pêche et les lois qui la regissent.* 1 vol., J-M., Douladoure Impr. Libr., Toulouse, 116 pp.

Drouin de Bouville R. (de), 1906. L'assainissement des rivières. *Bull. Soc. cent. Aquic. Pêche*, **18**, 3–15; 73–80; 97–104; 129–149; 161–175.

Ducret O., 1982. La gestion piscicole. *Pêcheur Fr.*, **51**, 8–9.

Ducret O., 1983. La gestion piscicole (suite). *Pêcheur Fr.*, **52**, 10.

Elliott J-M., 1987. Population regulation in contrasting populations of trout *Salmo trutta* in two lake district streams. *J. Anim. Ecol.*, **56**, 83–98.

Elliott J-M., 1988. Growth, size, biomass and production in contrasting population of trout *Salmo trutta* in two Lake District streams. *J. Anim. Ecol.*, **57**, 49–60.

Haime J., 1854. La pisciculture. *Rev. Deux Mondes*, **6**, 1006–1032.

Herbin de Halle P-E., 1835. *Recueil chronologique des règlements sur les forêts, la chasse et la pêche.* Tome 5. Arthus Bertrand Libr. Ed., Paris, 641 pp.

Heurtier, 1852. Rapport à M. le Ministre de l'Intérieur, de l'Agriculture et du Commerce sur les moyens de repeupler toutes les eaux de la France par l'éclosion artificielle de œufs de poisson. *Monit. Univ.*, **218**, 1191.

Hobbs D.F., 1937. Natural reproduction of quinnat salmon, brown trout and rainbow trout in certain New Zealand waters. *Fish. Bull., N.Z. Mar. Dep.*, **6**, 1–104.

Huet M., 1949. Appréciation de la valeur piscicole des eaux douces. *Trav. Stn. Rech. Groenendaal*, D, 10, 45 pp.

Larrere R., 1981. L'emphase forestière: adresse à l'Etat. In *Recherches*, no. 45, 113–154.

Leger L., 1907. Le laboratoire de pisciculture de l'Université de Grenoble. Campagne 1906–1907. Rapport présenté au conseil général (sesion d'août 1907). *Ann. Univ. Grenoble*, **19**, 719–725.

Leger L., 1909. Poissons et pisciculture dans le Dauphiné. *Trav. lab. piscic. Univ. Grenoble*, fasc. 2, 18–91.

Leger L., 1910. Principes de la méthode rationnelle du peuplement des cours d'eau à Salmonidés. *Bull. Soc. cent. Aquic. Pêche*, **22**, 241–269.

Lestidoubois, 1865. Exposé des motifs. Lois annotées, 29–30.

Maisse G., Baglinière J-L., Le Bail P-Y., 1987. Dynamique de la popultion de truite commune (*Salmo trutta*) d'un ruisseaux breton (France): les géniteurs sédentaires. *Hydrobiologia*, **148**, 123–130.

Paratre R., 1894. Du dépeuplement des cours d'eau de l'Indre. *Bull. Soc. cent. Aquic. Pêche*, **6**, 1–30.

Perau A., 1930. La statistique des pêche fluviales. *Bull. Fr. Piscic.*, **26**, 29–31.

Raumolin J., 1984. L'homme et la destruction des ressources naturelles: la Raubwirtschaft au tournant du siècle. *Ann. Econ. Soc. Civilis. Fr.*, **39**, 4, 798–819.

Roule L., Del Pere de Cardaillac de Saint-Paul G., 1902. Biologie et pisciculture; les causes réelles du dépeuplement des cours d'eau et les moyens d'y porter remède. *Bull. Soc. cent. Aquic. Pêche*, **14**, 73–84.

Thibault M., 1989. La redécouverte de la fecondation artificielle de la truite en France au milieu du XIXe siècle; les raisons de l'engouement et ses conséquences, pp. 205–231. In *Colloque Homme, Animal et Société*, 13–16 mai 1987. Tome 3, Histoire et Animal, des sociétes et des animaux, Inst. Et. Polit., Toulouse.

Vibert R., 1948. Dommages piscicoles des usines hydroélectriques, évaluation et limitation. *Bull. Fr. Piscic.*, **148**, 89–112.

Vibert R., 1961. *La recherche en biologie des pêches continentales. Conditions d'efficacité. Possibilités d'extension à l'échelon des fédérations et associations de pêche et de pisciculture.* Congrès de l'union des fédérations départementales des associations de pêche et de pisciculture des bassins de la Garonne, de l'Adour, de la Charente et de l'Aude. Auch, 19–22 mai 1961, 7 pp.

2. DEPARTMENTAL ARCHIVES

AD Ille-et-Vilaine, 7 Sc I et II, pêche maritime et fluviale.

A.D. Loir-et-Cher, 4 M 179, réglementation de la pêche, an IX 1865.

AD Tarn-et-Garonne, 108 S21, pêche fluviale, circulaires, instructions, 1831–1879.

A.D. Vienne 3 S 46, Circulaires et instructions; essais de pisciculture; repeuplement des cours d'eaux en écrevisses, construction d'une échelle à poissons au barrage de la manufacture d'armes de Chatellerault; réglementation annuelle, 1832–1880.

3. LIST OF LAWS, ORDINANCES, DECREES CONCERNING THE MANAGEMENT OF NATURAL POPULATIONS OF BROWN TROUT

Ordonnance de Louis XIV donnée à St Germain en Laye en août 1669 sur le fait des Eaux et Forêts. In Baudrillart, 1821, 41–92.

Loi relative à la pêche fluviale du 15 avril 1829. *Bull. Louis*, 87–159.

Ordonnance royale du 15 novembre 1830.

Circulaire du Ministre de l'Intérieur du 19 janvier 1852. Mesures à prendre pour prévenir les dégâts qui se commettent par contravention à la Loi sur la pêche fluviale, A.D. Loir-et-Cher, 4 M 179.

Décret impérial du 14 juillet 1853 portant règlement sur la pêche maritime côtière dans le premier arrondissement maritime (Cherbourg). Lois annotées, 108–116.

Décret impérial du 14 juillet 1853 portant règlement sur la pêche maritime côtière dans le deuxième arrondissement maritime (Brest). Lois annotées, 116–127.

Décret impérial du 14 juillet 1853 portant règlement sur la pêche maritime côtière dans le troisième arrondissement maritime (Lorient). Lois annotées, 127–132.

Décret impérial du 14 juillet 1853 portant règlement sur la pêche maritime côtière dans le quatrième arrondissement maritime (Rochefort). Lois annotées, 132–141.

Loi du 11 juin 1859 relative à l'exercice de la pêche dans le Bidassoa. *Bull. Lois Empire français*, 13, 931–938.

Décret impérial du 19 novembre 1859 sur la police de la pêche côtière dans le cinquième arrondissement maritime. Lois annotées, 117–127.

Décret impérial du 29 avril 1862 qui place dans les attributions du Ministre de l'Agriculture, du Commerce et des Travaux publics la surveillance, la police et l'exploitation de la pêche fluviale. *Bull. Lois*, 686–687.

Circulaire du 2 septembre 1862. Ministère de l'Agriculture, du Commerce et des Travaux Publics. Interdiction de la pêche du saumon et de la truite du 1 octobre au 15 janvier—demande d'avis. A.D. Ille-et-Vilaine, 7 Sc I et II.

Décret impérial du 19 octobre 1863 relatif à la pêche de la truite et du saumon dans la partie fluviale des cours d'eau navigables ou non navigables de l'Empire, à l'exception du Rhin et de la Bidassoa. *Bull. Lois*, 807.

Décret impérial du 24 octobre 1863 relatif à la pêche de la truite et du saumon, tant à la mer, le long des côtes que dans la partie des fleuves, rivières, étangs et canaux où les eaux sont salées. *Bull. Lois*, 807.

Loi relative à la pêche fluviale du 31 mai 1865. Lois annotées, 29–33.

Décret impérial du 26 août 1865 qui détermine le mode de verification de la dimension des mailles des filets et de l'espacement des verges des nasses autorisés pour la pêche de chaque espèce de poisson. *Bull. Off.*, 1336 no. 13659.

Décret impérial du 25 janvier 1868 qui désigne les parties des fleuves, rivières et canaux réservées pour la reproduction du poison dans les départements de Seine-et-Marne, de Seine-et-Oise, de la Seine, de l'Eure et de la Seine inférieure. *Bull. Lois*, no. 1568, 134–139.

Décret impérial du 25 janvier 1868 portant règlement sur la pêche fluviale. *Bull. Lois*, 158–161.

Décret impérial du 20 septembre 1868 qui désigne les parties des fleuves, rivières et canaux réservées pour la reproduction du poisson dans les départements de la Haute Garonne, de l'Ariège, du Tarn-et-Garonne, de Lot-et-Garonne, de la Gironde, de la Dordogne, de la Corrèze, du Lot, de l'Aveyron, du Cantal, du Tarn, des Landes, des Basses Pyrénees et des Haute Pyrénées. *Bull. Lois*, no. 1650, 767–782.

Décret impérial du 30 janvier 1869 qui désigne les parties des fleuves, rivières et canaux réservées pour la reproduction du poisson dans les départements de la Haute Loire, du Puy-de-Dôme, de la Loire, de Saône-et-Loire, de l'Allier, de la Nièvre, du Cher, du Loiret, de l'Indre, du Loir-et-Cher, d'Indre-et-Loire, de la Vienne, de la Sarthe, de la Mayenne, de Maine-et-Loire, de la Loire inférieure, d'Ille-et-Vilaine, du Morbihan, du Finistère et des Côtes-du-Nord. *Bull. Lois*, no. 1682, 117–130.

Décret impérial du 17 juillet 1869 qui désigne les parties des fleuves, rivières et canaux réservées pour la reproduction du poisson dans les département du Doubs, de la Haute Saône, de la Côte d'Or, du Jura, de l'Ain, de Saône-et-Loire, du Rhône, de la Haute Savoie, de la Loire, de l'Isère, de la Savoie, de l'Ardèche, de la Drôme, du Gard, du Vaucluse, des Hautes Alpes, des Basses Alpes et de l'Aude. *Bull. Lois*, no. 1741, 229–241.

Décret du 10 août 1875 portant règlement sur la pêche fluviale. Lois annotées, 760–761.

Décret du 20 novembre 1875 portant règlement sur la pêche maritime. *Bull. Off.* 289 no 4968.

Circulaire no 23 du 14 juin 1878. Ministère des travaux publics. Pêche fluviale modifications au décret du 10 août 1875. Envoi du décret du 18 mai 1878. A.D. Vienne 3S 46.

Rapport au Président de la République française suivi d'un décret du 2 juin 1886 portant interdiction de la pêche de la truite du 20 octobre au 31 mars de chaque année dan la partie maritime de la rivière la Liane (Département du Pas-de-Calais, quartier de Boulogne). *Bull. Off. Marine* no 254, 1026–1027.

Décret du 31 octobre 1886 qui prescrit la promulgation de la convention conclue, le 18 février 1886, entre la France et l'Espagne et relative à l'exercice de la pêche dans la Bidassoa. *Bull. Lois*, 472, 616–623.

Décret du 1 octobre 1888 qui prescrit la promulgation du protocole ayant pour objet de modifier la convention du 18 février 1886, relative à l'exercice de la pêche dans la Bidassoa, signée à Madrid, le 19 janvier 1888, entre la France et l'Espagne. *Bull. Lois*, 332, 402405.

Rapport au Président de la République et décret du 27 décembre 1889 modifiant la réglementation de la pêche dans les eaux douces. *J. Off.*, 31 décembre 1889, 6509–6511.

Rapport au Président de la République et décret du ler février 1890 réglementant la pêche maritime en ce qui concerne les espèces vivant alternativement dans les eaux douces et dans les eaux salées. *J. Off.*, 7 février 1890, 695.

Décret du 7 novembre 1896 relatif à la surveillance, à la police et à l'exploitation de la pêche fluviale. *J. Off.*, 11 novembre, 1896.

Décret du 5 septembre 1897 portant règlement général de la pêche fluviale. *Bull. Lois*, 1656–1661.

Loi du 20 janvier 1902 — loi complétant l'article 10 de la loi du 15 avril 1829 relative à la pêche fluviale. *J. Off.*, 22 janvier 1902.

Décret du 17 févier 1903 relatif à l'affermage par les sociétés de pêcheurs à la ligne de certains lots de pêche, sur les fleuves, rivières et canaux. *Bull. Lois*, 2446, 1389–1391.

Décret du 3 août 1904. Etablissement de pasages destinés à assurer la libre circulation des poissons sur les parties de cours d'eau du bassin de la Seine. *J. Off.*, 13 août 1904, 5093–5094.

Loi du 31 janvier 1905 portant approbation de la convention signée à Paris, le 9 mars 1904, entre la France et la Suisse pour réglementer la pêche dans les eaux frontières des deux pays *J. Off.*, 1 février 1905, 869.

Décret de promulgation du 4 février 1905 de la loi du 31 janvier 1905. *J. Off.*, 6 février 1905, 957–959.

Décret du 1 avril 1905. Etablisement de passages destinés à assurer la libre circulation des poisson sur les parties de cours d'eau du bassin de la Loire. *J. Off.*, 19 avril 1905, 2475–2477.

Loi due 18 juin 1909 portant modification des articles 1, 3, 4, 9 et 11 de la convention du 18 février 1886 et de l'acte additionnel du 19 janvier 1888, conclus entre la France et l'Espagne, reltivement à l'exercice de la pêche dans la Bidassoa. *J. Off.*, 20 juin 1909, 6613.

Décret du 19 août 1909 approuvant la déclaration relative à l'exercice de la pêche dans la Bidassoa. *J. Off.*, 20 août 1909, 8862.

Décret du 17 décembre 1909. Règlement pour l'exercice du droit de pêche dans les eaux frontières entre la France et la Suisse. *J. Off.*, 8 janvier 1910, 232.

La Convention signée à Paris le 9 mars 1904 entre la France et la Suisse pour réglementer la pêche dans les eaux frontières des deux pays est dénoncée par le Conseil Fédéral Suisse à la date du 31 décembre 1910. *J. Off.*, janvier 1911, 1–2.

Décret du 21 mars 1913 portant modification due 1er paragraphe de l'article 11 du décret du 5 septembre 1897 sur la pêche fluviale. *Bull. Lois.*, 567–568.

Circulaire no 824 du 15 octobre 1913. Direction générale des Eaux et Forêts, Ministère de l'Agriculture. Préparation des arrêtés préfectoraux annuels portant réglementation de la pêche, des déversements industriels et du rouissage. Annexes, 26 p.

Décret du 5 février 1921. Etablissement de passages destinés à assurer la libre circulation des poissons sur les parties de cours d'eau du bassin de la Canche. *J. Off.*, 16 février 1921, 2020.

Décret du 15 avril 1921. Etablissement de passages destinés à assurer la libre circulation du poisson sur les parties de cours d'eaux du bassin de l'Adour. *J. Off.*, 9 juin 1921, 6625.

Décret du 31 janvier 1922. Etablissement de passages destinés à assurer la libre circulation du poisson sur les parties de cours d'eaux côtiers de la Bretagne des départements des Côtes-du-Nord, du Finistère, de l'Ille-et-Vilaine, de la Manche et du Morbihan. *J. Off.*, 8 Février 1922, 1560–1561.

Décret du 2 février 1922. Etablissement de passages destinés à assurer la libre circulation des poissons sur la rivière l'Authie. *J. Off.*, 16 février 1922, 1985–1986.

Décret du 3 décembre 1923 portant relèvement de la taille marchande de l'esturgeon. Lois annotées 1924, 1570.

Décret du 23 février 1924. Etablissement de passages pour la libre circulation du poisson sur divers cours d'eau des départements du Calvados, de la Manche et de l'Orne, *J. Off.*, 29 septembre 1925, 9440.

Décret du 2 juin 1926. Promulgation de la déclaration relative à l'exercice de la pêche dans la Bidassoa, signée à Madrid le 2 juin 1924, portant modification de la convention franco-espagnole du 18 février 1886, modifiée par le protocole additionnel du 19 janvier 1888 et par la déclaration du 6 avril 1908. *Bull. Off. Mar. Marchande*, 6, 505–509.

Rapport au Président de la République et décret du 26 Juillet 1927 réglementant la pêche des poissons anadromes. *J. Off.* 28 juillet 1927, 7801–7802.

Décret du 15 mars 1929. Réglementation de la pêche dans le lac Léman. *J. Off.* 20 mars 1929, 3266–3267.

Décret du 7 décembre 1929. Règlement de la pêche dans les eaux françaises du Léman. *J. Off.*, 24 décembre 1929, 13728.

Rapport au Président de la République et décret du 23 novembre 1935 réglementant la pêche dans le estuaires en ce qui concerne les espèces vivant alternativement dans les eaux douces et dans les eaux salées. *J. Off.*, 24 novembre 1935, 12376–12378. Rectificatif *J. Off.*, 26 novembre 1935, 12416.

Décret du 25 janvier 1939. Réglementation de la pêche dans les eaux françaises du Léman. *J. Off.*, 2 février 1939, 1530–1531.

Décret du 29 août 1939. Réglementation de la pêche fluviale. *J. Off.* du 31 août 19390, 10916–10919. rectificatif *J. Off.*, 30 mai 1941, 2254.

Arrêté du 17 juillet 1941. Cours d'eaux et canaux. Etat de classement des cours d'eau. *J. Off.*, 25 juillet 1941, 3111–3121, rectificatif *J. Off.*, 28 septembre 1941, 4177.
Arrêté due 15 avril 1941. Etat de classement des cours d'eau. *J. Off.*, 12 mai 1942, 1749–1750.
Décret no 1167 du 29 avril 1943 portant modification au décret du 29 août 1939 sur la pêche fluviale. *J. Off.*, 4 mai 1943, 1237.
Arrêté du 14 août 1943. Protection de la truite commune. *J. Off.*, septembre 1943, 2329. Rectificatif *J. Off.*, 3 octobre 1943, 2586.
Arrêté du 14 août 1943. Mesure des mailles des filets. *J. Off.*, 3 septembre 1943, 2329.
Arrêté du 13 juin 1944. Emploi de certains filets ou engins dans les cours d'eau de 1 catégorie (protection du brochet). *J. Off.*, 9 juillet 1944, 1756.
Décret no 45-208 du 9 février 1945 portant modification du décret du 29 août 1939 concernant la réglementation de la pêche fluviale. *J. Off.*, 11 février 1945, 704.
Décret no 45-2191 du 1 septembre 1943 portant modification du décret du 29 août 1939 concernant la réglementation de la pêche fluviale. *J. Off.*, 28 septembre 1945, 6100.
Décret no 46-1757 du 5 août 1946 modifiant les articles 5, 9, 15, 17 20 et 31 du décret du 29 août 1939 concernant la réglementation de la pêche fluviale. *J. Off.*, 8 août 1946, 7019.
Décret no 47-1350 du 18 juillet 1947 modifiant l'article 15 du décret du 29 août 1939 concernant la réglementation de la pêche fluviale. *J. Off.*, 22 juillet 1947, 7068.
Arrêté du 17 juin 1948. Modification de l'état de classement des cours d'eau en première et deuxième catégorie. *J. Off.*, 1 juillet 1948, 6365–6366.
Décret du 27 août 1948 modifiant l'article 31 du décret du 29 août 1939 concernant la réglementation de la pêche fluviale. *J. Off.*, 31 août 1948, 8595.
Arrêté du 29 décembre 1949. Interdiction de l'exercice de toute espèce de pêche jusqu'au 1 janvier 1955 dans la partie martime de certaines rivières. *J. Off.*, 5 janvier 1950, 184. Rectificatif *J. Off.*, 10 janvier 1950, 331.
Arrêté du 6 février 1950. Réglementation de la pêche au sud de la réserve de l'estuaire de la Penzé. *J. Off.*, 12 février 1950, 1714.
Décret no 50-1126 du 14 septembre 1950 modifiant les articles 1, 3, 4, 9, 12, 15 et 23 du décret du 29 août 1939 concernant la réglementation de la pêche fluviale. *J. Off.*, 15 septembre 1950, 9801.
Arrêté du 17 janvier 1952. Modification à l'arrêté du 17 juillet 1941 à l'état de classement des cours d'eau en première et deuxième catégorie. *J. Off.*, 25 janvier 1952, 1102–1103. Rectificatif *J. Off.*, 19 février 1952, 2055.
Décret no 52-315 du 17 mars 1952 modifiant les articles 1, 2, 9, 12, 14, 15, 17, 19, 24, 25 et 30 du décret du 29 août 1939 concernant la réglementation de la pêche fluviale. *J. Off.*, 19 mars 1952, 3105–3107.
Décret no 52-381 du 8 avril 1952 modifiant les articles 1, 4, 12, 22 et 26 du décret du 29 août 1939 concernant la réglementation de la pêche fluviale. *J. Off.*, 10 avril 1952, 3770–3771.
Arrêté du 29 avril 1952. Classement des cours d'eau en catégories. *J. Off.*, 7 mai 1952, 4677–4681.

Arrêté du 23 juin 1952. Classement des cours d'eau en catégories. *J. Off.*, 2 juillet 1952, 6598–6600.

Arrêté du 25 juin 1952. Conditions de vérification des dimensions des mailles des filets. *J. Off.*, juillet 1952, 6849.

Décret no 52-1348 du 15 décembre 1952 portant réglementation de la pêche dans les estuaires en ce qui concerne les espèces vivant alternativement dans les eaux douces et dans les eaux salées. *J. Off.*, 19 décembre 1952, 11675–11676.

Arrêté du 16 juillet 1953. Destruction des poissons des espèces reconnues esentiellement nuisibles. *J. Off.*, 28 juillet 1953, 6632.

Décret no 54-99 du 23 janvier 1954 modifiant certains articles du décret du 29 août 1939 sur la pêche fluviale. *J. Off.*, 28 jnvier 1954, 1011–1012.

Arrêté du 3 mars 1954. Classement des cours d'eau en catégories. *J. Off.*, 17b mars 1954, 2554–2555. Rectificatif *J. Off.*, 25 avril 1954, 4023.

Arrêté du 11 juin 1954. Destruction des poissons des espèces reconnues comme particulièrement nuisibles. *J. Off.*, 22 juin 1954, 5946–5947.

Décret no 54-1176 du 23 novembre 1954 modifiant les articles 9, 12, 15 et 19 du décret du 29 août 1939 sur la pêche fluviale. *J. Off.*, 27 novembre 1954, 11112–11113.

Arrêté du 28 avril 1955. Classement des cours d'eau en catégories. Modification de l'état de classement des cours d'eau en première et deuxième catégorie. *J. Off.*, 4 mai 1955, 4438–4440. Rectificatif *J. Off.*, 17 juillet 1955, 7154.

Arrêté du 27 juillet 1955. Classement des cours d'eau en catégories. *J. Off.*, 10 août 1955, 8045–8047.

Arrêté du 15 décembre 1955. Interdiction de toute espèce de pêche dans la partie maritime des rivières Penzé et du Dossen (quartier de Morlaix). *J. Off.*, 23 décembre 1955, 12533. Rectificatif, *J. Off.*, 12 janvier 1956, 508.

Arrêté du 29 décembre 1955. Classement en catégorie de certains cours d'eau. *J. Off.*, 19 janvier 1956, 711–715. Rectificatif *J. Off.*, 16 février 1956, 1802.

Arrêté du 22 juin 1956. Modification du classement en catégories de certains cours d'eau. *J. Off.*, 8 juillet 1956, 6353–6356. Rectificatif *J. Off.*, 26 juillet 1956, 6974.

Décret no 58-368 du 2 avril 1958 relatif à la réglementation de la pêche dans les eaux françaises du lac Léman. *J. Off.*, 6 avril 1958, 3373–3375.

Décret no 58-873 du 16 septembre 1958 déterminant le classement des cours d'eau en deux catégories. *J. Off.*, 25 septembre 1958, 8789–8802.

Décret no 58-874 du 16 septembre 1958 relatif à la pêche fluviale. *J. Off.*, 25 septembre 1958, 8802–8806.

Arrêté du 16 septembre 1958. Autorisation d'employer divers filets, engins ou lignes dans certaines eaux de la première catégorie. *J. Off.*, 25 septembre 1958, 8806. Etat des eaux de la première catégories dans lesquelles outre l'emploi de la ligne (flottante ou plombée ordinaire) de la vermée, de la bosselle à anguilles et de la balance à écrevisses sont autorisés certains filets, engins ou lignes. *J. Off.*, 25 septembre 1958, 8806–8808.

Arrêté du 16 septembre 1958. Liste des départments où les truites et saumons de fontaine peuvent être pêchés à partir d'une longueur de 18 centimètres. *J. Off.*, 25 septembre 1958, 8809.

Décret du 9 janvier 1960 modifiant le décret no 58–874 du 16 septembre 1958 relatif à la pêche fluviale. *J. Off.*, 16 Janvier 1960, 502.

Arrêté du 26 juillet 1960. Modification de l'arrêté du 16 septembre 1958 autorisant l'emploi de divers filets, engins ou lignes dans certaines eaux de la première catégorie. *J. Off.*, 6 août 1960, 7341–7343.

Décret no 60-827 (rectificatif au *J. Off.*, du 2 août 1960) portant modification du décret no 58-368 du 2 avril 1958 relatif à la réglementation de la pêche dans les eaux françaises du lac Léman. *J. Off.*, 19 octobre 1960, 9519.

Arrêté du 11 février 1961. Interdiction de l'exercice de toute espèce de pêche dans la partie martime des rivières de Penzé et du Dossen ou rivière de Morlaix. *J. Off.*, 18 février 1961, 1824.

Arrêté du 15 mars 1961. Conditions de vérification des mailles des filets et des engins. *J. Off.*, mars 1961, 3145.

Arrêté du 6 avril 1961. Modification de l'arrêté du 16 septembre 1958 autorisant l'emploi de divers filets, engins ou lignes dans certaines eaux de la première catégorie. *J. Off.*, 21 avril 1961, 3794–3795.

Décret no 61-616 du 5 juin 1961 portant modification du décret du 16 septembre 1958 déterminant le classement des cours d'eau en deux catégories. *J. Off.*, 16 juin 1961, 5428–5433. Rectificatif *J. Off.*, 6 août 1961, 7344.

Arrêté du 31 octobre 1961. Création de réserves de pêche dans la zone martime des rivières Yères, Arques, Scie, Saane (quartier de Dieppe) et Durdent (quartier de Fécamp). *J. Off.*, 9 novem bre 1961, 10323–10324.

Loi no 61-1243 du 21 novembre 1961 tendant à interdire la vente des Salmonidés sauvages. *J. Off.*, 22 novembre 1961, 10716.

— Sénat–1 session ordinaire de 1961–1962. Annexe au procès verbal de la séance du 25 octobre 1961. No 39. Raport fait au nom de la commission des Affaires économiques et du plan sur la proposition de Loi, adoptée par l'Assemblée nationale tendant à interdire la ventre des Salmonidés sauvages, 1–5.

— Documents de l'Assemblée nationale:

- annexe no 902—1 session ordinaire de 1960–1961. Séance du 27 octobre 1960. Proposition de Loi tendant à interdire la ventre des Salmonidés sauvages présentée par MM. Guillon, Chazelle, Dalainzy, Schaffner et Jean Valentin députés, 884–885.
- annexe no 1188—deuxième session ordinaire de 1960–1961. Séance du 17 mai 1961. Rapport fait au nom de la commission de la production et des échange sur la proposition de Loi (no 902) de M. Guillon et plusieurs de ses collègues tendant à interdire la ventre des Salmonidés sauvages par M. Grasset-Morel, député, 136–138.
- annexe no 1380—deuxième session ordinaire de 1960-1961. Séance du 18 juillet 1961. Avis présenté au nom de la commission des lois constitutionnelles sur la proposition de Loi (no 902) de M. Guillon et plusieurs de ses collègues par M. Carous, député, 428–429.

Décret du 22 janvier 1962 portant modification du décret du 16 septembre 1958 déterminant le classement des cours d'eau en deux catégories. *J. Off.*, 27 janvier 1962, 944–945.

Arrêté du 26 janvier 1962. Modification de l'arrêté du 16 septembre 1958 autorisant l'emploi de divers filets, engins ou lignes dans certaines eaux de la première catégories. *J. Off.*, 9 février 1962, 1434–1435.

Arrêté du 15 mai 1962. Interdiction de la pêche des saumons et des truites dans un secteur de la partie martime de la Bresle (quartier de Dieppe). *J. Off.*, 23 mai 1962, 5032.

Décret no 62-654 du 8 juin 1962 modifiant les articles 3 et 6 du décret du 16 septembre 1958 relatif à la pêche fluviale. *J. Off.*, 9 juin 1962, 5565.

Décret no 62-813 du 16 juillet 1962 portant règlement d'administration publique pour l'application de l'article 439-2 du code rural. *J. Off.*, 19 juillet 1962, 7128.

Décret o 62-1018 du 24 août 1962 portant modification du décret no 58-873 du 16 septembre 1958 déterminant le classement des cours d'eau en deux catégories. *J. Off.*, 29 août 1962, 8496–8497. Rectificatif *J. Off.*, 22 septembre 1962, 9249.

Décret no 63-98 du 11 février 1963 modifiant les articles 2, 3, 9, 18, 21, 22 et 33 du décret no 58-874 du 16 septembre 1958 relatif à la pêche fluviale. *J. Off.*, 12 février 1963, 1452–1453.

Arrêté ministériel du 25 février 1963. Plans d'eau, cours d'eau ou parties de cours d'eau dans lesquels les truites et les saumons de fontaine peuvent être pêchés pour la consommation familiale. *J. Off.*, 12 mars 1963, 2410–2412.

Arrêté du 12 novembre 1963. Modification de l'arrêté du 16 septembre 1958 autorisant l'emploi de divers filets, engins ou ligne dans certaines eaux de la première catégorie. *J. Off.*, 29 novembre 1963, 10647.

Arrêté du 24 février 1964. Plans d'eau, cours d'eau ou parties de cours d'eau dans lesquels les truites et les saumons de fontaine peuvent être pêchés pour la consommation familiale. *J. Off.*, 7 mars 1964, 2188.

Décret no 64-826 du 28 juillet 1964 portant modification du décret no 58-873 du 16 septembre 1958 déterminant le classement des cours d'eau en deux catégories. *J. Off.*, 8 août 1964, 7324–7328. Rectificatif *J. Off.*, 13 octobre 1964, 9170.

Arrêté du 19 octobre 1964. Taille marchande des poissons et crustacés (pêche maritime). *J. Off.*, 3 novembre 1964, 9849–9850.

Décret no 64-1263 du 19 décembre 1964 modifiant le décret no 58-874 du 16 septembre 1958 modifié relatif à la pêche fluviale. *J. Off.*, 22 décembre 1964, 11367–11368.

Arrêté du 7 janvier 1965. Interdiction de la pêche du saumon, des truites et des écrevisses sur certains cours d'eau ou portions de cours d'eau pendant l'année 1965. *J. Off.*, 30 janvier 1965, 850–851.

Décret no 65–53 du 14 janvier 1965 portant modification du décret no 58-873 du 16 septembre 1958 déterminant le classement des cours d'eau en deux catégories. *J. Off.*, 24 janvier 1965, 625–626.

Décret no 65-173 du 4 mars 1965 portant publication de la convention entre la France et l'Espagne relative à la pêche en Bidassoa et baie du Figuier du 14 juillet 1959. *J. Off.*, 9 mars 1965, 1893–1897.

Arrêté du 7 janvier 1966. Interdiction de la pêche du saumon, des truites, des écrevisses sur certains cours d'eau ou sections de cours d'eau pendant l'année 1966. *J. Off.*, 4 février 1966, 1032.

Arrêté du 13 mai 1966. Interdiction de la pêche des écrevisses, du saumon, de la truite et des autres espèces dans certains cours d'eau ou sections de cours d'eau juqu'au 31 décembre 1966. *J. Off.*, 9 juin 1966, 4624.

Décret no 66-597 du 27 juillet 1966 réglementant la pêche dans la section du Doubs qui forme frontière avec la Suisse. *J. Off.*, 10 août 1966, 7003–7004.

Arrêté du 13 janvier 1967. Création de réserves dans la zone maritime des rivières Yères, Arques, Scie, Saane (quartier de Dieppe) et Durdent (quartier de Fécamp). *J. Off.*, 12 février 1967, 1558.

Arrêté du 17 janvier 1967. Interdiction de la pêche du saumon, de certains autres Salmonidés et des écrevisses sur divers cours d'eau ou section de cours d'eau pendant l'année 1967. *J. Off.*, 2 février 1967, 1211–1212.

Décret no 67-281 du 15 mars 1967 portant modification de décret no 58-873 du 16 septembre 1958 déterminant le classement des cours d'eau en deux catégorie. *J. Off.*, 2 avril 1967, 3236–3237. Rectificatif *J. Off.*, 12 mai 1967, 4726.

Arrêté du 22 mars 1967. Emploi de divers filets, engins ou lignes dans certaines eaux de la première catégories. *J. Off.*, 15 avril 1967, 3836–3837.

Arrêté du 27 juillet 1967. Interdiction de la pêche dans la partie maritime des rivières Penzé et du Dossen. *J. Off.*, août 1967, 8321.

Arrêté du 27 décembre 1967. Interdiction de la pêche du saumon, de certains autres Salmonidés et des écrevisses sur divers cours d'eau ou sections de cours d'eau pendant l'année 1968. *J. Off.*, 17 janvier 1968, 698–699.

Décret no 68-33 du 10 janvier 1968 portant modification des articles 2 et 3 du décret no 58-874 du 16 septembre 1958 modifié relatif à la pêche fluviale. *J. Off.*, 13 janvier 1968, 545.

Arrêté du 15 janvier 1968. Modification dans certains départements, pour l'année 1968 des périodes d'interdiction génerale de la pêche et des périodes d'interdiction spécifique de la pêche du saumon et de l'ombre commun. *J. Off.*, 31 janvier 1968, 1122.

Décret no 68-1032 du 22 novembre 1968 portant modification de l'article 1er du décret no 58-368 du 2 avril 1958 modifié relatif à la réglementation de la pêche dans le eaux française du lac Léman. *J. Off.*, 27 novembre 1968, 11150.

Arrêté du 30 décembre 1968. Modification dans certains départements pour l'année 1969 des périodes d'interdiction générale de la pêche et des périodes spécifiques de la pêche du saumon. *J. Off.*, 1er février 1969, 1149–1150.

Arrêté du 30 décembre 1968. Interdiction de la pêche du saumon, de certains autres Salmonidés et des écrevisses sur divers cours d'eau ou sections de cours d'eau pendant l'année 1969. *J. Off.*, 1 février 1969, 1150–1151.

Arrêté du 2 avril 1969. Modification de l'arrêté du 16 septembre 1958 autorisant l'emploi de divers filets, engins ou lignes dans certaines eaux de la première catégorie. *J. Off.*, 24 avril 1969, 4131–4132.

Décret no 69-438 du 3 mai 1969 portant modification du décret no 58–873 du 16 septembre 1958 déterminant le classement des cours d'eau en deux catégories. *J. Off.*, 20 mai 1969, 5024–5027.

Arrêté du 8 janvier 1970. Interdiction de la pêche du saumon, de certain autre Salmonidés et des écrevisses sur divers cours d'eau ou sections de cours d'eau pendant l'année 1970. *J. Off.*, 23 janvier 1970, 858–859.

Arrêté du 8 janvier 1970. Modification dans certains départements, pour l'année 1970, des périodes d'interdiction générale de la pêche et des périodes d'interdiction spécifique de la pêche du saumon. *J. Off.*, 23 janvier 1970, 860.

Arrêté du 17 février 1970. Plans d'eau ou parties de cours d'eau dans lesquels les truites et les saumons de fontaine peuvent être pêchés pour la consommation familiale. *J. Off.*, 12 mars 1970, 2436.

Décret no 70-179 du 4 mars 1970 portant modification du décret no 58-873 du 16 septembre 1958 déterminant le classement des cours d'eau en deux catégories. *J. Off.*, 10 mars 1970, 2365–2366.

Décret no 70-496 du 5 juin 1970 relatif à la réglementation de la pêche dans les eaux française du lac Léman. *J. Off.*, 13 juin 1970, 5486–5487.

Arrêté du 7 juillet 1970. Modification de l'arrêté du 16 septembere 1958 autorisant l'emploi de divers filets, engins ou lignes dans certaines eaux de la première catégorie. *J. Off.*, 2 août 1970, 7253–7254.

Décret no 70-888 du 25 septembre 1970 portant modification du décret no 58-873 du 16 septembre 1958 déterminant le classement des cours d'eau en deux catégories. *J. Off.*, 2 octobre 1970, 9179–9180.

Décret no 70-958 du 15 octobre 1970 portant modification dess articles 3, 4, 12, 22 et 25 du décret no 58-874 du 16 septembre modifié relatif à la pêche fluviale. *J. Off.*, 23 octobre 1960, 9846.

Arrêté du 20 novembre 1970. Emploi de divers filets, engins ou lignes dans certaines eaux de la première catégorie. *J. Off.*, 17 décembre 1970, 11600.

Arrêté du 5 janvier 1971. Interdiction de la pêche du saumon, de certains autres Salmonidés et des écrevisses sur divers cours d'eau ou sections de cours d'eau pendant l'année 1971. *J. Off.*, 28 janvier 1971, 929–930.

Arrêté du 5 janvier 1971. Modification, sur certains cours d'eau ou sections de cours d'eau de la période d'interdiction générale et prolongation, dans tous les départements, de la période d'interdiction spécifique de la pêche du saumon bécart ou saumon de descente l'année 1971. *J. Off.*, 28 janvier 1971, 930.

Décret no 71-115 du 3 février 1971 portant modification du décret no 58-873 du 16 septembre 1958 déterminant le classement des cours d'eau en deux catégories. *J. Off.*, février 1971, 1395.

Arrêté du 14 avril 1971. Plans d'eau, cours d'eau ou parties de cours d'eau dans lesquels les truites et les saumons de fontaine peuvent être pêchés à partir d'une longueur de 18 cm pour la consommation familiale. *J. Off.*, 2 mai 1971, 4202.

Décret no 71-1048 du 17 décembre 1971 portant modification du décret no 58-873 du 16 septembre 1958 déterminant le classement des cours d'eau en deux catégories. *J. Off.*, 27 décembre 1971, 12783.

Arrêté du 18 janvier 1972. Interdiction de l'exercice de toute pêche dans certaines parties martimes des rivières Penzé et du Dossen. *J. Off.*, 4 février 1972, 1345.

Arrêté du 4 février 1972. Modification de la période d'interdiction générale de la pêche sur certains cours d'eau ou sections de cours d'eau du département du Calvados et prolongation dans tous les départements de la périoide d'interdiction spécifique de la pêche du saumon bécart ou saumon de descente pendant l'année 1972. *J. Off.*, 17 février 1972, 1764.

Arrêté du 4 février 1972. Interdiction de la pêche dans certains ruisseaux du département de l'Yonne ainsi que de la pêche du saumon, de certains autre Salmonidés et des écrevisses sur diver cours d'eau ou sections de cours d'eau pendant l'année 1972. *J. Off.*, 17 février 1972, 1764–1766.
Arrêté du 13 juin 1972. Interdiction de la pêche sur le lac de retenue du Tech (Hautes Pyréneess) pendant l'année 1972. *J. Off.*, 21 juin 1972, 6301.
Arrêté du 16 janvier 1973. Modification exceptionnelle pour l'année 1973 des périodes d'interdiction générale de la pêche dans les cours d'eau de 1 et de 2 catégorie dans certains départements. *J. Off.*, 18 janvier 1973, 734.
Arrêté du 14 février 1973. Interdiction de la pêche dans certains cours d'eau ou sections de cours d'eau des départements des Pyrénées atlantiques, des Hautes Pyrénées, de l'Essonne et de l'Yonne ainssi que de la pêche du saumon, de certains autres Salmonidés et des écrevisses sur divers cours d'eau ou sections de cours d'eau pendant l'année 1973. *J. Off.*, 10 mars 1973, 2645–2647.
Arrêté du 14 juin 1973. Plans d'eau, cours d'eau ou parties de cours d'eau dans lesquels les truites et les saumons de fontaine peuvent être pêchés à partir d'une longueur de 18 cm pour la consommation familiale. *J. Off.*, 24 juin 1973, 6684.
Arrêté du 12 octobre 1973. Emploi de divers filets, engins ou lignes dans certaines eaux de la 1er catégorie. *J. Off.*, 20 novembre 1973, 12306.
Arrêté du 23 janvier 1974. Interdiction totale de la pêche dans certain cours d'eau ou sections de cours d'eau des départements des Pyrénées atlantiques, de l'Essonne et de l'Yonne, du Gard et de la Vienne, et interdsant la pêche du saumon, de certains autre Salmonidés et des écrevisses sur divers cours d'eau ou sections de cours d'eau pendant l'année 1974. *J. Off.*, 3 février 1974, 1290–1292.
Arrêté du 5 février 1974. Interdiction de la pêche, par quelque mode que ce soit, dans certain cours d'eau ou sections de cours d'eau des départements du Finistère et du Morbihan pendant l'année 1974. *J. Off.*, 9 février 1974, 1513.
Décret no 74-177 du 7 février 1974 portant modification du décret no 58-873 du 16 septembre 1958 déterminant le classement des cours d'eau en deux catégories. *J. Off.*, 1 mars 1974, 2387–2389.
Arrêté du 26 février 1974. Plans d'eau, cours d'eau ou parties de cours d'eau dans lesquels les truites et les saumons de fontaine peuvent être pêchés à partir d'une longueur de 18 centimètres. *J. Off.*, 5 mars 1974, 2506.
Arrêté du 21 juin 1974. Modification de l'arrêté ministériel du 25 février 1963 fixant les plans d'eau, cours d'eau ou parties de cours d'eau dans lesquels le truites et les saumons de fontaine peuvent être pêchés à partir d'une longueur de 18 cm pour la consommation familiale. *J. Off.*, 18 juillet 1974, 7500.
Arrêté du 3 juillet 1974. Emploi de divers filets, engins ou lignes dans certaines eaux de la première catégorie. *J. Off.*, 22 septembre 1974, 9796.
Arrêté du 7 octobre 1974. Interdiction de la pêche sur une section de la Maronne (Cantal). *J. Off.*, 12 octobre 1974, 10475.
Décret no 74-956 du 9 octobre 1974 portant modification du décret no 58-873 du 16 septembre 1958 déterminant le classement des cours d'eau en deux catégorie. *J. Off.*, novembre 1974, 11645.
Arrêté du 24 février 1975. Prolongation dans tous les départements de la période

d'interdiction spécifique de la pêche du saumon bécart ou saumon de descente et interdiction de la pêche du saumon, de certains autres Salmonidés ou de tous poissons sur divers cours d'eau ou sections de cours d'eau pendant l'année 1975. *J. Off.*, 25 mars 1975, 3214.

Arrêté du 13 juin 1975. Interdiction de pêche dans de lac de retenue de Saint-Peyres (Tarn) pendant l'année 1975. *J. Off.*, 21 juin 1975, 6187.

Décret no 75-1093 du 21 novembre 1975 portant modification du décret no 58-874 du 16 septembre 1958 modifié relatif à la pêche fluviale. *J. Off.*, 26 novembre 1975, 12134.

Décret no 75-1144 du 1 décembre 1975 portant modification du décret no 58-873 du 16 septembre 1958 déterminant le classement des cours d'eau en deux catégories. *J. Off.*, 14 décembre 1975, 12798–12799.

Arrêté du 29 janvier 1976. Prolongation dans tous les départements de la période d'interdiction spécifique de la pêche du saumon bécart ou saumon de descente et interdiction de la pêche du saumon, de certains autres Salmonidés ou de tous poissons sur divers cours d'eau ou sections de cours d'eau pendant l'année 1976. *J. Off.*, 14 février 1976, 1060–1062.

Arrêté du 13 avril 1976. Interdiction de la pêche des Salmonidés dans l'estuaire du Scorff, du Trieux et du Jaudy. *J. Off.*, 16 mai 1976, 2950.

Arrêté du 9 juin 1976. Interdiction de la pêche dans une section de la Sénouire et certains de ses affluents (Haute-Loire) pendant l'année 1976. *J. Off.*, 16 juillet 1976, 4265.

Arrêté du 13 janvier 1977. Prolongation dans tous les départements de la période d'interdiction spécifique de la pêche du saumon bécart ou saumon de descente et interdiction de la pêche du saumon, de certains autres Salmonidés et de tous poisson sur divers cours d'eau ou sections de cours d'eau pendant l'année 1977. *J. Off.*, 16 février 1977, 939–941.

Décret no 77-192 du 23 février 1977 portant modification du décret no 58-873 du 16 septembre 1958 déterminant le classement des cours d'eau en deux catégories. *J. Off.*, 4 mars 1977, 1219–1220.

Arrêté du 21 avril 1977. Interdiction de la pêche des Salmonidés dans l'estuaire du Scorff, du Trieux et du Jaudy. *J. Off.*, 3 mai 1977, 2537.

Arrêté de 23 juin 1977. Modification de l'arrêté du 16 septembre 1958 autorisant l'emploi de divers filets, engins ou lignes dans certaines eaux de la première catégorie. *J. Off.*, 9 juillet 1977, 3615.

Arrêté du 23 juin 1977. Interdiction totale de la pêche ou interdiction de la pêche des écrevisses dans certains cours d'eau ou sections de cours d'eau pendant l'année 1977. *J. Off.*, 9 juillet 197, 3616.

Arrêté du 12 janvier 1978. Interdiction de la pêche des Salmonidés (estuaires du Trieux, du Jaudy et du Scorff). *J. Off.*, 23 février 1978, 1544 NC.

Arrêté du 22 février 1978. Prolongation dans tous les départements de la période d'interdiction spécifique de la pêche du saumon bécart ou saumon de descente et interdiction de la pêche du saumon, de certains autres Salmonidés ou de tous poissons sur divers cours d'eau ou sections de cours d'eau pendant l'année 1978. *J. Off.*, 2 mars 1978, 873–875.

Arrêté du 22 février 1978. Interdiction totale de la pêche ou interdiction de la pêche des écrevisses dans certains cours d'eau ou sections de cours d'eau pendant l'année 1978. *J. Off.*, 2 mars 1978, 876.

Décret no 78-287 du 22 février 1978 portant modification du décret no 58-874 du 16 septembre 1958 modifié relatif à la pêche fluviale. *J. Off.*, 12 mars 1978, 1063.

Décret no 78-845 du 9 août 1978 portant modification du décret no 58-873 du 16 septembre 1958 déterminant le classement des cours d'eau en deux catégories. *J. Off.*, août 1978, 3053 et 6405 NC.

Arrêté du 6 novembre 1978. Modification de l'arrêté du 25 février 1963 fixant les plans d'eau, cours d'eau ou parties de cours d'eau dans lesquels les truites et saumons de fontaine peuvent être pêchés à partir d'une longueur de 18 centimètres pour la consommation familiale. *J. Off.*, 17 bnovembre 1978, 8724 NC.

Arrêté du 8 janvier 1979. Prolongation dans tous les départements de la période d'interdiction spécifique de la pêche du saumon bécart ou saumon de descente et interdiction de la pêche du saumon, de certains autres Salmonidés ou de tous poissons sur divers cours d'eau ou sections de cours d'eau pendant l'année 1979. *J. Off.*, 17 janvier 1979, 593–595 NC.

Arrêté du 26 janvier 1979. Modification en 1979 dans certains départements de la période d'interdiction générale de la pêche afférente aux eaux de la 1 catégorie. *J. Off.*, 2 février 1979, 307–308.

Arrêté du 19 février 1979. Modification de l'état annexé à l'arrêté du 8 janvier 1979 prolongeant dans tous les cours d'eau la période d'interdiction spécifique de la pêche du saumon bécart ou saumon de descente et interdisant la pêche du saumon, de certains autres Salmonidés ou de tous poissons sur divers cours d'eau ou sections de cours d'eau pendant l'année 1979. *J. Off.*, 7 mars 1979, 2070 NC.

Arrêté du 25 avril 1979. Interdiction de la pêche du saumon et de toutes espèces de truites à l'embouchure de la Bresle (port du Tréport). *J. Off.*, 13 mai 1979, 4033 NC.

Décret no 79-993 du 23 novembre 1979 portant modification du décret no 58-874 du 16 septembre 1958 modifié relatif à la pêche fluviale. *J. Off.*, 25 novembre 1979, 2926.

Arrêté du 16 janvier 1980. Prolongation dans tous les départements de la période d'interdiction spécifique de la pêche du saumon bécart ou saumon de descente et interdiction de la pêche du saumon, de certains autres Salmonidés ou de tous poissons sur divers cours d'eau ou sections de cours d'eau pendant l'année 1979. *J. Off.*, 21 février 1980, 1911–1914 NC.

Arrêté du 21 janvier 1980. Interdiction de la pêche des Salmonidés dans les estuaires du Trieux, du Leff, du Jaudy et du Scorff. *J. Off.*, 13 février 1980, 1673 NC.

Arrêté du 15 avril 1980. Interdiction de la pêche du saumon et de toutes espèces de truites à l'embouchure de la Bresle (port du Tréport). *J. Off.*, 6 mai 1980, 4065 NC.

Décret no 80-296 du 22 avril 1980 portant modification du décret no 58-873 du 16 septembre 1958 déterminant le classement des cours d'eau en deux catégories. *J. Off.*, 26 avil, 1078 et 3075–3876 NC.

Arrêté du 26 décembre 1980. Interdiction de la pêche des Salmonidés dans les estuaires du Trieux, du Leff, du Jaudy et du Scorff. *J. Off.*, 18 janvier 1981, 714 NC.

Arrêté du 14 janvier 1981. Interdiction de la pêche des Salmonidés dans les estuaires du Gouët et du Gouessant. *J. Off.*, 4 février 1981, 1198 NC.

Arrêté du 15 janvier 1981. Prolongation dans tous les départements de la période d'interdiction spécifique de la pêche du saumon bécart ou saumon de descente et interdiction de la pêche du saumon, de certains autres Salmonidés ou de tous poissons sur divers cours d'eau ou sections de cours d'eau pendant l'année 1981. *J. Off.*, 22 février 1981, 1851–1854 NC.

Décret no 81-201 du 3 mars 1981 portant modification du décret no 58-874 du 16 septembre 1958 modifié relatif à la pêche fluviale. *J. Off.*, 5 mars 1981, 696.

Arrêté du 16 mars 1981. Interdiction de la pêche du saumon et de toutes espèces de truites à l'embouchure de la Bresle (port du Trépot). *J. Off.*, 19 avril 1981, 3934 NC.

Arrêté du 8 juillet 1981. Interdiction de la pêche des Salmonidés dans l'estuaire de l'Aberwrac'h. *J. Off.*, 22 juillet 1981, 6632 NC.

Arrêté du 3 août 1981. Interdiction de la pêche des Salmonidés dans une partie du port de Fécamp. *J. Off.*, 18 août 1981, 7358 NC.

Décret no 81-800 du 14 août 1981 portant modification du décret no 58-873 du 16 septembre 1958 déterminant le classement des cours d'eau en deux catégories. *J. Off.*, 22 août 1981, 2293–2294.

Arrêté du 21 janvier 1982. Interdiction de la pêche des Salmonidés dans les estuaires du Gouet, du Gouessant, du Trieux, du Trieux, du Jaudy, du Leff, de l'Aberwrac'h et du Scorff. *J. Off.*, février 1982, 1497 NC.

Arrêté du 11 février 1982. Prolongation dans tous les départements de la période d'interdictioin spécific de la pêche du saumon bécart ou saumon de descente et interdiction de la pêche du saumon, de certains autres Salmonidés ou de tous poissons sur divers cours d'eau ou sections de cours d'eau pendant l'année 1982. *J. Off.*, 3 mars 1982, 2297–2300 NC.

Arrêté du 30 mars 1982. Interdiction de la pêche du saumon et de toutes espèces de truites à l'embouchure de la Bresle (Port du Tréport). *J. Off.*, avril 1982, 3537 NC.

Arrêté du 27 avril 1982. Interdiction de la pêche des Salmonidés dans une partie du port de Fécamp. *J. Off.*, 13 mai 1982, 4515 NC.

Arrêté du 30 avril 1982. Interdiction de la pêche des Salmonidés dan l'estuaire de l'Elorn. *J. Off.*, 22 mai 1982, 4843–4844 NC.

Décret no 82-781 du ler septembre 1982 portant publication de l'accord entre le gouvernement de la République française et le Conseil fédéral Suisse concernant la pêche dans le lac Léman (ensemble une annexe et un règlement d'application signé à Berne le 20 novembre 1980). *J. Off.*, 16 septembre 1982, 2788–2792.

Décret no 82-911 du 15 octobre 1982 portant modification du décret no 58-874 du 16 septembre 1958 modifié relatif à la pêche fluviale. *J. Off.*, octobre 1982, 3223.

Décret no 82-978 du 17 novembre 1982 relatif à la réglementation de la pêche dans les eaux françaises du lac Léman. *J. Off.*, 20 novembre 1982, 3497–3499.

Arrêté du 7 janvier 1983. Interdiction de la pêche des Salmonidés dans les estuaires du Gouet, du Gouessant, du Trieux, du Jaudy, du Leff, de l'Elorn, de l'Aberwrac'h et du Scorff. *J. Off.*, janvier 1983, 1027 NC.

Arrêté du 31 janvier 1983. Prolongation dans tous les départements de la période d'interdiction spécifique de la pêche du saumon bécart ou saumon de descente et interdiction de la pêche du saumon, de certains autres Salmonidés ou de tous poissons sur divers cours d'eau ou section de cours d'eau pendant l'année 1983. *J. Off.*, 3 mars 1983, 2333–2337 NC.

Décret no 83-81 du 8 février 1983 portant modification du décret no 58-873 du 16 septembre 1958 déterminant le classement des cours d'eau en deux catégories. *J. Off.*, 10 février 1983, 517–518.

Arrêté du 19 juillet 1983. Interdiction de la pêche du saumon et de toutes espèces de truites à l'embouchure de la Bresle (port du Tréport). *J. Off.*, août 1983, 7441 NC.

Arrêté du 178 janvier 1984. Interdiction de la pêche des Salmonidés dans une partie du port de Fécamp. *J. Off.*, 26 janvier 1984, 967 NC.

Arrêté du 19 janvier 1984. Interdiction de la pêche des Salmonidés dans les estuaires du Gouet, du Gouessant, du Trieux, du Jaudy, du Leff, du Léguer, de l'Elorn, de l'Aberwrac'h et du Scorff pour l'année 1984. *J. Off.*, 1 février 1984, 1149–1150 NC.

Décret no 84-90 du 7 février 1984 portant modification du décret no 58-873 du 16 septembre 1958 déterminant le classement des cours d'eau en deux catégories. *J. Off.*, 10 février 1984, 554.

Arrêté du 14 février 1984. Prolongation de la période d'interdiction spécifique de la pêche du saumon bécart ou saumon de descente et interdiction de la pêche du saumon, de certains autres Salmonidés ou de tous poissons sur divers cours d'eau ou sections de cours d'eau pendant l'année 1984. *J. Off.*, 1 mars 1984, 2097–2102 NC.

Arrêté du 18 mi 1984. Création de réserves dans la zone maritime des rivières Yères, Scie, Saane, Durdent, Le Dun et dans une partie des ports de Fécamp, de Dieppe et du Tréport. *J. Off.*, 30 mai 1984, 4810 NC.

Loi no 84-512 du 29 juin 1984 relative à la pêche en eau douce et à la gestion des ressources piscicoles. *J. Off.*, 30 juin 1984, 2039–2045.

Décret no 85-16 du 3 janvier 1985 portant modification du décret no 58-874 du 16 septembre 1958 modifié relatif à la pêche fluviale. *J. Off.*, 4 janvier 1985, 118.

Arrêté du 18 février 1985 portant interdiction de la pêche des Salmonidés dans les estuaires du Gouët, du Gouessant, du Trieux, du Leff, du Jaudy, du Léguer, de l'Elorn, de l'Aberwrac'h, du Goyen, de la Laïta, du Scorff et du Blavet. *J. Off.*, 24 février 1985, 2439–2440.

Décret no 85-942 du 30 août 1985 portant modification du décret no 58-573 du 16 septembre 1958 déterminant le classement des cours d'eau en deux catégories. *J. Off.*, 6 septembre 1985, 10330–10331.

Arrêté du 4 octobre 1985 relatif à la protection de certains poissons d'eau douce. *J. Off.*, 27 octobre 1985, 12487.

Arrêté du 31 octobre 1985 portant interdiction de la pêche des Salmonidés dans les estuaires du Gouet, du Gouessant, du Trieux, du Leff, du Jaudy, du Léguer, de l'Elorn, de l'Aberwrac'h, du Goyen, de la Laïta, du Scorff et du Blavet en 1986. *J. Off.*, 7 décembre 1985, 14249.

Décret no 85-1369 du 20 décembre 1985 pris en application de l'article 435 du code rural et fixant les conditions dans lesquelles la pêche est interdite en vue de la protection du poisson. *J. Off.*, 24 décembre 1985, 15070–15071.

Décret no 85-1375 duu 23 décembre 1985 portant modification du décret no 58-873 du 16 septembre 1958 déterminant le classement des cours d'eau en deux catégories. *J. Off.*, 26 décembre 1985, 15127–15128.

Décret no 85-1385 du 23 décembre 1985 pris pour l'application de l'article 437 du code rural et réglementant la pêche en eau douce. *J. Off.*, 28 décembre 1985, 15242–15246.

Arrêté du 2 janvier 1986 fixant la liste des espèces migratrices présentes dans certains cours d'eau classés au titre de l'article 411 de la Loi du 29 juin 1984 sur la pêche en eau douce et la gestion des ressources piscicoles. *J. Off.*, 4 février 1986, 1959–1966.

Décret no 86-223 du 13 février 1986 portant publication de l'échange de notes en date du 16 décembre 1985 entre le gouvernement de la République française et le Conseil fédéral Suise relatif l'accord du 20 novembre 1980 concernant la pêche dan le lac Léman (ensemble deux annexes). *J. Off.*, 19 février 19086, 2748–2750.

Arrêté du 21 février 1986 relatif aux périodes d'ouverture de la pêche de la truite de mer durant l'année 1986. *J. Off.*, 1 mars 1986, 3230.

Arrêté du 21 février 1986 fixant la liste des cours d'eau ou parties de cours d'eau classés comme cours d'eau à truite de mer. *J. Off.*, 2 mars 1986, 3297–3300.

Arrêté du 21 février fixant la liste des cours d'eau, parties de cours d'eau et plans d'eau où la taille minimum de capture des truites et de l'omble de fontaine est ramenée à 0.18 mètre. *J. Off.*, 7 mars 1986, 3557–3558.

Arrêté du 26 mai 1986 relatif à la procédure de contrôle des filets, engins et hameçons utilisés pour la pêche en eau douce. *J. Off.*, 31 mai 1986, 6972.

Arrêté du 4 juillet 1986 relatif aux périodes d'ouverture de la pêche de la truite de mer durant l'année 1986. *J. Off.*, 10 juillet 1986, 8598.

Arrêté du 10 juillet 1986 modifiant la liste annexée à l'arrêté du 21 février 1986 fixant la liste des cours d'eau, parties de cours d'eau et plans d'eau où la taille minimum des truites et de l'omble de fontaine est ramenée à 0.18 mètre. *J. Off.*, 19 juillet 1986, 8957.

Arrêté du 12 décembre 1986 relatif aux périodes d'ouverture de la pêche de la truite de mer durant l'année 1987. *J. Off.*, 18 décembre 1986, 15160–15161.

Conclusion: Current ideas on the biological basis for management of trout populations (*Salmo trutta* L.)

G. Maisse, J. L. Baglinière

The decision to produce this work, dedicated to the biology and ecology of the trout (*Salmo trutta* L.) was made at the Colloquium organized at Paraclet in September 1988. The scientists expressed, by this decision, a desire to review the diverse studies on trout and to transmit the acquired knowledge in a synthesized form, which could be understood by those with little biological knowledge.

The very great plasticity of the trout has long posed taxonomic problems: 'there is no other fish which varies more in appearance with locality, which can lead taxonomists to create different species which are only varieties' (La Blanchere, 1926). Now we speak of one species, *Salmo trutta* L., with three ecotypes or forms; brown trout, sea trout and lake trout.

However, while the idea of monospecificity no longer needs to be questioned, the concept of ecotype is not as clear as it sounds. There are more differences between two populations of brown trout, one from Corsica and the other from Normandy, than between a brown trout and a sea trout hatched in the same Normandy river.

I. BIOLOGICAL INFORMATION AND SIGNIFICANT PROGRESS IN METHODOLOGY

As a result of regional studies; Brittany for brown trout, Lake Léman for lake trout, Normandy and Picardy for sea trout, knowledge of life cycles has progressed most, benefiting from long-term studies. This concentration of research effort has allowed the input of disciplines such as physiology and genetics, complementing biometry, on which most studies of wild fish have traditionally been based.

The 1980s have thus seen major progress in the classical study techniques, both at the level of data collection (the effectiveness of trapping was improved with the installation of electrically operated barriers (Gosset, 1989)) and at that of analysis (the interpretation of sea trout scales has been improved, in particular the analysis of 'double bands' characterizing the finnock stage (Richard and Baglinière, 1990). Other methods have been introduced, both at the population study level (echo-sounding carried out in Lake Léman, allowing the estimation of population sizes according to species (Gerdeaux, personal communication)) and that of individuals (characterization of sexual maturity stage from blood parameters (Lebail and Fostier, 1984).

Often, this methodological contribution results in the sophistication of techniques which can then only be carried out by qualified personnel. However, certain techniques have been developed intended for easy use on the field; this is the case for morphological sexing; using the sexual dimorphism of the upper jaw, demonstrated in the sea trout by Lebail (1981). Richard and Baglinière (1988) proposed keys for sex determination, with reference to the time of capture and individual characteristics, for the populations of Lower-Normandy.

Parallel to this technical progress, a development of type of approach has occurred, favoured by the variety of original disciplines of researchers collaborating on one programme. For example, the effects of sediment and water quality on incubating brown trout embryos in streams have been studied both by physiologists, ecologists and soil specialists (Massa *et al.*, 1998). For a long time, the biologist has tended to research simple cause and effect relationships. Interactions between environmental factors and the complexity of fish responses to these factors did not allow the hypothesis stage to be passed. In more recent times, different, complementary approaches of scientists, engaged in a single study, are exploited by utilizing multifactorial analyses, which are simple to do since the appearance of computing science. Baglinière *et al.* (1987) thus showed that mature trout show variable sensitivity, according to sex, to external factors initiating spawning migration. Finally, the long-term collection of data permits the modellization of the dynamics of the populations according to several biological and mesological parameters (Sabaton *et al.*, 1997; Gouraud *et al.*, 1998).

II. THE DEFICIENCIES IN ECOLOGY

The current work shows that there is a clear gap between studies carried out on the biology of fisheries and those relevant to ecology. Regional studies providing the description of biological cycles of the three ecotypes fall into the first category, consisting of the counting and characterization of individuals of trout species, the description of biotic and abiotic parameters of the environment remaining short. There are also the problems of behaviour (Part I, Chapter 4), habitat (Part I, Chapter 2) and feeding strategies (Part I, Chapter 3).

This division of research does not allow the satisfactory inclusion of equally important points, such as trout growth during the first year of life. This subject, closely linked to the concept of 'biogenic capacity of the streams', described by Léger (1910) would require the establishment of multidisciplinary studies, carried out on sections of rivers which best represent the diverse situations in France (varying in relation to hydraulic and temperature regimes, substrate type, slope, width, hardness of the water, river-bank vegetation). The relationship between fish and their habitat must be analysed on different spatial, temporal and biological scales. This concept was well developed during the international symposium 'Fish and their habitat' in 1994 (Gaudin *et al.*, 1995). Only Cuinat (1971) approached the problem of growth in trout, in relation to slope, width of the water course and calcium concentration of the water, at a national level. Twenty years on, while knowledge of the biological cycles of the three ecotypes has reached a satisfactory level, it would be particularly interesting and important to follow up the studies of Cuinat (1971) with more modern means of investigation. Such a step would allow the answering, at least in part, of questions posed by the coexistence of several ecotypes in the one basin: to what extent are these ecotypes genetically differentiated? What part does the environment play in the determination of migrations towards this or that feeding area? What is the response of the species to changes in the environment? Sound knowledge of the factors acting on growth of early stages of the trout is probably one of the keys to the answer to these questions.

III. MANAGEMENT *A PRIORI*

Facing the increase in knowledge and understanding, fisheries management appears to be founded uniquely on the idea of catches becoming reduced. With the exception of examples from populations of sea trout in the Bresle and the Adour (Prouzet *et al.*, 1988) and trout in Lake Léman (Gerdeaux *et al.*, 1989) and Lake Annecy (Gerdeaux, 1988), there is little or no information on the exploitation of trout in France. Proper management should be based on the continuous observation of control fisheries. Only reliable statistics on catch per unit effort and estimation of populations on site can allow the demonstration of fluctuations in exploitable stocks.

However, it remains a fact that, despite the real progress in techniques of water quality improvement and installation of fish passes (Larinier, 1987), numerous water courses have become unsuitable for the maintenance of wild trout populations. Such a situation is the result of the disappearance either of the spawning zones or of water quality being insufficient for the ecological requirements of the trout or of the seasonal drying up of the stream.

It is clear that the protection of areas for spawning and juvenile production must be a priority for fisheries management efforts. Part I provides the basic elements for the establishment of such a policy, adapted to the particularities of each ecotype; these are the heads of basins (drainage order <3) which play a primordial role in the recruitment of juvenile brown trout, while for sea trout or lake trout, it is the higher order sectors of the river which are more favourable.

In the area of fisheries management, in the true sense, lawyers have put various tools at the manager's disposal; open and closed seasons for fishing, legal size, creation of reserves in sectors of river. The biological justification of a particular measure is not always easy to establish, in particular if it concerns the choice of reserve zones and their true value, other than the proximity of obstacles to migration. In certain cases, the knowledge acquired can provide the biological justifications for the implementation of such measures. Thus, taking into account the observations carried out in Brittany on the brown trout (Part I, Chapter 1), it would be interesting to put the spawning tributaries into a reserve: this measure would aim at protecting not only the breeding fish coming from the main river, but also and above all, the breeding fish residing in the streams; the protection of the latter would assure an important reproductive potential, the source of the majority of juveniles, allowing the renewal of the population of catchable trout in the river.

IV. RESTOCKING, A LITTLE-STUDIED PRACTICE

The Paraclet Colloquium showed that few studies had been carried out in the area of restocking, although, since the rediscovery of artificial fertilization in the mid-nineteenth century, this has become common practice (Thibault, 1989).

The prerequisite question for all restocking problems is the definition of the concept of 'stocking capacity of the environment'. Roule (1914) mentioned the problem: 'the important question, in the problem of restocking, is that of feeding equilibrium and the limits which it imposes. ... Also it is useless to dream about increasing the normal population, if one cannot increase the feeding richness of the waters, and improve all conditions favourable to life, at the same time'. The latter remark alludes to habitat characteristics and goes further than the idea of 'biogenic capacity' of Léger (1910). In other words, a water course is not a fish farm and the failures of all attempts to artificially feed in streams attest to this (most data unfortunately unpublished!).

The concept of 'stocking capacity of the environment' remains to be determined; this would be one of the important themes to research, particularly in the areas of habitat and intraspecific competition. The practice of restocking is one of the means at the disposal of managers but, as stated by Roule (1914) 'these methods have as their aim, after the causes of depopulation have been stopped, to increase the population of the waters up to a level determined by the feeding equilibrium'.

In other words, restocking is only used when the receiving environment has been judged to be in a satisfactory state. Unfortunately, the term 'restocking' is often taken to mean a policy of systematic releases, without preliminary analysis and without control of its effectiveness. Such a practice can have repercussions for the wild populations, since incidences of genetic introgression between these stocks and released fish can be observed. Nevertheless, systematic restocking is now contested by the French Council for Fisheries.

V. CONCLUSION

This work shows that the minimum biological bases necessary are available to define the main lines for the rational management of the three trout ecotypes. But are these bases sufficient? The answer must be certainly not. Numerous question marks surround the relationships between the diverse ecotypes (should they be managed separately?) in relation to the importance of habitat (the concept of 'biogenic capacity' in relation to that of 'stocking capacity' must be defined) and exploitation by fishing. Only direct collaboration between researchers and fisheries managers, seeking mutual benefit, would allow the conditions vital for the resolution of these problems to develop. We leave the final word to Roule (1914): 'One often moans about the depopulation of water courses. Amateurs and professionals frequently express their grievances. Lawyers and administrators are employed to fix things: the preferable methods are indicated to them. Each recommends his own, according to his observations, temperament or interests. Sentiments differ, as do views. Often the most rational is forgotten; this consists of searching firstly for the primary cause of such a decrease, to try to check its action. ... It is vital therefore to envisage the problem in its entirety, and to give it a global solution, in order to effectively to treat each of its parts'.

BIBLIOGRAPHY

Baglinière J. L., Maisse G., Lebail P. Y., Prevost E., 1987. Dynamique de la population de truite commune (*Salmo trutta* L.) d'un ruisseau breton (France). II. Les géniteurs migrants. *Acta Œcol. Œcolo. Appl.*, **8**, 201–215.

Cuinat T., 1971. Principaux caractères démographiques observés sur 50 rivières à truites françaises. Influence de la pente et du calcium. *Ann. Hydrobiol.*, **2**, 187–207.

Gaudin P., Souchon Y., Orth D. J., Vigneux E., 1995. Colloque 'Habitat-Poissons', *Bull. Fr. Pêche Piscic.*, **337/338/339**, 418 pp.

Gerdeaux D., 1988. Synthèse des connaissances actuelles sur le peuplement piscicole du lac d'Annecy, Octobre 1988. Bilan piscicole et halieutique. St. Hydrobiol. Lac., INRA Thonon les Bains, 43 pp.

Gerdeaux D., Buttiker B., Pattay D., 1989. La pêche et les recherches piscicoles en 1988 sur le Léman. Rapport annuel 1988, CIPEL, 7 pp.

Gosset C., 1989. Etude sur l'installation d'écrans électriques à poissons. Convention d'étude Région Midi-Pyrénées-INRA. Rapp. tech. INRA—Saint-Pée-sur-Nivelle, 55 pp.

Gouraud V., Baglinière J. L., Sabaton C., Ombredane D., 1998. Application d'un modèle de dynamique de population de truite commune (*Salmo trutta*) sur un bassin de Basse Normandie: premières simulations. *Bull. Fr. Pêche Piscic.*, **350/351**, 675–691.

La Blanchere H. (de), 1926. *La pêche et les poissons. Dictionnaire général des pêches.* De la Grave, Paris, 842 pp.

Larinier M., 1987. Les passes à poissons: méthodes et techniques générales. *La Houille blanche*, **1-2**, 51–57.

Lebail P. Y., Fostier A., 1984. Techniques d'identification du sexe et d'estimation de la maturité sexuelle chez les poissons vivants. In G. Barnabé and R. Billard (Eds), *L'Aquaculture du Bar et des Sparidés*, INRA Publ., Paris, 45–52.

Lebail P. Y., 1981. Identification du sexe en fonction de l'état de maturité chez les poissons. Thèse de Docteur-Ingénieur, Université de Rennes I, 71 pp.

Leger L., 1910. Principe de l améthode rationnelle du peuplement des cours d'eau. *Ann. Univ. Grenoble*, **22**, 533–568.

Massa F., Grimaldi C., Baglinière J. L., Prunet P., 1998. Evolution des caractéristiques physicochimiques de deux zones de frayères à sédimentation contrastée et premiers résultats de survie embryo-larvaire de truite commune (*Salmo trutta*). *Bull. Fr. Pêche Piscic.*, **350/351**, 359–376.

Prouzet P., Martinet J. P., Casaubon S., 1988. Rapport sur la pêche des marins pêcheurs dans l'estuaire de l'Adour en 1988. Rap. IFREMER/DRV/RH/St Pée-sur-Nivelle, 15 pp.

Richard A., Baglinière E., 1988. Le sexage morphologique des truites de mer (*Salmo trutta* L.) des rivières Orne et Touques (Basse-Normandie). *Colloque sur la truite* (*Salmo trutta*). Le Paraclet, 6–8 September 1988.

Richard A., Baglinière J. L., 1990. Description et interprétation des écailles de truite de mer (*Salmo trutta* L.) des deux rivières de Basse-Normandie, l'Orne et la Touques. *Bull. Fr. Pêche Piscic.*, **319**, 239–257.

Roule L., 1914. Traité raisonné de la Pisciculture et des Pêches, Baillière, Paris, 734 pp.

Sabaton C., Siegler L., Gouraud V., Manne S., Baglinière J. L., 1997. Presentation and first application of a dynamic population model for brown trout, *Salmo trutta* L. Aid to river management. *Fish Mngmt. Ecol.*, **4**, 425–438.

Thibault M., 1989. La redécouverte de la fécondation artificielle de la truite en France au milieu du XIXXe siècle; les raisons de l'engouement et ses conséquences. In Colloque *Hommes, Animal et Sociéte*, 13–16 May 1987. Vol. 3, Histoire et Animal, des Sociétés et des Animaux, Inst. Et. Polit., Toulouse, 205–211.

Index